T0292698

5G Physical Layer

Principles, Models and Technology Components

5G Physical Layer

Principles, Models and Technology Components

Ali Zaidi

Fredrik Athley

Jonas Medbo

Ulf Gustavsson

Giuseppe Durisi

Xiaoming Chen

Academic Press is an imprint of Elsevier
125 London Wall, London EC2Y 5AS, United Kingdom
525 B Street, Suite 1650, San Diego, CA 92101, United States
50 Hampshire Street, 5th Floor, Cambridge, MA 02139, United States
The Boulevard, Langford Lane, Kidlington, Oxford OX5 1GB, United Kingdom

Notices

Knowledge and best practice in this field are constantly changing. As new research and experience broaden our understanding, changes in research methods, professional practices, or medical treatment may become necessary.

Practitioners and researchers must always rely on their own experience and knowledge in evaluating and using any information, methods, compounds, or experiments described herein. In using such information or methods they should be mindful of their own safety and the safety of others, including parties for whom they have a professional responsibility.

To the fullest extent of the law, neither the Publisher nor the authors, contributors, or editors, assume any liability for any injury and/or damage to persons or property as a matter of products liability, negligence or otherwise, or from any use or operation of any methods, products, instructions, or ideas contained in the material herein.

Library of Congress Cataloging-in-Publication Data
A catalog record for this book is available from the Library of Congress

British Library Cataloguing-in-Publication Data
A catalogue record for this book is available from the British Library

ISBN: 978-0-12-814578-4

For information on all Academic Press publications
visit our website at https://www.elsevier.com/books-and-journals

Publisher: Mara Conner
Acquisition Editor: Tim Pitts
Editorial Project Manager: Leticia Lima
Production Project Manager: R. Vijay Bharath
Designer: Christian J. Bilbow

Typeset by VTeX

To my loving wife, Marzieh, for bringing new inspirations.

(Ali Zaidi)

To Linnea, Simone, and my parents for constant support and patience.

(Ulf Gustavsson)

Contents

Acknowledgments

We thank the following colleagues for their continuous support, collaboration, and inspiration: Robert Baldemair (Ericsson), Vicent Molés-Cases (former Ericsson, now UPV), Markus Ringström (Ericsson), Joakim Sorelius (Ericsson), Marie Hogan (Ericsson), Gianluigi Liva (German Aerospace Center), Hua Wang (Keysight), Kittipong Kittichokechai (Ericsson), Mattias Andersson (Ericsson), Erik Dahlman (Ericsson), Stefan Parkvall (Ericsson), Kristoffer Andersson (Ericsson), Sven Mattisson (Ericsson), Lars Sundström (Ericsson), Per Landin (Ericsson), Sven Jacobsson (Ericsson), Thomas Eriksson (Chalmers University), Christian Fager (Chalmers University), Erik G. Larsson (Linköping University), Christopher Mollén (former Linköping University), Katharina Hausmair (former Chalmers University, now Qamcom), Christoph Studer (Cornell University), Jian Luo (Huawei), Jaakko Vihriälla (Nokia), Andreas Wolfgang (Qamcom), Robin Gerzaguet (ENSSAT), Yinan Qi (Samsung), Ning He (Ericsson), Karl Werner (Ericsson), Sebastian Faxér (Ericsson), Shehzad Ali Ashraf (Ericsson), Eleftherios Karipidis (Ericsson), Peter von Wrycza (Ericsson), Miurel Tercero (Ericsson), Håkan Björkegren (Ericsson), Mikael Wahlén (Ericsson), Joakim Hallin (Ericsson), Arne Simonsson (Ericsson), Kjell Larsson (Ericsson), Göran Klang (Ericsson), Dennis Sundman (Ericsson), Henrik Asplund (Ericsson), and Satyam Dwivedi (Ericsson).

We acknowledge mmMAGIC (a European collaborative research project), 3GPP (a collaboration project between standard development organizations), and Ericsson Research for their tremendous contributions to the development of 5G New Radio. We are also grateful to Qamcom Research & Technology AB for publishing the source code that is part of this book.

List of Acronyms

3GPP third generation partnership project
2G second generation
3G third generation
4G fourth generation
5G fifth generation
ACEPR adjacent channel error power ratio
ACLR adjacent channel leakage ratio
ACE active constellation extension
ACLR adjacent channel leakage ratio
ADC analog-to-digital converter
ADSL asymmetric digital subscriber line
AMPS advanced mobile phone system
AMAM amplitude-to-amplitude modulation
AWGN additive white Gaussian noise
BICM bit-interleaved coded modulation
BP belief propagation
BEC binary erasure channel
BER bit error rate
BLER block error rate
BS base station
CCDF complementary CDF
CDD cyclic delay diversity
CDF cumulative distribution function
CDMA code division multiple access
CFO carrier frequency offset
CN check node
CPE common phase error
CTF channel transfer function
CIR channel impulse response
CoMP coordinated multipoint
CP cyclic prefix
CP-OFDM cyclic prefix OFDM
CPRI common public radio interface
CQI channel quality indicator
CRI CSI-RS resource indicator
CRC cyclic-redundancy check
CRS cell-specific reference signal
CSI channel state information
CSI-IM CSI interference measurement
CSIR channel state information at the receiver
CSI-RS channel state information reference signal
CSIT channel state information at the transmitter
CW continuous wave
D2D device-to-device
DAB digital audio broadcast
DAC digital-to-analog converter
DFT discrete Fourier transform

DFTS-OFDM DFT-spread OFDM
DLL delay-locked loop
DPD digital pre-distortion
DM-RS demodulation reference signal
DPC dirty paper coding
DVB-T digital video broadcast-terrestrial
EDGE enhanced data rates for GSM evolution
EGT equal gain transmission
EMF electromagnetic field
EPDCCH enhanced physical downlink control channel
eNB evolved Node B
eMBB enhanced MBB
EVM error vector magnitude
FBMC filter bank multicarrier
FCC federal communications commission
FDD frequency division duplex
FDM frequency division multiplexing
FD-MIMO full-dimension MIMO
FEC forward error correction
FFT fast Fourier transform
FIR finite impulse response
F-OFDM filtered OFDM
FS frequency spreading
FSTD frequency-switched transmit diversity
GaN Gallium-Nitride
GI guard interval
GMP generalized memory polynomial
GMSK Gaussian minimum shift keying
gNB the next generation Node B
GoB grid-of-beams
GPRS general packet radio service
GSM global system for mobile communications
HARQ hybrid ARQ
HetNet heterogeneous network
HPBW half-power beam width
HSPA high speed packet access
ICI intercarrier interference
ICT information and communication technologies
ID identity
IFDMA interleaved frequency division multiple access
IFFT inverse FFT
IID independent identically distributed
IMT international mobile telecommunications
IoT internet of things
IP internet protocol
ISI intersymbol interference
ITU international telecommunications union
ITU-R ITU-radiocommunications sector
KPIs key performance indicators
L1-RSRP layer-1 RSRP
LBT listen-before-talk
LDPC low density parity check
LFDMA localized frequency division multiple access

LLR log-likelihood ratio
LNA low-noise amplifier
LoS line-of-sight
LS least-square
LTE long term evolution
LUT look-up table
MBB mobile broadband
MBSFN multicast-broadcast single-frequency network
MAC medium access control
MCS modulation and coding scheme
MIMO multiple-input multiple-output
ML maximum likelihood
MP memory polynomial
MMS multimedia message service
MMSE minimum mean square error
mmMAGIC millimeter magic project
mMTC massive machine type communications
MRC maximum ratio combining
MRT maximum ratio transmission
MSE mean squared error
MSK minimum shift keying
MU multi user
MU-MIMO multiuser MIMO
NB Node B
NB-IoT narrow band-IoT
NMSE normalized mean squared error
NMT nordic mobile telephone
NLoS non line-of-sight
NR new radio
NSA non-stand-alone
NZP CSI-RS non-zero-power CSI-RS
OFDM orthogonal frequency division multiplexing
OSTBC orthogonal space–time block code
OOB out-of-band
OQAM offset QAM
OSD ordered-statistic decoding
PA power amplifier
PAPR peak-to-average power ratio
PBCH physical broadcast channel
PCM parity check matrix
PDCCH physical downlink control channel
PDCP packet data convergence protocol
PDSCH physical downlink shared channel
PHD polyharmonic distortion
PHY physical
PLL phase-locked loop
PN phase-noise
PMI precoding matrix indicator
PPN polyphase network
PQI PDSCH rate matching and quasicolocation indicator
PRACH physical random access channel
PSD power spectral density
PSK phase shift keying

PSS primary synchronization signal
PTS partial transmit sequence
PT-RS phase tracking reference signal
PUCCH physical uplink control channel
PUSCH physical uplink shared channel
QAM quadrature amplitude modulation
QoS Quality of Service
QPSK quadrature phase shift keying
RAN radio access network
RB resource block
RCFO residual CFO
RF radio frequency
RFICs radio frequency integrated circuits
RI rank indicator
RLC radio link control
RLF radio link failure
RMa rural macro
RMS root mean square
RRC radio resource control
RSRP reference signal received power
Rx receive
SA stand-alone
SAR specific absorption rate
SC-FDMA single-carrier frequency division multiple access
SDAP service data adaptation protocol
SDMA space division multiple access
SER symbol error rate
SFO sampling frequency offset
SFBC space frequency block coding
SFTD space frequency transmit diversity
SIC successive interference cancellation
SINDR signal-to-interference, distortion and noise ratio
SINR signal-to-interference-and-noise ratio
SIR signal-to-interference-ratio
SIMO single-input multiple-output
SISO single-input single-output
SLI strongest layer indicator
SLM selective mapping
SMS short message service
SNR signal-to-noise ratio
SORTD spatial orthogonal-resource transmit diversity
SQNR signal to quantization noise ratio
SRI SRS resource indicator
SRS sounding reference signal
SSB synchronization signal block
SSS secondary synchronization signal
SSBRI SSB resource indicator
STTD space–time transmit diversity
SU-MIMO single-user MIMO
SVD singular value decomposition
TB tail-biting
TCI transmission configuration indicator
TDD time division duplex

TDM time division multiplexing
TI tone injection
TM transmission mode
TPMI transmit precoder matrix indicator
TR tone reservation
TRI transmit rank indicator
TRS tracking reference signal
TS technical specification
TSG technical specification group
TTI transmit time interval
Tx transmit
QCL quasicolocation
UE user equipment
UF-OFDM universally filtered OFDM
ULA uniform linear array
UMa urban macro
UMTS universal mobile telecommunications system
UPA uniform planar array
URLLC ultra-reliable low-latency communications
V2I vehicle-to-infrastructure
V2N vehicle-to-network
V2P vehicle-to-pedestrian
V2X vehicle-to-anything
V2V vehicle-to-vehicle
VCO voltage controlled oscillator
VN variable node
WAP wireless application protocol
WAVA wraparound Viterbi algorithm
WiMAX worldwide inter-operability for microwave access
W-OFDM windowed OFDM
WRC world radiocommunication conference
WSS wide-sense stationary
VNA vector network analyzer
VS-GMP vector switched GMP
WCDMA wideband CDMA
WG working group
WLAN wireless local area network
ZF zero-forcing
ZP CSI-RS zero-power CSI-RS

CHAPTER

INTRODUCTION: 5G RADIO ACCESS 1

Information and communication technologies (ICT) have sparked innovations in almost every society. The ever-growing capabilities for instantly transferring and processing information are transforming our societies in many ways—online shopping, social interactions, professional networking, media distribution, e-learning, instant information access, remotely watching live events, audio and video communication, virtual offices and workforce, and so on. Various industries and corporations have also been evolving their processes and businesses based on technological advancements in information and communication technologies.

Fifth generation (5G) mobile communication is expected to enormously expand the capabilities of mobile networks. New technologies and functionalities are being introduced for 5G systems in various domains—wireless access, transport, cloud, application, and management systems [6]. These advancements are targeting traditional mobile broadband users as well as emerging machine-type users, so that new and superior services can be enabled for both consumers and industries at large, unleashing the potential of the internet of things (IoT), and virtual and augmented reality. According to a recent survey performed across 10 different industries [4], the global revenues driven by 5G technologies will be as high as 1.3 trillion USD by 2026 (see Fig. 1.1 for revenues per industrial segment). It is estimated that by 2023, there will be around 3.5 billion cellular IoT connections [24].

The backbone of any mobile communication system is its wireless access technology, which connects devices with radio base stations. As almost every society and industry is looking forward to the 5G revolution with its specific set of requirements, the design of the 5G wireless access is challenging. A 5G wireless access technology is expected to provide extreme data rates, ubiquitous coverage, ultra-reliability, very low latency, high energy efficiency, and a massive number of heterogeneous connections. The human-centric emerging applications are augmented reality, virtual reality, and online gaming—these demand extreme throughput and low latency. For machine-type communication there are two main segments: massive IoT and critical IoT. Massive IoT is characterized by a high number of low cost device connections, supporting small volumes of data per device with long battery life and deep coverage (for example, for underground and remote areas). The applications are in smart buildings, utilities, transport logistics, agriculture, and fleet management. The critical IoT is characterized by ultra-reliability and very low-latency connectivity, for example, to support autonomous vehicles, smart power grids, robotic surgery, traffic safety, and industrial control.

This book is about the upcoming 5G wireless access technology, and it focuses on its physical layer. A preview of the book is provided in Section 1.4. This chapter gives a holistic view of the 5G wireless access technology and its global development. We start with a brief history of mobile access technologies in Section 1.1 and introduce the 5G mobile access technology in Section 1.2. In Section 1.3, we provide a global picture of 5G wireless access—spectrum allocations, standardization, use cases and their requirements, field trials, and future commercial deployments.

5G Physical Layer. https://doi.org/10.1016/B978-0-12-814578-4.00006-0

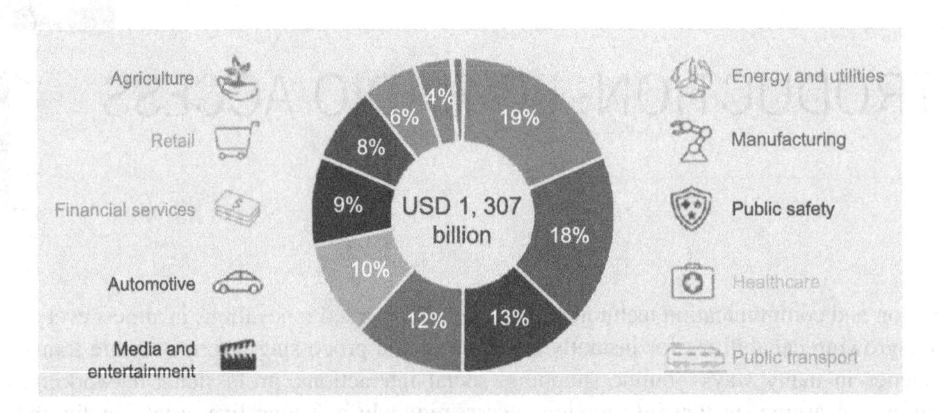

FIGURE 1.1

5G enabled industry digitalization revenues for ICT players, 2026 (Source: Ericsson [4]).

1.1 EVOLUTION OF MOBILE COMMUNICATION

In 1946, the US federal communications commission (FCC) approved a first mobile telephony service to be operated by AT&T in 1947. At this time, the equipment were bulky and had to be installed in a vehicle due to the weight and its excessive power consumption. From this point on, more than three decades of cellular communication technology evolution has led to a shift from analog to digital formats of communication, going from what was mainly voice to high-speed data communication.

Leading up from the mid-1980s, the first generation (1G) of cellular communication, which mainly carried voice, grew up using formats such as advanced mobile phone system (AMPS) in the USA and nordic mobile telephone (NMT) in Scandinavia. These analog formats were later replaced moving towards 2G with the first digital communication schemes around the mid to late-1990s—global system for mobile communications (GSM) in Europe and digital-AMPS for the USA. At this point, the short message service (SMS) was introduced, being one of the first widely used non-voice applications for cellular communication. Enhancement for 2.5G using enhanced data rates for GSM evolution (EDGE), general packet radio service (GPRS) and code division multiple access (CDMA) sparked the use of mobile data communication and early cellular internet connectivity in the early-2000s. This was an early enabler, which did, however, require a specific protocol, known as wireless application protocol (WAP).

Moving forward from 2G into 3G, in order to meet the increasing demand for cellular access data rates, universal mobile telecommunications system (UMTS) based on wideband CDMA (WCDMA) technology was introduced by third generation partnership project (3GPP) just around 2000. With advances in mobile user equipment technology, this enabled the user to not only communicate via multimedia message service (MMS), but also stream video content. Transitioning to 4G, long term evolution (LTE) was introduced, which does not only imply major changes on the air interface, but was moving from code division multiplexing to orthogonal frequency division multiplexing (OFDM) and time division duplex (TDD) or frequency division duplex (FDD).

Entering the era of 4G, there were mainly two competing technologies at an early stage. These were worldwide inter-operability for microwave access (WiMAX), based on IEEE 802.16m, and LTE Advanced, which is an extension of LTE. LTE-A introduced technology components such as carrier aggregation and improved support for coordinated multipoint (CoMP) transmission and heterogeneous network (HetNet) deployments for improving Quality of Service (QoS) in hot-spots and coverage for cell-edge users. LTE-A prevailed as the dominant cellular access technology today and has served as the basis of the transition to 5G mobile communications. The transition from 4G to 5G is inspired by new human-centric and machine-centric services across multiple industries.

1.2 5G NEW RADIO ACCESS TECHNOLOGY

5G wireless access is envisioned to enable a networked society, where information can be accessed and shared anywhere and anytime, by anyone and anything [2]. 5G shall provide wireless connectivity for anything that can benefit from being connected. To enable a truly networked society, there are three major challenges:

- A massive growth in the number of connected devices.
- A massive growth in traffic volume.
- A wide range of applications with diverse requirements and characteristics.

To address these challenges, 5G wireless access not only requires new functionalities but also substantially more spectrum and wider frequency bands.

Fig. 1.2 illustrates the operational frequency ranges of existing (2G, 3G, 4G) and future (5G) mobile communication systems. The current cellular systems operate below 6 GHz. A large amount of spectrum is available in the millimeter-wave frequency band (30–300 GHz); however, there is no commercial mobile communication system operating in the millimeter-wave frequencies today. 4G LTE is designed only for frequencies below 6 GHz. There are some local area networks and (mostly) indoor communication systems based on the IEEE 802.11ad and 802.15.3c standards that operate in the unlicensed 60 GHz band. IEEE 802.11ay, a follow-up of 802.11ad, is under development. 3GPP is currently developing a global standard for new radio access technology, 5G new radio (NR), which will operate in frequencies from below 1 GHz up to 100 GHz. 5G NR shall unleash new frequencies and new functionalities to support ever-growing human-centric and machine-centric applications.

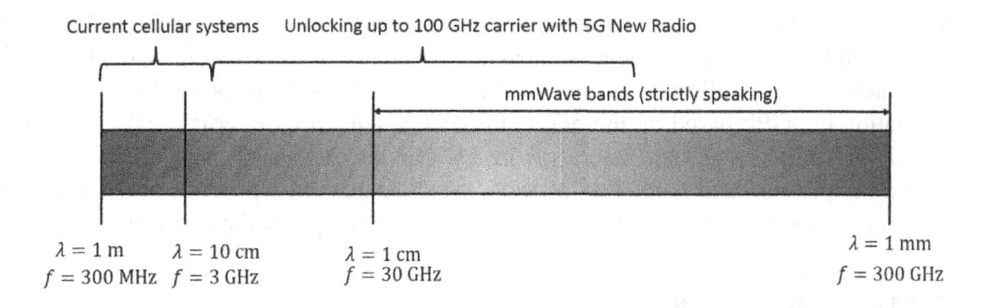

FIGURE 1.2

Frequency ranges of current and future mobile communication systems.

The vision of 5G wireless access is shown in Fig. 1.3. 5G wireless access comprises both 5G NR and LTE evolution. LTE is continuously evolving to meet a growing part of the 5G requirements. The evolution of LTE towards 5G is referred to as the LTE Evolution [13]. LTE will operate below 6 GHz and NR will operate from sub-1 GHz up to 100 GHz. 5G NR is optimized for superior performance; it is not backwards compatible to LTE, meaning that the legacy LTE devices do not need to be able to access the 5G NR carrier. However, a tight integration of NR and LTE evolution will be required to efficiently aggregate NR and LTE traffic.

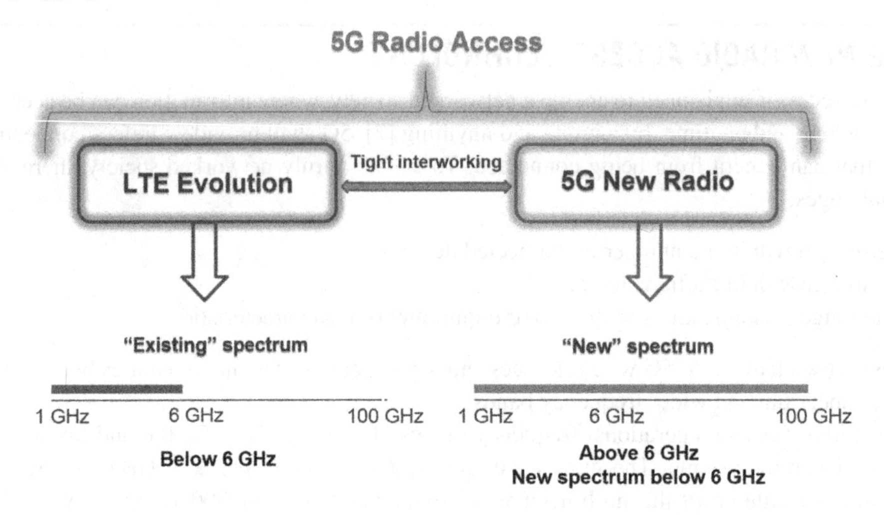

FIGURE 1.3

5G radio access vision.

1.3 5G NR GLOBAL VIEW

The research and concept development of the 5G wireless access technology (named 5G NR since 2016) started almost a decade ago with new inspiring applications and business cases in mind. The research efforts led to development of 5G test-beds in universities as well as industry. Like previous generations of cellular systems, the development of 5G NR is a well-coordinated global effort—addressing the new spectrum allocations on global and regional levels for 5G NR and the global 5G NR standardization in 3GPP based on the 5G requirements defined by international telecommunications union (ITU). To give a global picture of the 5G NR developments, we will in the following discuss spectrum allocations and regulatory aspects, the standardization process, major use cases and their requirements, some precommercial trials, and expected commercial deployments.

1.3.1 5G STANDARDIZATION

The specifications for NR and LTE are developed by the 3GPP which is a collaboration between seven regional and national standard development organizations from Asia, Europe and North America:

ARIB, ATIS, CCSA, ETSI, TSDSI, TTA, and TTC. 3GPP produces technical specifications which are transposed into standards by the standardization bodies. 3GPP was started in 1998 with the initial goal of developing globally applicable specifications for the third generation (3G) of mobile communications. The scope has since then been widened and now includes the development and maintenance of specifications for the second generation (2G) GSM, 3G WCDMA/HSPA, fourth generation (4G) LTE, and 5G NR/LTE evolution.

The international radio frequency (RF) spectrum is managed by the ITU-radiocommunications sector (ITU-R). ITU-R has also a responsibility to turn technical specifications from, e.g., 3GPP into global standards, including also countries not covered by the standardization bodies in 3GPP. ITU-R also defines the spectrum for the so-called international mobile telecommunications (IMT) systems. The IMT systems correspond in practice to the different generations of mobile communications, from 3G onwards. The 3G and 4G technologies are included in the IMT-2000 and IMT-Advanced recommendations, respectively. A new ITU-R recommendation for 5G, called IMT-2020, is planned to be developed in 2019–20.

ITU-R does not produce the detailed technical specifications but defines IMT in cooperation with regional standardization bodies by specifying requirements that an IMT technology should fulfill. The actual technology is developed by others, e.g. 3GPP, and submitted to ITU-R as a candidate IMT technology. The technology is evaluated against the specified requirements and may then be approved as an IMT technology. ITU-R gives recommendations of radio interface technologies in radio interface specifications for a particular IMT system and provides references to the corresponding detailed specifications which are maintained by the corresponding standardization bodies. The IMT-2000 radio interface specifications include six different radio interface technologies while IMT-Advanced includes two. Unlike previous generations, for 5G it is not expected that competing technologies will be submitted as candidates for IMT-2020; only 3GPP-based technologies are anticipated. 3GPP will submit LTE evolution and NR together as their candidate system for IMT-2020.

Different generations of mobile communications have appeared around every ten years. However, the individual systems are continuously evolving with new features. The 3GPP specifications are divided into releases, where each release consists of a complete and self-contained set of specifications. This means that a particular release contains all components needed to build a complete cellular network, not just the newly added features. When a release is completed, the features are frozen and ready for implementation. When a release has been frozen, only essential corrections are permitted. Further functionalities will have to go into the next release. The work on different releases has some overlap, so that the work on a new release starts before the completion of the current release. The releases should be backwards compatible so that a user equipment (UE) developed for one release can also work in a cell that has implemented a previous release. The first version of LTE was part of release 8 of the 3GPP specifications. LTE release 10 was named *LTE-Advanced* since it was approved as an IMT-Advanced technology by ITU-R. In release 13, the marketing name for LTE changed to *LTE-Advanced Pro*.

The everyday 3GPP work is divided into study items and work items. Study items are feasibility studies of concepts, where the results are documented in a technical report (TR). The details of agreed concepts are worked out in the work items, where the features are defined and end up in a technical specification (TS). The 3GPP technical specifications (TSs) are organized in series and are numbered TS XX.YYY, where XX defines the series. The radio aspects of NR are defined in the 38-series. All 3GPP specifications are publicly available at www.3gpp.org.

Organizationally, 3GPP consists of three technical specification groups (TSGs) where TSG radio access network (RAN) is responsible for the radio access specifications. TSG RAN in turn consists of six working groups (WGs), where RAN WG1 deals with the physical layer specifications. The WGs meet regularly and come together four times a year in TSG plenary meetings. The decisions in 3GPP are based on consensus among the members.

Fig. 1.4 illustrates the ITU and 3GPP time lines as well as the expected development of commercial equipment for 5G. The standardization work of 5G NR started in 3GPP in April 2016, with the aim of making it commercially available before 2020. 3GPP is taking a phased approach in defining the 5G specifications. A first standardization phase, with limited NR functionality, NR release 15, was completed for non-stand-alone (NSA) operation at the end of 2017 and for stand-alone (SA) operation in mid-2018. The second standardization phase, NR release 16, is expected to meet all the requirements of IMT-2020 and be completed by 2019.

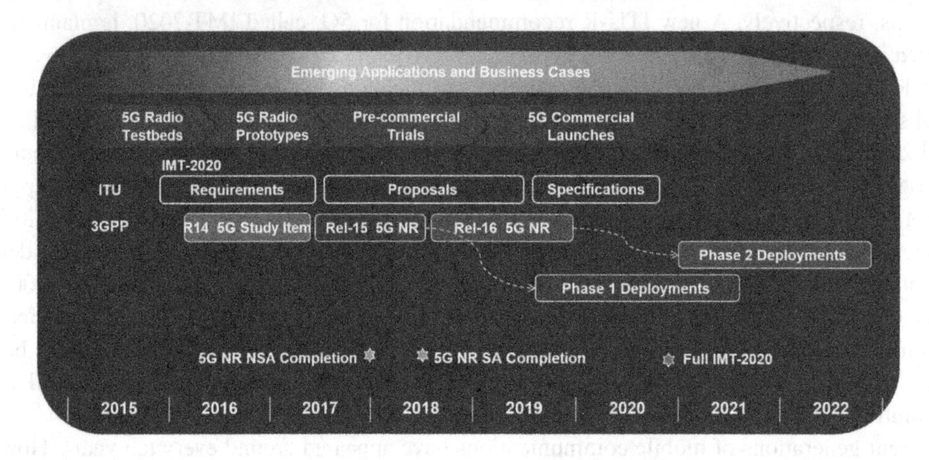

FIGURE 1.4

5G global view.

With 5G standardization in acceleration, 5G precommercial trials are also taking place worldwide. Based on the release 15 specifications, 3GPP-compliant base stations and devices are already under development. The commercial deployments are then expected in two phases. The first phase of NR commercial deployments is expected in 2019; it is based on the release 15 specifications. The second phase of NR commercial deployments is expected to start in the 2021 time frame, based on the release 16 specifications. It is likely that the NR specifications will continue to evolve in 3GPP beyond 2020, with a sequence of releases including additional features and functionalities.

1.3.2 SPECTRUM FOR 5G

One of the main changes when going from previous generations of mobile communications to the fifth generation is the spectrum use at radically higher frequencies in the millimeter-wave range. 3GPP has decided to support the range from below 1 GHz up to 52.6 GHz already from the first releases of NR [12]. A main reason for this change is the availability of large amounts of spectrum with very large

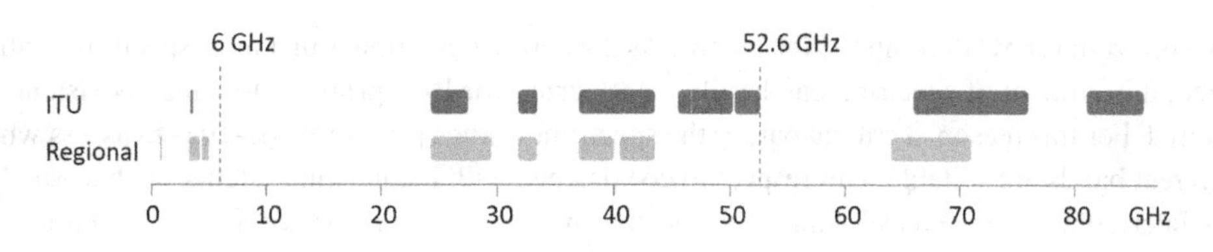

FIGURE 1.5

Global (ITU) and regional spectrum identifications for 5G.

bandwidths in the range of several GHz. Though the millimeter-wave spectrum seems very attractive, there are also many challenges:

- The transmission loss goes up substantially if multiantenna and beam-forming techniques are not used.
- RF hardware performance regarding e.g. phase noise and output power is degraded.
- There is no non-utilized spectrum, meaning that coexistence, with e.g. satellite systems, will be required where acceptable interference levels have to be guaranteed.

As a consequence, 5G is designed by 3GPP (NR) for flexible frequency use over the full frequency range. Joint operation at lower frequencies and higher frequencies is supported in order to provide both reliable coverage (utilizing e.g. frequencies below 6 GHz) and very high capacity and bitrates when possible (where millimeter-wave coverage is available).

An overview of the spectrum identified for 5G is given in Fig. 1.5. The 3300–3600 MHz band was identified as a global IMT band by ITU-R at world radiocommunication conference (WRC)-15.[1] The corresponding global allocation above 6 GHz is subject for WRC-19. There is a substantial additional regional spectrum identified below 6 GHz. Above 6 GHz the regional spectrum is largely overlapping with the global bands except for around 28 GHz and 65 GHz. It is evident that the amount of spectrum below 6 GHz is scarce and that spectrum above 6 GHz will be needed for fulfilling the requirements of 5G. For early deployments of 5G, bands below 6 GHz at 600–700 MHz, 3300–4200 MHz, and 4400–5000 MHz have drawn attention [11]. Except the 3300–3600 MHz band which was identified as a global IMT band at WRC15 these bands are subject to regional regulations.

In Europe the focus for early deployments is put on the 3600–3800 MHz band [3]. Also the United States, Japan, South Korea and China use different allocations in the 3300–4200 MHz band in their planning for early 5G systems, and 3GPP is developing a specification for this band.

The 4400–5000 MHz band is mainly promoted by China and Japan but could potentially also be adopted by other countries in the Asia–Pacific region.

Though the bandwidth is limited in the 600–700 MHz band, it is useful for providing coverage as the transmission losses are low without requiring any advanced multiantenna transmission techniques. There are already allocations for LTE at 700 MHz which could be migrated to 5G in 2020 or later. Moreover, the United States has identified 614–698 MHz for mobile use.

In the process of allocation of IMT frequencies above 6 GHz it was decided at WRC-15 to study 11 bands in the range 24.5 GHz to 86 GHz for decisions at WRC-19. ITU-R has estimated the maximum

[1]A WRC is held every 3–4 years where regulations are revised for the RF spectrum use across the world.

need of spectrum for 5G to be up to 20 GHz [8]. As the current situation is that most spectrum is already allocated, this amount of spectrum can hardly be allocated for IMT primary use, and coexistence will be required. For this reason, a crucial part of the spectrum studies prior to WRC-19 is to assess whether the different bands are suitable with respect to coexistence with incumbent systems, such as satellites.

The interference between IMT and other systems on earth, in air or in space, is mainly of long range. For this reason, mainly the aggregation of multiple IMT transmitters is of interest. In this case the specific pointing directions and antenna gain is averaged out, providing an effective omni-directional radiation. For base stations, however, the radiation in elevation angle is focused in down-tilted directions.

In order to accurately determine interference signals between IMT and other systems, proper propagation models are required. Such models are provided by the ITU-R study group 3 recommendations. There are earlier recommendations (P.619, P.2041, P.1409, P.452, P.2001, etc.) which are valid for atmospheric propagation paths and paths over the earth profile and for the full range of frequencies under consideration by WRC-19. However, two important cases have been missing for modeling of additional losses due to clutter (vegetation and man made structures like buildings) and penetration into buildings. In March 2017 ITU-R study group 3 succeeded in providing proper models for these cases [10], [9]. The corresponding urban clutter loss is up to 50 dB (median) for low elevation angles. For higher elevation angles the clutter loss may be very low, down to 0 dB, as no buildings obstruct the sky. However, for these cases the building penetration loss for indoor transmitters is in the range 40 dB to 60 dB (median) at 30 GHz. As 5G deployments are expected to be highly concentrated in urban environments below roof-top and indoors, the urban clutter and building penetrations losses greatly improve the prospects for IMT coexistence with incumbent systems.

In addition to the global ITU process there have been a lot of regional efforts for finding a suitable spectrum for 5G. The United States, Korea, Canada, Japan, Singapore, and Sweden are focusing on bands in the 26.5–29.5 GHz range. The United States is also targeting bands in the 37–40 GHz range and Canada is considering the very high end of frequencies in the 64–71 GHz band. Both Europe and China have identified bands in the range 24.25–27.5 GHz. In addition China is looking at 37–42.5 GHz, whereas Europe considers the ranges 31.8–33.4 GHz and 40.5–43.5 GHz. The 5G spectrum availability around the world is summarized in Fig. 1.6.

Except the obvious need for contiguous bands of large bandwidths, the allocation of spectrum for 5G will crucially depend on the frequency dependent performance of RF hardware, propagation channel and multiantenna techniques. All these aspects are addressed in depth in Chapter 3, Chapter 4, and Chapter 7.

1.3.3 USE CASES FOR 5G

Differently from its predecessors, 5G has the goal of providing optimized support for a number of drastically different types of services and user requirements. Conforming to the ITU nomenclature for international mobile communications for 2020 and beyond (IMT-2020) [7], 5G will target three use case families with very distinct features: enhanced mobile broadband (eMBB), massive machine-type communications (mMTC), and ultra-reliable low-latency communications (URLLC). We will next describe the features of these three use case categories, which are depicted in Fig. 1.7

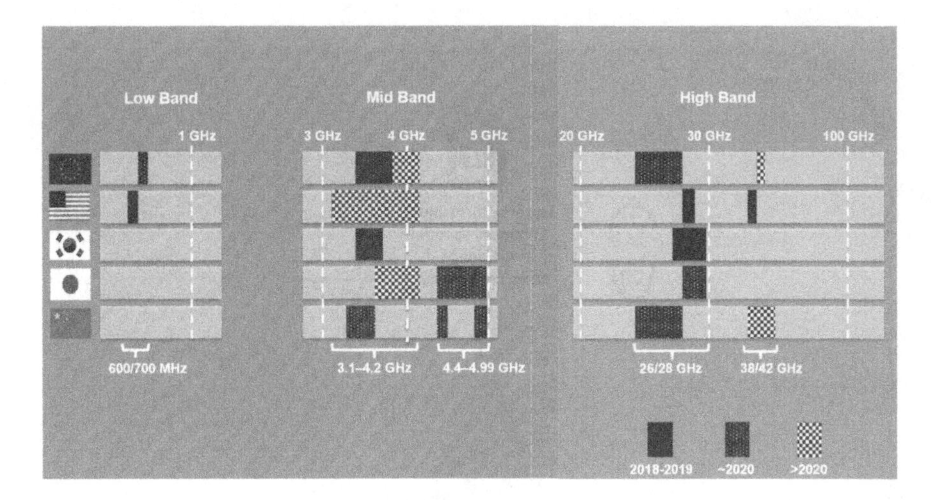

FIGURE 1.6

5G spectrum availability around the world.

eMBB:

This use case category is the natural extension of the classic mobile broadband connectivity scenario under current mobile telecommunication standards. It addresses human-centric connectivity, including access to multimedia content, services, and data. This is done by providing the high data rates that are required to support future multimedia services and the increasing traffic volume generated by these services. The eMBB use case covers a range of scenarios including:

- *Hotspot connectivity*, which is characterized by a high user density and extremely high data rates, and low mobility.
- *Wide-area coverage*, where the user density and data rates are lower, but the mobility is higher.

URLLC:

Stringent requirements on both latency and reliability are the distinctive features of this use case category, which targets mainly machine-type communications (MTC). Envisaged applications include wireless control of industrial manufacturing and production processes, remote medical surgery, driverless and/or remotely driven vehicles, and distribution automation in smart grids.

mMTC:

The growth of the Internet of Things (IoT) is causing a proliferation of wirelessly connected devices carrying MTC traffic. Indeed, the number of such devices is expected to exceed soon the number of devices carrying human-generated traffic. The focus of the mMTC is on providing connectivity to a massive number of devices, which are assumed to transmit sporadically a low amount of traffic, which is not delay critical. The mMTC devices are expected to have a very long battery lifetime to allow for remote deployments. A unique feature of this use case is that the MTC devices will be extremely heterogeneous in terms of capabilities, cost, energy consumption, and transmission power.

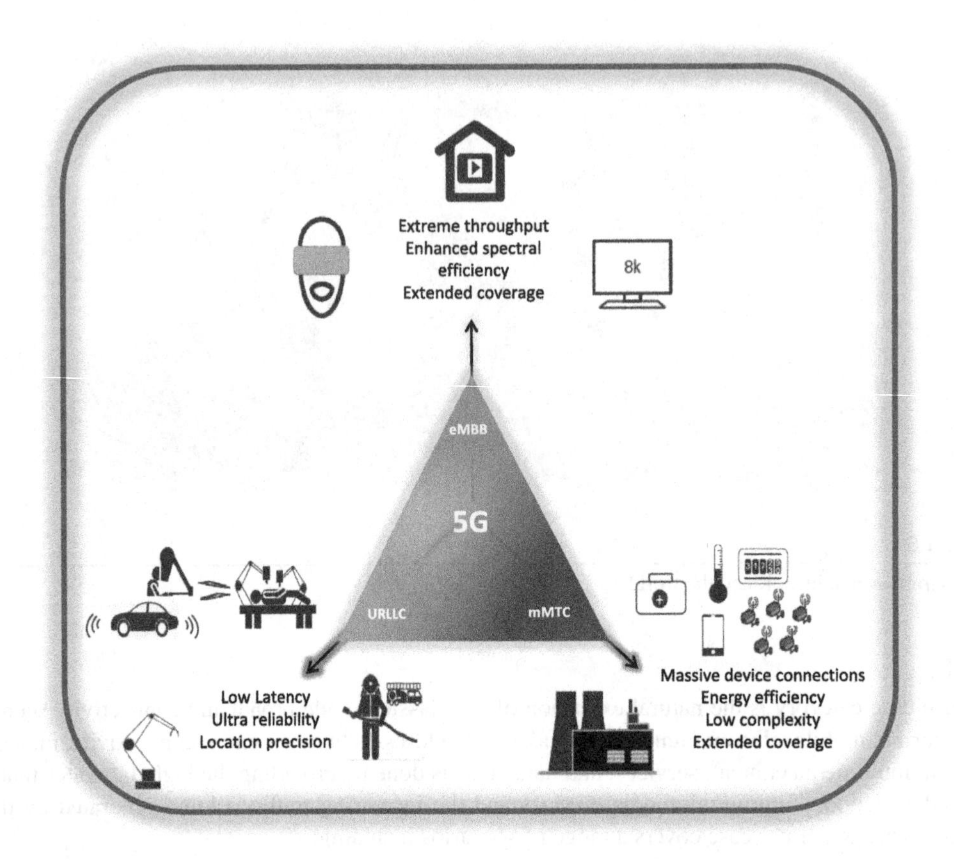

FIGURE 1.7

The three 5G use cases, according to the ITU nomenclature, and their main features.

Fig. 1.8 provides a graphical summary of some of the applications targeted by 5G, and their relation to the IMT-2020 use cases just reviewed.

The ITU has set key performance requirements for IMT-2020 to be able to address satisfactorily the specific needs of eMBB, URLLC, and mMTC traffic. Such requirements are summarized in Fig. 1.9, where a comparison with the key capabilities of the previous mobile generation—IMT-advanced—is provided. The requirements include a peak data rate of 20 Gbit/s, a latency below 1 ms, and the capability to support a connection density of 10^6 devices per square kilometer. As exemplified in Tables 1.1–1.2, NR is being designed to satisfy (in some cases, with large margins) all these requirements [1].

1.3.4 5G FIELD TRIALS

In the following, we will briefly present some of the world's most exciting 5G trials, which show the potential of 5G to unlock new and exciting opportunities. Some pictures from these trials are shown in Fig. 1.10.

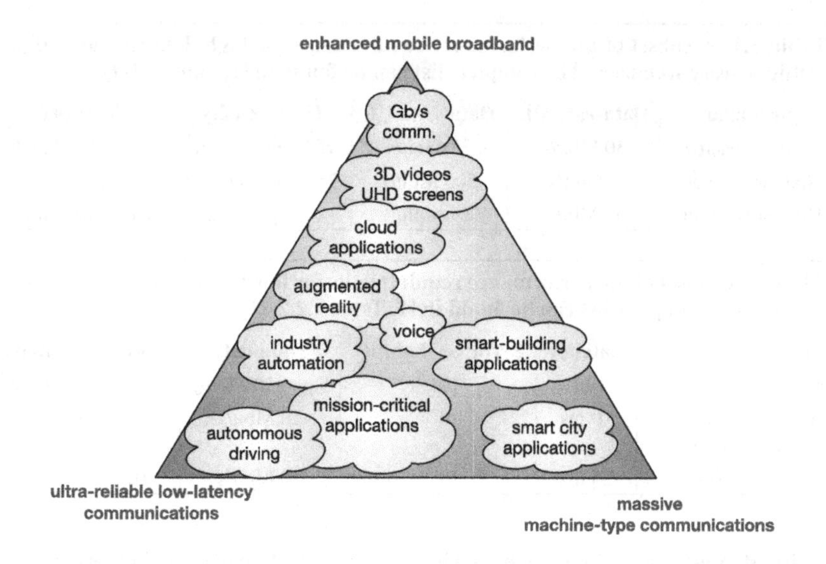

FIGURE 1.8

The three IMT-2020 use cases and some targeted applications.

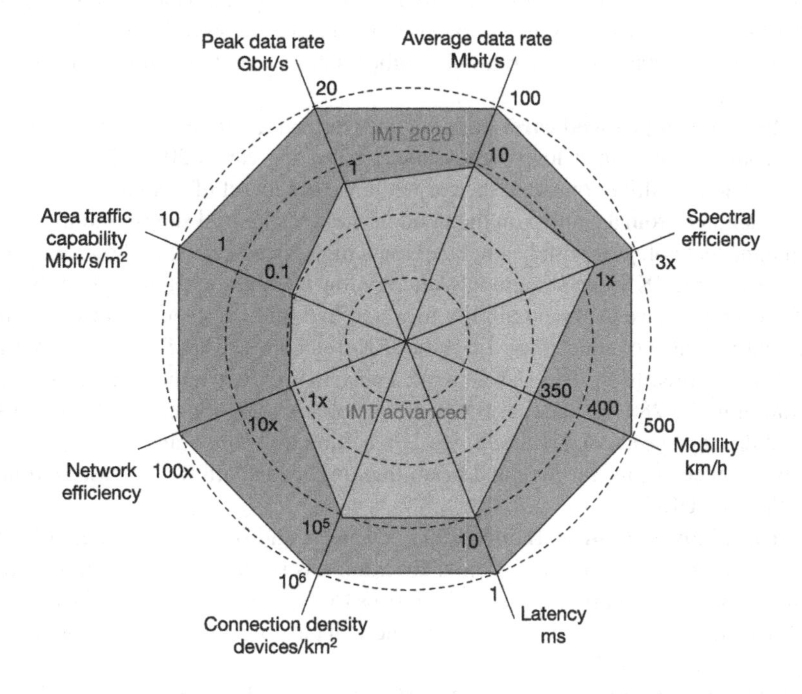

FIGURE 1.9

The key capabilities of IMT-2020 in relation to the ones of its predecessor, IMT-advanced.

Table 1.1 A subset of the performance requirements for high-data-rate and high-traffic-density scenarios. The complete list can be found in [1, Table 7.1-1]

Scenario	Data-rate DL	Data-rate UL	User density	Mobility
Urban macro	50 Mbit/s	25 Mbit/s	10^5 UEs per km^2	Up to 120 km/h
Indoor hotspot	1 Gbit/s	500 Mbit/s	25×10^4 per km^2	Pedestrian
High-speed train	50 Mbit/s	25 Mbit/s	10^3 per train	Up to 400 km/h

Table 1.2 A subset of the performance requirements for low-latency and high-reliability scenarios. The complete list can be found in [1, Table 7.2.2-1]

Scenario	Latency	Error probability	Data rate	Connection density
Motion control	1 ms	10^{-6}	Up to 10 Mbit/s	10^5 UEs per km^2
Transport system infrastructure backhaul	10 ms	10^{-6}	10 Mbit/s	10^3 UEs per km^2
Electricity distribution	25 ms	10^{-3}	10 Mbit/s	10^3 UEs per km^2

5G for flying drones: Ericsson and China Mobile successfully performed the world's first 5G trial for flying drones on a commercial network in 2016, with 5G enabled technologies on a cellular network [14]. A drone (an unmanned aerial vehicle) was flown with handovers across multiple sites. In order to demonstrate the concept's validity in a real-world setting, the handovers were performed between sites that were simultaneously in use by commercial mobile phone users. The applications of drones are emerging, for example, in agriculture, public safety, search and rescue, inventory management, and goods delivery.

5G connectivity for high-speed cars: Ericsson and Verizon tested the limits of 5G by demonstrating 6+ Gbps throughput and super-low latency inside a moving car in 2017 [20]. The car was completely backed out and the driver operated the car while wearing a set of virtual reality glasses, relying solely on video captured from a camera on the hood of the car. The trial showcased the massive potential of multiantenna technologies with beams tracking a high-speed car. This shows that 5G is capable of supporting 360-degree 4K video streaming with very low latencies, even in high mobility scenarios. This would, for example, give us the possibility to watch live sporting events with virtual reality.

A similar trial was also conducted by Ericsson, SK Telecom and BMW Group on a racetrack in Yeonjong-do, South Korea [22]. The test network consisted of four radio transmission points from Ericsson operating in the 28 GHz band at BMW Group Korea's driving center. Under this setup, test results achieved driving speeds of 170 km/h, while reaching downlink data speeds of 3.6 Gbps. This was enabled by advanced beam-forming and beam tracking, which allows the base station to transmit signals that follow the UEs.

5G connecting ships and ports: Tallink, Telia, Ericsson, and Intel have created a 5G test and exploration area at the Port of Tallinn in Estonia. Ericsson has set up a 5G base station in the port of Tallinn and Intel has placed equipment inside the ferries to receive signals. The 5G trial network is delivering high-speed internet connectivity to the commercial passenger cruise ships and their passengers while in the port area since 2017 [19].

5G delivering tens of Gbps worldwide: In 2016, Ericsson and Korea Telecom (KT) demonstrated the world's first 5G achieving 25.3 Gbps throughput [16]. The trial was performed in millimeter-wave frequency bands for 5G mobile communications. (KT was aiming to provide 5G trial services for the

FIGURE 1.10

5G trials. (A) 5G connectivity for high-speed cars (Ericsson, Verizon); (B) 5G connectivity for high-speed cars (Ericsson, SK Telecom); (C) 5G for flying drones (Ericsson, China Mobile); (D) 5G connecting ships and ports (Ericsson, Telia, Tallink Grupp, Intel); (E) 5G for remotely driving cars (Ericsson, Telefonica); (F) 5G delivering tens of Gbps (Ericsson, Korea Telecom).

PyeongChang 2018 winter games.) In the same year (2016), Ericsson and Telstra conducted Australia's first live 5G trial, achieving 20+ Gbps and a latency of half of what is seen in Telstra's 4G networks [15]. The trial used 800 MHz of spectrum, which is 10 times more than what Telstra currently uses for its 4G networks.

Ericsson and Mobile Telesystems (MTS) have built a prototype 5G network in Moscow and successfully achieved 25 Gbps in 2017, with the use of advanced antenna systems on 5G base stations [21].

In January 2018, Mobile Telephony Network Group (MTN) and Ericsson performed the first 5G trial in Africa and achieved a throughput of more than 20 Gbps with less than 5-ms latency. This was the best performance ever achieved on a mobile network in Africa. The 5G trial was based on Ericsson's 5G prototype radios [23].

5G making remote driving a reality: Telefonica, Ericsson, the Royal Institute of Technology (KTH), and Applus Idiada developed a revolutionary demo to showcase the world's first 5G remote driving of a car on Applus Idiada race track during the Mobile World Congress 2017 [17,18]. A 5G base station provided connectivity to a car at 15 GHz carrier frequency. The driver in a remote location gets an in-car experience with 4K video streaming in the uplink. In the downlink, the 5G connectivity provides ultra-low latency and high reliability to communicate driving decisions to the car. One possible application of remote driving is remote parking, where a driver can leave a car in a drop-off zone and request assistance from a remote driver to ensure safe and efficient parking.

1.3.5 5G COMMERCIAL DEPLOYMENTS

The deployment of a 5G NR network should start from an existing 4G LTE network, which already has good coverage [5]. As shown in Fig. 1.3, NR can coexist and interwork with LTE, which reduces the time to market for NR. Commercial deployment of the 5G NR network is expected to occur in two phases:

1. The first phase is a non-standalone (NSA) mode, with tight interworking between 5G NR and LTE, as shown in Fig. 1.3. This option uses LTE as the control-plane[2] anchor for NR, and it uses either LTE or NR for user traffic (user-plane). The first 5G specification (3GPP Release 15, NSA) was finalized at the end of 2017. Fully standard-compliant radio systems are likely to be available at the end of 2018. We expect to see 5G NR to start being deployed during 2019.
2. The second phase is a standalone (SA) mode, with both control-plane and user-plane existing over 5G NR. The 5G specification for the SA mode was released in mid-2018. The commercial deployments for the SA mode are expected beyond 2020.

NR will need to coexist and interwork with LTE for many years to come, not only as a way of reducing the time to market but also ensuring good coverage and mobility. With 5G NR deployments, new use cases will pick up and 5G devices will come to market. We see the deployment of standalone NR beginning when devices are more widely available, new use cases (e.g., ultra-reliable and low-latency communications, industrial IoT) start to gain momentum, and NR has greater access to both new and existing spectrum. Ultimately, 5G NR is likely to become the mainstream cellular technology used to address multiple use cases across multiple industries. The evolution path of 5G NR deployments is sketched in Fig. 1.11.

There are three main 5G deployment scenarios considering that LTE operates below 6 GHz (i.e., low or mid frequency band) and NR can operate up to 100 GHz (i.e., in low, mid, and high bands) [5]:

1. **NR NSA in mid/low with LTE in low/mid band** In this deployment scenario, NR and LTE will have similar coverage per base station, since they are both deployed in similar frequencies. The

[2] The control-plane is mainly responsible for control signaling for connection setup, mobility (handovers), and security.

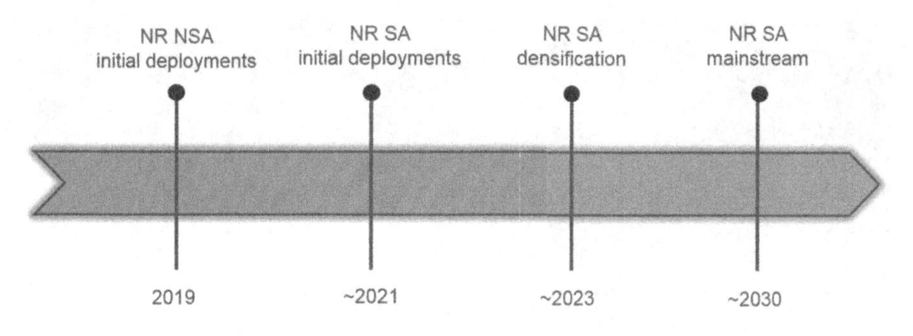

FIGURE 1.11

The evolution path of 5G NR deployments.

greater capacity and throughput enabled by the 5G NR spectrum implies that this is a more likely way to deployment for high user density areas, typically in city and urban areas.

2. **NR NSA in high band with LTE in low/mid band** In this deployment scenario, NR coverage area per base station will be smaller than LTE coverage area, due to higher transmission losses at higher frequencies. The main driver for this deployment is to enable high capacity and extreme throughput cells with wider channel bandwidths available in millimeter-wave frequencies. In addition to eMBB, fixed wireless access is an emerging use case for this deployment category.
3. **NR SA in mid/low/high band** NR standalone will be deployed in all frequency bands, depending on the spectrum availability. In addition to eMBB, standalone NR deployments are very likely to be used for private or enterprise networks to support industrial applications (for example, manufacturing). Although NR SA deployments do not rely on LTE for its operation, many NR SA deployments are likely to reuse the existing 4G physical infrastructure and transport network, thereby reducing the cost of deployment. NR NSA in low/mid bands will be suitable for massive IoT applications. Deployment of 5G NR in low bands is of interest to boost coverage. Deployments in high bands are more likely to be used for high-traffic areas and for private (industrial) IoT networks.

Although deployment for cost-efficient and well-performing 5G NR networks is crucial for 5G success, it is the availability of NR capable devices and their market uptake that will eventually determine growth driven by 5G NR connectivity. Fig. 1.12 shows approximate timing of NR capable devices. Early fixed wireless access devices are already available in some regions (e.g., in USA) for operation in high bands. The first 3GPP-compliant 5G smart-phones and tablets are likely to be launched in 2019. New NR capable devices for IoT are expected to be introduced in 2020 and beyond. For many IoT use cases, the cost of NR capable devices might have to significantly go down below eMBB devices in order to allow mass adoption. However, for critical IoT applications requiring high-reliability and very low-latency connections, device cost is not expected to be a barrier.

The United States will be among the first to experience 5G commercial services [24]. The major operators in the US have already announced that they will start providing 5G services between late 2018 and mid-2019. South Korea, Japan, and China are expected to be among early 5G service providers with significant volumes of 5G subscriptions. Ericsson forecasts over 1 billion 5G mobile broadband subscriptions by the end of 2023, accounting for 12 percent of all mobile subscriptions.

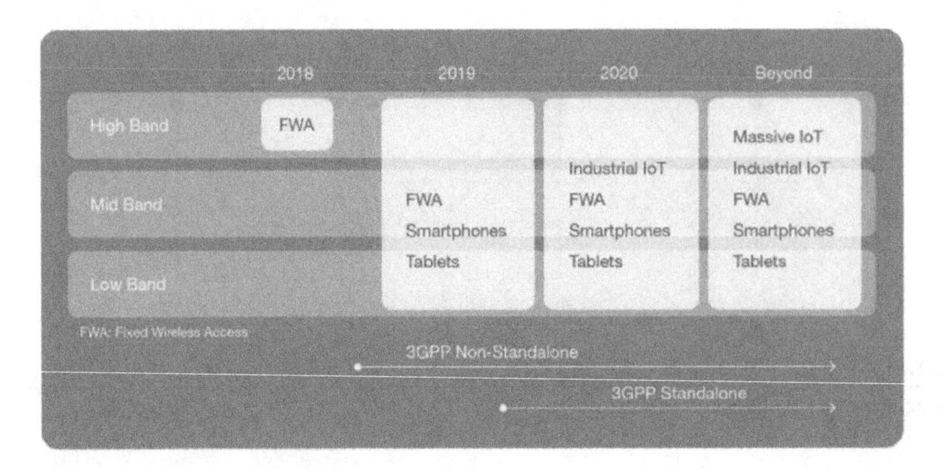

FIGURE 1.12

5G device availability (Source: Ericsson [5]).

1.4 PREVIEW OF THE BOOK

This book is written for researchers and system designers in the field of mobile radio communication, interested in understanding principles, models, and technology components for the 5G NR physical layer. Although the focus is on 5G NR, many concepts presented in the book are of a fundamental nature and are applicable beyond 5G. We assume that the reader has a basic understanding of digital wireless communications and signal processing; however, familiarity with cellular technologies (for example, 4G LTE standard) is not required. We will introduce relevant standard related concepts and terminologies in the book.

A preview of the book is sketched in Fig. 1.13, highlighting key aspects covered in various chapters. The book is composed of nine chapters covering various aspects of 5G NR—a global picture of its development, the physical layer overview based on the first 5G NR release in 3GPP, the physical limitations imposed by radio wave propagation and hardware impairments, the key physical layer technologies, and an open-source link-level simulator. In the following, we briefly outline the content of each chapter.

Chapter 1 introduces 5G NR and discusses global efforts in the development of 5G NR and its future impact on industry and society. We provide a holistic view on 5G use cases and their requirements, spectrum allocations, standardization, field trials, and future commercial deployments.

Chapter 2 provides an overview of the 5G NR physical layer based on the first 3GPP NR release. We will see that the physical layer components of NR are flexible, ultra-lean and forward-compatible. Moreover, we provide an overview of radio wave propagation and hardware impairment related challenges associated with enabling a high performing NR.

Chapter 3 presents state-of-the-art insights on radio wave propagation along with description of fundamental concepts and propagation characteristics. We focus on the frequency dependency of the channel properties for the full range of frequencies envisioned for 5G NR, with experimental examples.

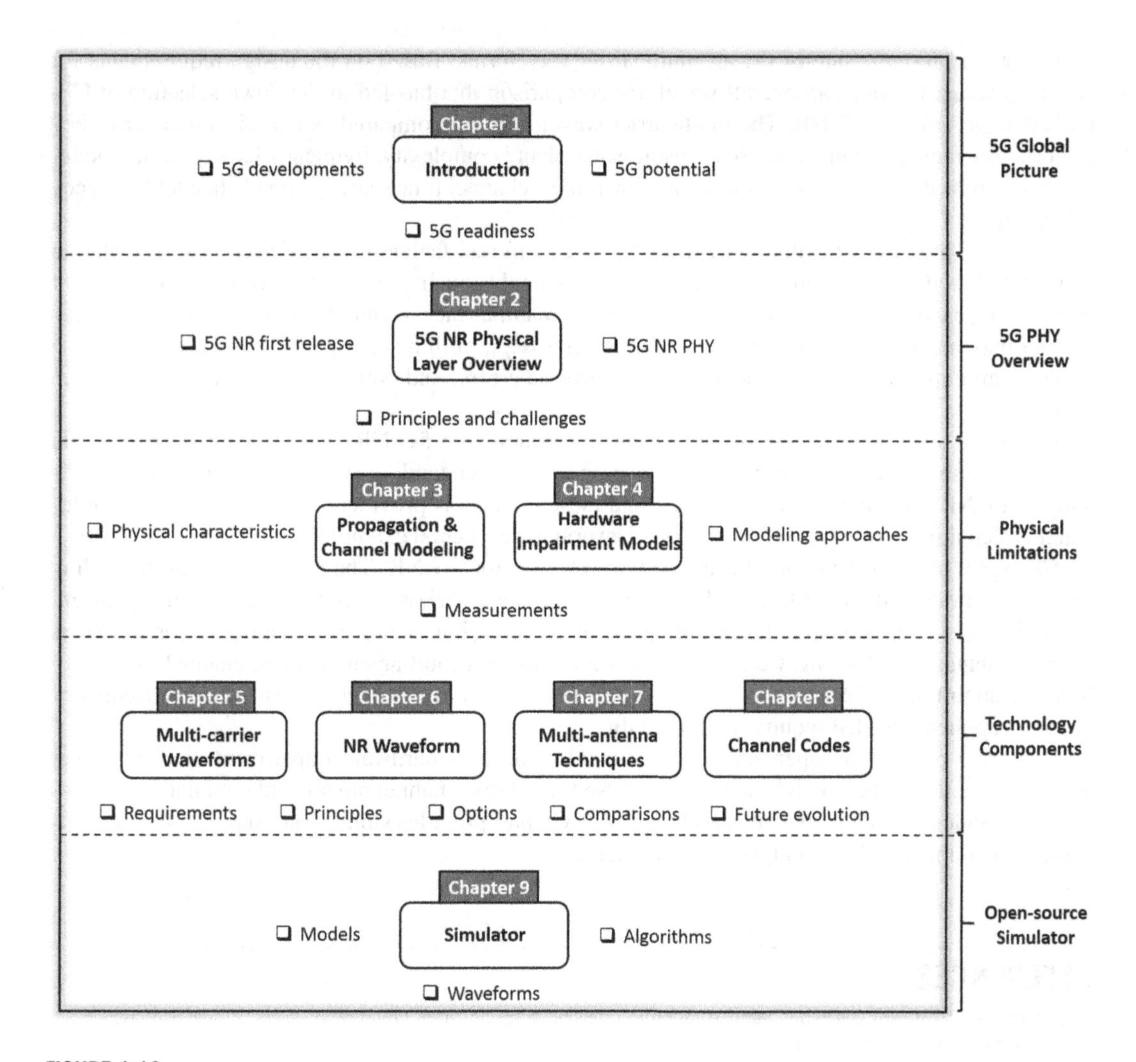

FIGURE 1.13

Preview of the book.

The channel modeling for 5G NR is discussed. Moreover, we point out both validated and non-validated (or deficient) aspects of the current 5G channel models defined by 3GPP and ITU-R.

Chapter 4 covers some of the traditional behavioral models for power amplifiers, local oscillators and data converters. These models may accurately predict the input–output relation of analog and mixed-signal components. Furthermore, a novel modeling approach is presented that provides the second order statistics of the errors caused by non-ideal components. This stochastic modeling framework provides a powerful tool for link-level evaluations and aids in making sound choices in terms of the radio performance versus energy efficiency trade-offs.

Chapter 5 presents state-of-the art multicarrier waveforms. Based on the design requirements for NR, the chapter provides an overall waveform comparison that has led to the down selection of CP-OFDM waveform for 5G NR. The multicarrier waveforms are compared as regards a number of key performance indicators: phase-noise robustness, baseband complexity, frequency localization, time localization, robustness to power amplifier nonlinearities, channel time selectivity and channel frequency selectivity.

Chapter 6 presents a flexible OFDM for 5G NR. Different factors involved in the implementation of OFDM-based NR modems are discussed, for example, quality of service requirements, type of deployment, carrier frequency, user mobility, hardware impairments, and implementation aspects. This chapter puts special focus on high carrier frequencies (e.g., millimeter-wave band), where robustness to hardware impairments (phase noise, synchronization errors) and power efficiency of the waveform is crucial.

Chapter 7 discusses the role of multiantenna techniques in 5G NR and the features included in the first release of the NR specifications. To provide an understanding and motivation of the features adopted by NR, the fundamental theory behind these features is provided. The viability of the multiantenna techniques presented is illustrated by several experimental examples.

Chapter 8 presents different channel coding schemes for 5G NR. The performance of the coding schemes is evaluated for different blocklength values. We review recently developed information-theoretic tools to benchmark the performance of these coding schemes. Looking beyond what is currently standardized in NR, we consider transmissions over multiantenna fading channels and highlight the importance of exploiting frequency and spatial diversity through the use of space-frequency codes, in applications that require high reliability.

Chapter 9 presents an open-source simulator that includes hardware impairment models (power amplifier, oscillator phase noise), a geometry-based stochastic channel model, and modulation/demodulation modules of state-of-the art waveforms. The chapter provides simulation exercises with various waveforms subject to different types of impairments.

REFERENCES

[1] 3GPP TS 22.261, 3rd generation partnership project: technical specification group services and system aspects; service requirements for the 5G systems; stage 1; (release 16), V16.2.0, 2017.

[2] E. Dahlman, G. Mildh, J. Peisa, J. Sachs, Y. Selén, S. Parkvall, 5G radio access, Ericsson Technology Review (2014, July), https://www.ericsson.com/assets/local/publications/ericsson-technology-review/docs/2014/er-5g-radio-access.pdf.

[3] ECC, Operational guidelines for spectrum sharing to support the implementation of the current ECC framework in the 3600–3800 MHz range, ECC Report 254, 2016.

[4] Ericsson, The 5G business potential, 2017, Oct.

[5] Ericsson, 5G deployment considerations, 2018.

[6] Ericsson White Paper, 5G systems, 2017, Jan.

[7] ITU-R, Recommendation ITU-R M.2083-0: IMT vision–framework and overall objectives of the future development of IMT for 2020 and beyond, 2015.

[8] ITU-R, Liaison statement to task group 5/1-spectrum needs and characteristics for the terrestrial component of IMT in the frequency range between 24.25 GHz and 86 GHz, ITU-R, WP 5D, Doc. TG5.1/36, 2017.

[9] ITU-R, Prediction of building entry loss, Recommendation ITU-R P.2109, 2017.

[10] ITU-R, Prediction of clutter loss, Recommendation ITU-R P.2108, 2018.

[11] J. Lee, E. Tejedor, K. Ranta-aho, H. Wang, K.T. Lee, E. Semaan, E. Mohyeldin, J. Song, C. Bergljung, S. Jung, Spectrum for 5G: global status, challenges, and enabling technologies, IEEE Communications Magazine 56 (3) (2018) 12–18.

[12] S. Parkvall, E. Dahlman, A. Furuskär, M. Frenne, NR: the new 5G radio access technology, IEEE Communications Standards Magazine (2017, Dec.) 24–30.
[13] O. Teyeb, G. Wikström, M. Stattin, T. Cheng, S. Fáxer, H. Do, Evolving LTE to fit the 5G future, Ericsson Technology Review (2017, Jan.), https://www.ericsson.com/assets/local/publications/ericsson-technology-review/docs/2017/etr_evolving_lte_to_fit_the_5g_future.pdf.
[14] Ericsson and China Mobile conduct world's first 5G drone prototype field trial, Ericsson, 2016, https://www.ericsson.com/en/news/2016/8/ericsson-and-china-mobile-conduct-worlds-first-5g-drone-prototype-field-trial-.
[15] Ericsson and Telstra conduct the first live 5G trial in Australia, Ericsson, 2016, https://www.ericsson.com/en/news/2016/9/ericsson-and-telstra-conduct-the-first-live-5g-trial-in-australia.
[16] KT demonstrates world's first 5G performing at 25 Gbps, Business Korea, 2016, http://www.businesskorea.co.kr/news/articleView.html?idxno=13896.
[17] 5G can make remote driving a reality, Telefónica and Ericsson demonstrate at MWC, Telefonica, 2017, https://www.telefonica.com/en/web/press-office/-/5g-can-make-remote-driving-a-reality-telefonica-and-ericsson-demostrate-at-mwc.
[18] 5G can make remote driving a reality, Telefónica and Ericsson demonstrate at MWC, Ericsson, 2017, https://www.ericsson.com/en/press-releases/latin-america/2017/2/5g-can-make-remote-driving-a-reality-telefonica-and-ericsson-demonstrate-at-mwc.
[19] 5G goes live in the Port of Tallinn, Ericsson, 2017, https://www.ericsson.com/en/news/2017/9/5g-goes-live-in-the-port-of-tallinn.
[20] The 5G race is on, Ericsson, 2017, https://www.ericsson.com/en/news/2017/5/ericsson-and-verizon-test-the-limits-of-5g.
[21] Ericsson and MTS test new 5G features, Ericsson, 2017, https://www.ericsson.com/en/press-releases/2017/4/ericsson-and-mts-test-new-5g-features.
[22] Ericsson, SK Telecom and BMW Group Korea reach new world record speed with 5G, Ericsson, 2017, https://www.ericsson.com/en/news/2017/2/ericsson-sk-telecom-and-bmw-group-korea-reach-new-world-record-speed-with-5g.
[23] MTN and Ericsson first in Africa to trial 5G technology, Ericsson, 2018, https://www.ericsson.com/en/news/2018/1/5g-in-south-africa.
[24] Ericsson Mobility Report, June 2018, https://www.ericsson.com/assets/local/mobility-report/documents/2018/ericsson-mobility-report-june-2018.pdf.

CHAPTER

NR PHYSICAL LAYER: OVERVIEW

2

Like for any wireless technology, the physical layer forms the backbone of 5G NR. The NR physical layer has to support a wide range of frequencies (from sub-1 GHz to 100 GHz) and various deployment options (pico cells, micro cells, macro cells). There are human-centric and machine-centric use cases with extreme and sometimes contradictory requirements. There may also be unforeseen applications with new requirements in the future. To successfully address these challenges, 3GPP is developing a flexible physical layer for NR. The flexible components can be properly optimized with an accurate understanding of radio wave propagation and hardware imperfections in networks and devices. This is a challenge, because these characteristics are less understood. NR is the first ever mobile radio access technology going into millimeter-wave frequency range (with frequencies as high as 100 GHz), targeting channel bandwidths in the GHz range, and enabling massive multiantenna systems.

The first release of NR (3GPP NR Release 15) was completed in June 2018. Any future releases of NR will be backwards compatible to its first release. This is usually referred to as forward compatibility of NR, that is, NR will be developed in such a manner that any future releases of NR will be backwards compatible to its initial release(s). For the reader interested in understanding what an NR physical layer is, in this chapter we provide an overview of the NR physical layer (based on the first NR release) and discuss the radio wave propagation and hardware impairment related challenges associated with enabling the NR physical layer. Our focus in the following chapters will be on fundamental principles, models, and technology components for the NR physical layer. If the reader is interested in the details of the NR specification, we strongly recommend [6].

The chapter is organized as follows. In Section 2.1, we briefly describe the protocol stack of the NR radio interface and the role of the physical layer therein. Due to brevity, this section may be hard to fully grasp if the reader is not familiar with 3GPP-based cellular technologies (e.g., 4G LTE [5]). The rest of the chapter focuses only on physical layer aspects. Section 2.2 gives an overview of the key physical layer technology components of NR—modulation, waveform, multiantenna, and channel coding schemes. These technology components are explained in detail in Chapters 5–8. Section 2.2 introduces the physical time-frequency resource structure of NR. Sections 2.4 and 2.5 describe how the time-frequency resources are allocated to different types of signals. Sections 2.6 and 2.7 explain flexible duplexing schemes and a flexible transmission structure (frame structure) available in NR. Finally, Section 2.9 briefly summarizes some challenges associated with radio wave propagation and hardware impairments. This section motivates Chapters 3 and 4, which, respectively, cover radio wave propagation and hardware impairments in depth.

2.1 RADIO PROTOCOL ARCHITECTURE

In 3GPP terminology, a base station is an implementation of a logical radio access network node. For example, in 3G UMTS and 4G LTE, the network node is termed Node B (NB) and evolved Node B

5G Physical Layer. https://doi.org/10.1016/B978-0-12-814578-4.00007-2

(eNB), respectively. A 5G NR radio access network node has been named the next generation Node B (gNB) by 3GPP. It is important to stress that gNB is a logical entity and not a physical implementation of a base station. A base station can be realized in different ways based on a standardized gNB protocol. Similarly, any device is referred to as a UE in 3GPP specifications.

The radio protocol architecture for NR can be separated into control-plane architecture and user-plane architecture. The user-plane delivers user data, whereas the control-plane is mainly responsible for connection setup, mobility, and security. Fig. 2.1 illustrates the user-plane protocol stack of NR. The protocol is split into the following layers: physical (PHY) layer, medium access control (MAC) layer, radio link control (RLC) layer, packet data convergence protocol (PDCP) layer, and service data adaptation protocol (SDAP) layer. The main functionalities of these layers are briefly described now.

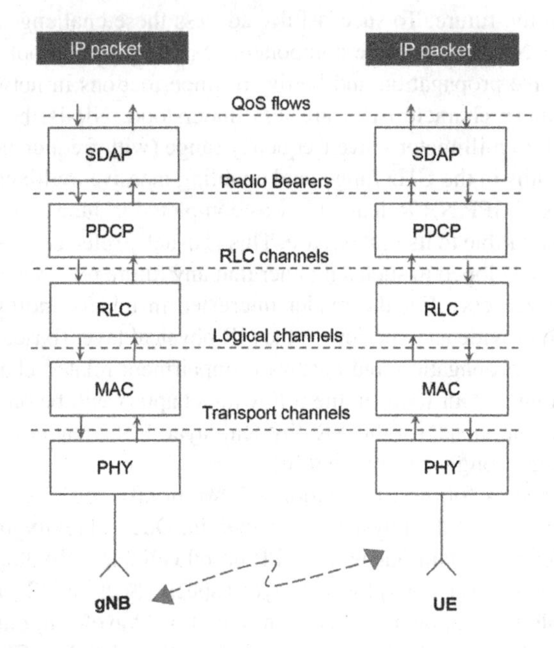

FIGURE 2.1

NR user-plane protocol stack.

- The SDAP layer handles the mapping between quality of service (QoS) flow and radio bearers.[1] IP packets are mapped to radio bearers according to their QoS requirements.
- The PDCP layer is primarily responsible for IP header compression/decompression, reordering and duplicate detection, ciphering/deciphering and integrity protection. The header compression mechanism reduces the number of bits to transmit over radio interface. The ciphering protects from

[1] A radio bearer can be seen as a pipe that carries internet protocol (IP) packets through a network and gets prioritized according to the specified QoS requirement.

eavesdropping and ensures message integrity. The reordering and duplication detection mechanisms allow in-sequence delivery of data units and removes duplicate data units.

- The RLC layer mainly performs error correction through an automatic repeat request[2] (ARQ) mechanism, segmentation/resegmentation of (header compressed) IP packets, and in-sequence delivery of data units to higher layers.
- The MAC layer is mainly responsible for error correction through hybrid ARQ[3] (HARQ) mechanism and uplink and downlink scheduling. The scheduler controls the assignment of uplink and downlink physical time-frequency resources for transmission. The MAC layer also takes care of multiplexing data across multiple component carriers when carrier aggregation is employed.
- The PHY layer handles coding/decoding, modulation/demodulation, multiantenna processing, and mapping of signals to physical time-frequency resources.

The control-plane is mainly responsible for control signaling for connection setup, mobility, and security. Control signaling originates either from a core network or from a radio resource control (RRC) layer in gNB. The main services from a RRC layer include broadcasting of system information, transmission of paging messages, security management including key management, handovers, cell-selection/reselection, QoS management, and detection of and recovery from radio link failures. The RRC messages are transmitted using the same PDCP, RLC, MAC, and PHY layers as for the user-plane. Therefore, from a physical layer perspective, there is not at all a fundamental technological difference in providing services to higher layers in control-plane and user-plane protocol stacks.

2.2 NR PHY: KEY TECHNOLOGY COMPONENTS

The key technology components of the NR physical layer are modulation, waveform, multiantenna transmission, and channel coding. In the following, we provide a brief overview of these physical layer components.

2.2.1 MODULATION

NR supports quadrature phase shift keying (QPSK), 16 quadrature amplitude modulation (QAM), 64 QAM and 256 QAM modulation formats for both uplink and downlink, as in LTE. Moreover, $\pi/2$-BPSK is supported in uplink to enable a further reduced peak-to-average power ratio and enhanced power amplifier efficiency at lower data rates, which is important for mMTC services. Since NR will cover a wide range of use cases, it is likely that the set of supported modulation schemes may be expanded. For example, 1024 QAM may become part of the NR specification, since fixed point-to-point backhaul already uses modulation orders higher than 256 QAM. Different modulation schemes for different UE categories may also be included in the NR specification.

[2] Automatic repeat request (ARQ) is an error-control method for data transmission that uses acknowledgments and timeouts to achieve reliable data transmission.

[3] Hybrid ARQ is a combination of forward error-correction coding and ARQ error-control.

Table 2.1 Scalable OFDM numerology for 5G NR (3GPP Release 15)

OFDM numerology	15 kHz	30 kHz	60 kHz	120 kHz
Frequency band	0.45–6 GHz	0.45–6 GHz	0.45–6 GHz 24–52.6 GHz	24–52.6 GHz
OFDM symbol duration	66.67 μs	33.33 μs	16.67 μs	8.33 μs
Cyclic prefix duration	4.69 μs	2.34 μs	1.17 μs	0.59 μs
OFDM symbol with CP	71.35 μs	35.68 μs	17.84 μs	8.91 μs
Maximum bandwidth	50 MHz	100 MHz	200 MHz	400 MHz

2.2.2 WAVEFORM

NR employs cyclic prefix OFDM (CP-OFDM) in both uplink and downlink up to at least 52.6 GHz. This is in contrast to LTE, where CP-OFDM is only used for downlink transmissions and DFT-Spread OFDM (DFTS-OFDM) is used for uplink transmissions. Having the same waveform in both directions simplifies the overall design, especially with respect to wireless backhauling and device-to-device (D2D) communications. Additionally, there is an option for DFT-spread OFDM (DFTS-OFDM) in uplink for coverage-limited scenarios, with single stream transmissions (that is, without spatial multiplexing). In practice, a gNB can select the uplink waveform (either CP-OFDM or DFTS-OFDM) and a UE should be able to support both OFDM and DFTS-OFDM. Any operation that is transparent to a receiver can be applied on top of the NR waveform, such as windowing/filtering to improve spectrum confinement.

NR has a scalable OFDM numerology to enable diverse services on a wide range of frequencies and deployments. The subcarrier spacing is scalable and specified as 15×2^n kHz, where n is an integer and 15 kHz is the subcarrier spacing used in LTE. In Chapter 6, we provide details on numerology design for NR. In 3GPP Release 15, four subcarrier spacings are specified: 15 kHz, 30 kHz, 60 kHz, and 120 kHz (i.e., $n = 1/2/3/4$), all with 7% CP overhead as in LTE (see Table 2.1). For the 60 kHz numerology, an extended CP is also defined. Different numerologies are specified for different frequency bands. Currently, there is no spectrum identified for NR between 6 and 24 GHz; therefore, the corresponding numerology has not been specified. When new frequencies will become available, the corresponding numerologies can be specified according to 15×2^n kHz. For all numerologies, the number of active subcarriers is 3300. Considering 3300 active subcarriers, the maximum bandwidths enabled by different numerologies are given in Table 2.1. To support even larger channel bandwidths, carrier aggregation can be employed. In Release 15, up to 16 component carriers are supported, where each component carrier can have up to 3300 active subcarriers.

The spectrum of an OFDM signal decays rather slowly outside the transmission bandwidth. In order to limit out-of-band emission, the spectrum utilization for LTE is 90 percent. For NR, it has been agreed that the spectrum utilization will be 94 to 99 percent. Windowing and filtering operations are viable ways to confine the OFDM signal in the frequency domain. In Chapters 5 and 6, we provide details on the spectrum confinement techniques for multicarrier waveforms.

2.2.3 MULTIPLE ANTENNAS

Multiantenna techniques were important already in LTE, but in NR they have a more fundamental role in the system design. The extension of the spectrum for mobile communication to include also the millimeter-wave bands has led to a beam-centric design of NR in order to support analog beam-forming for achieving sufficient coverage. Furthermore, multiantenna techniques are crucial for fulfilling the performance requirements for 5G also for the traditional cellular frequency bands.

For low frequencies, multiantenna techniques are mainly enhancements of features developed in the later releases of LTE. The purpose with these enhancements is improved spectral efficiency driven by the ever-increasing quest for higher data rates and capacity in a congested spectrum. Advances in active array antenna technology have made it possible to have digital control over a large number of antenna elements, sometimes referred to as massive multiple-input multiple-output (MIMO). This enables higher spatial resolution in the multiantenna processing which can give higher spectral efficiencies. To this end, NR provides better support for multiuser MIMO (MU-MIMO) and reciprocity-based operation. A new framework for acquiring channel state information (CSI) has been developed to allow for more flexibility in the transmission of reference signals and to enable CSI with higher spatial resolution. This framework also provides a leaner system design and makes it easier to adapt to diverse use cases and to introduce new features in future releases of NR.

For high frequencies, obtaining coverage is the main challenge rather than obtaining high spectral efficiency. The reason for this is that transmission losses, when using legacy transmission techniques, are considerably higher, while there is a large amount of bandwidth available in the millimeter-wave spectrum. To overcome higher transmission losses and provide sufficient coverage, beam-forming is useful, particularly under LoS conditions and possibly both at the gNB and UE. With current hardware technology, analog beam-forming is expected to be prevalent at millimeter-wave frequencies. Therefore, procedures for supporting analog beam-forming in both the gNB and the UE have been developed in NR. Unlike previous generations of mobile communication systems, NR supports beam-forming not only for the data transmission but also for initial access and broadcast signals.

NR multiantenna techniques for both the gNB and the UE are discussed in detail in Chapter 7.

2.2.4 CHANNEL CODING

NR employs low density parity check (LDPC) codes for the data transmission for mobile broadband (MBB) services and polar codes for the control signaling. LDPC codes are attractive from an implementation perspective, especially at multigigabits-per-second data rates. Unlike the LDPC codes implemented in other wireless technologies, the LDPC codes considered for NR use a rate-compatible structure. This allows for transmission at different code rates and for HARQ operation using an incremental redundancy.

For the physical layer control signaling where the information blocks are small compared to data transmission and HARQ is not used, NR employs polar codes. By concatenating the polar code with an outer code and by performing successive cancellation list decoding, good performance is achieved at shorter block lengths. For the smallest control payloads, Reed–Muller codes are used.

NR LDPC and polar codes are discussed in detail in Chapter 8. For URLLC services, channel codes have not been agreed in 3GPP yet. In Chapter 8, a selection of coding schemes that exhibit a favorable performance/complexity tradeoff in the short block length regime is discussed. These schemes may be candidates for the future NR releases in addition to LDPC and polar codes.

2.3 PHYSICAL TIME-FREQUENCY RESOURCES

Physical time-frequency resources correspond to OFDM symbols and subcarriers within the OFDM symbols. The smallest physical time-frequency resource consists of one subcarrier in one OFDM symbol, known as a resource element. The transmissions are scheduled in group(s) of 12 subcarriers, known as physical resource blocks (PRBs). An example of NR physical time-resource structure is shown in Fig. 2.2.

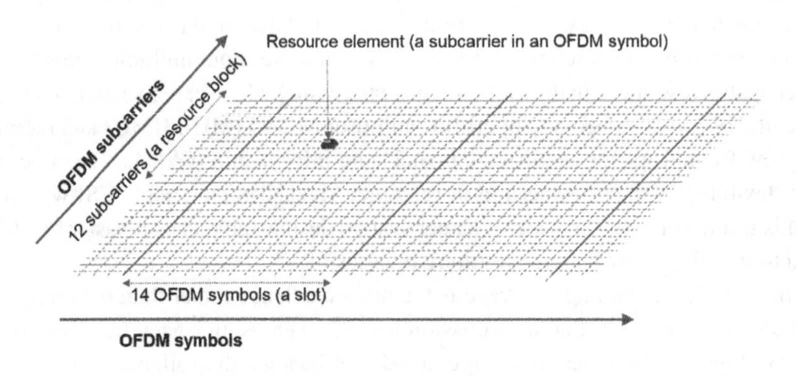

FIGURE 2.2

NR physical time-frequency structure.

In the time domain, the radio transmissions are organized into *radio frames*, *subframes*, *slots*, and *mini-slots*. As illustrated in Fig. 2.3, each radio frame has a duration of 10 ms and consists of 10 subframes with a subframe duration of 1 ms. A subframe is formed by one or multiple adjacent slots, each slot having 14 adjacent OFDM symbols. A mini-slot, in principle, can be as short as one OFDM symbol, but in Release 15 mini-slots are restricted to 2, 4, and 7 OFDM symbols. The time duration of a slot/mini-slot scales with the chosen numerology (subcarrier spacing) since the duration of an OFDM symbol is inversely proportional to its subcarrier spacing.

Physical layer uses time-frequency resources for transmission. In NR, the time-frequency resources (resource elements) represent either *physical channels* or *physical signals*, as in LTE. In the 3GPP terminology, a physical channel corresponds to a set of resource elements carrying information originating from higher layers, whereas a physical channel corresponds to a set of resource elements (used by physical layer) that do not contain information from higher layers. In the following sections we discuss both.

2.4 PHYSICAL CHANNELS

The time-frequency resources carrying information from higher layers (layers above PHY) are termed physical channels [1]. There are a number of physical channels specified for uplink and downlink:

- Physical downlink shared channel (PDSCH), used for downlink data transmission.

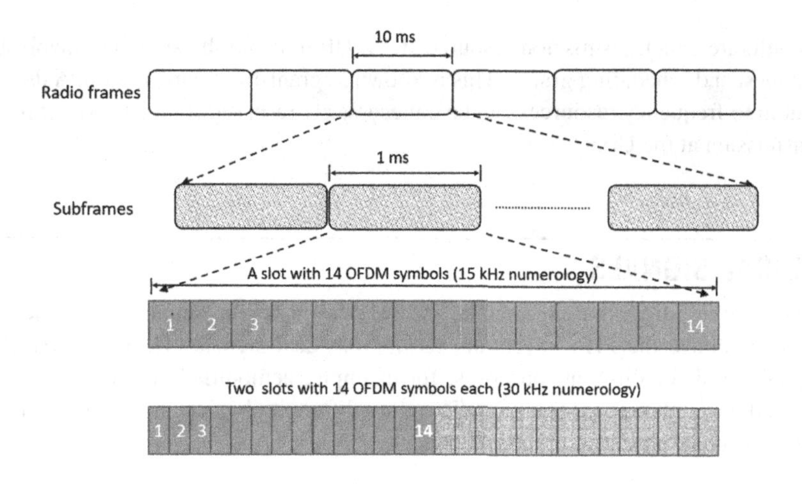

FIGURE 2.3

NR frame structure.

- Physical downlink control channel (PDCCH), used for downlink control information, which includes scheduling decisions required for downlink data (PDSCH) reception and for scheduling grants giving permission for uplink data (PUSCH) transmission by a UE.
- Physical broadcast channel (PBCH), used for broadcasting system information required by a UE to access the network.
- Physical uplink shared channel (PUSCH), used for uplink data transmission (by a UE).
- Physical uplink control channel (PUCCH), used for uplink control information, which includes: HARQ feedback acknowledgments (indicating whether a downlink transmission was successful or not), scheduling request (requesting time-frequency resources from network for uplink transmissions), and downlink channel-state information for link adaptation.
- Physical random access channel (PRACH), used by a UE to request connection setup referred to as random access.

On a high level, the downlink and uplink transmissions between gNB and UE work as follows. In downlink, a UE monitors the PDCCH, typically once per slot (PDCCH can also be configured more than once to enable ultralow-latency transmissions). Upon detection of a valid PDCCH, the UE receives one unit of data (named a transport block) on the PDSCH following the scheduling decision of gNB. Afterwards, the UE responds with a hybrid ARQ acknowledgment indicating whether the data was successfully decoded or not. In the case of unsuccessful decoding, retransmission(s) is scheduled. For uplink data transmission, a UE first requests gNB for physical time-frequency resources for the data awaiting transmission. This is termed a scheduling request and is sent over the PUCCH. In response, gNB sends a scheduling grant (over the PDCCH) which gives permission to a UE to use certain time-frequency resources for transmission. Following the scheduling grant, the UE schedules its data transmission over the PUSCH. gNB receives uplink data and sends hybrid ARQ acknowledgment indicating whether the uplink data transmission was successfully decoded or not. In the case of a decoding failure, retransmission(s) is scheduled. To enable ultralow-latency communication, the network can also preconfigure data transmission resources for a UE to avoid the signaling involved in sending scheduling request and scheduling grants. This is known as grant-free transmission. A drawback of this scheme is that time-frequency resources can be unnecessarily reserved for a UE even if there is no data awaiting transmission at the UE.

2.5 PHYSICAL SIGNALS

The time-frequency resources that are used by the PHY layer but do contain information from higher layers (i.e., layers above the PHY layer) are termed physical signals [1]. The physical signals are reference signals used for different purposes, for example, demodulation, channel estimation, synchronization, and channel-state information. There are different physical signals in the uplink and the downlink. The downlink physical signals include:

- demodulation reference signal (DM-RS)
- phase tracking reference signal (PT-RS)
- channel state information reference signal (CSI-RS)
- primary synchronization signal (PSS)
- secondary synchronization signal (SSS)

Furthermore, the following uplink physical signals are defined:

- demodulation reference signal (DM-RS)
- phase tracking reference signal (PT-RS)
- sounding reference signal (SRS)

NR has an ultralean design, which minimizes always-on transmissions to enhance network energy efficiency, reduce interference, and ensure forward compatibility. In contrast to the setup in LTE, the reference signals in NR are transmitted only when necessary. Next, we briefly discuss four main reference signals: DM-RS, PT-RS, CSI-RS, and SRS.

DM-RS is used to estimate the radio channel for demodulation. DM-RS is UE-specific, can be beam-formed, confined in a scheduled resource, and transmitted only when necessary, both in the downlink and the uplink. The DM-RS design takes into account the early decoding requirement to support low-latency applications. For this reason, the DM-RS is placed in the beginning of a slot (known as a front-loaded DM-RS). For low-speed scenarios, DM-RS uses low density in the time domain (i.e., fewer OFDM symbols in a slot contain DM-RS). For high-speed scenarios, the time density of DM-RS is increased to track fast changes in the radio channel.

PT-RS is introduced in NR to enable compensation of the oscillator phase noise. Typically, phase noise increases as a function of the oscillator carrier frequency. PT-RS can therefore be utilized at high carrier frequencies (such as by millimeter wave) to mitigate phase noise. One of the main degradations caused by phase noise in an OFDM signal is an identical phase rotation of all the subcarriers, known as common phase error (CPE). (This is discussed in detail in Chapter 6 and Chapter 7.) PT-RS is designed so that it is sparse in the frequency domain and has a high density in the time domain and the reason is as follows. The phase rotation produced by CPE is identical for all subcarriers within an OFDM symbol, but there is low correlation of phase noise across OFDM symbols. The density of PT-RS in frequency domain is one subcarrier in every PRB, every second PRB, or every fourth PRB. The density in time domain is every OFDM symbol, every second OFDM symbol, or every fourth OFDM symbol. Fig. 2.4 shows an example of DM-RS and PT-RS time-frequency structure. Like DM-RS, PT-RS is also UE-specific, it is confined in a scheduled resource, and it can be beam-formed. PT-RS is configurable depending on the quality of the oscillators, the carrier frequency, the OFDM subcarrier spacing, and on the modulation and coding schemes used for transmission.

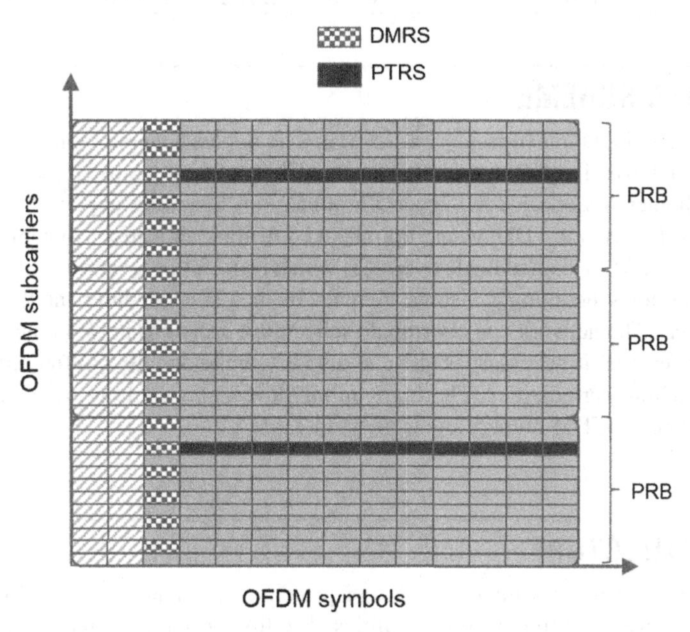

FIGURE 2.4

An example of time-frequency structure DM-RS and PT-RS.

CSI-RS is a downlink reference signal which is used mainly for CSI acquisition, beam management, time/frequency tracking and uplink power control. It has a very flexible design in order to support diverse use cases. CSI-RS for CSI acquisition is used for determining CSI parameters such as channel quality indicator (CQI), rank indicator (RI), and precoding matrix indicator (PMI), which are needed for link adaptation and for determining precoders. Furthermore, a so-called CSI interference measurement (CSI-IM) resource, which is a zero-power CSI-RS (ZP CSI-RS) resource, can be configured for interference measurements in the UE. For beam management, CSI-RS is used for evaluating candidate transmission beams by measuring reference signal received power (RSRP) for each beam. It can also be used for beam recovery purposes. A CSI-RS configured for time/frequency tracking is called a tracking reference signal (TRS). TRS can be used for fine time and frequency synchronization, and Doppler and delay spread estimation. This is needed for channel estimation and demodulation.

The SRS is transmitted in uplink to perform CSI measurements mainly for scheduling and link adaptation. For NR, it is expected that the SRS will also be utilized for reciprocity-based precoder design for massive MIMO and uplink beam management. SRS has a modular and flexible design to support different procedures and UE capabilities.

Table 2.2 Semi-static TDD configurations

OFDM numerology	Uplink/downlink switching periodicities [ms]
15 kHz	0.5, 1, 2, 5, 10
30 kHz	0.5, 1, 2, 2.5, 5, 10
60 kHz	0.5, 1, 1.25, 2, 2.5, 5, 10
120 kHz	0.5, 0.625, 1, 1.25, 2, 2.5, 5, 10

2.6 DUPLEXING SCHEME

NR supports TDD and FDD transmissions, as in LTE. The duplex scheme typically depends on spectrum allocation. At lower frequencies, the spectrum allocations are mostly paired, implying FDD transmission. At higher frequencies the spectrum allocations are often unpaired, implying TDD. In addition, NR supports dynamic TDD, where uplink and downlink allocations dynamically change over time. This is one of the key enhancements over LTE which is useful in scenarios with rapid traffic variations. The transmission scheduling decisions are made by the gNB scheduler and the UEs follow these scheduling decisions. The network can coordinate scheduling decisions between neighboring network sites to avoid interference, if necessary. TDD can also be semi-statically configured with certain uplink/downlink switching periodicities. There are uplink/downlink switching periodicities specified in Release 15 for different OFDM numerologies given in Table 2.2.

2.7 FRAME STRUCTURE

The NR frame structure follows three key design principles to enhance forward compatibility and reduce interactions between different functionalities. The first principle is that transmissions are self-contained. Data in a slot and in a beam is decodable on its own without dependency on other slots and beams. This implies that reference signals required for the demodulation of data are included in a given slot and a given beam. The second principle is that transmissions are well confined in time and frequency. Keeping transmissions confined makes it easier to introduce new types of transmissions in parallel with legacy transmissions in the future. The NR frame structure avoids the mapping of control channels across the full system bandwidth. The third principle is to avoid static and/or strict timing relations across slots and across different transmission directions. For example, asynchronous HARQ is used instead of a predefined retransmission time.

The NR frame structure supports TDD and FDD transmissions and operation in both the licensed and the unlicensed spectrum. It enables very low latency, fast HARQ acknowledgments, dynamic TDD, coexistence with LTE and transmissions of variable length (for example, short duration for URLLC and long duration for enhanced MBB (eMBB)). Considering the TDD operation, Fig. 2.5 provides examples of NR frame structure for different scenarios.

NR can also employ mini-slots to support transmissions with a flexible start position and a duration shorter than a regular slot duration. In principle, a mini-slot can be as short as one OFDM symbol and can start at any time. In Release 15, mini-slots are limited to 2, 4, and 7 OFDM symbols. Mini-slots can be useful in various scenarios, including low-latency transmissions, transmissions in unlicensed spectrum and transmissions in the millimeter-wave spectrum. In low-latency scenarios, transmission needs

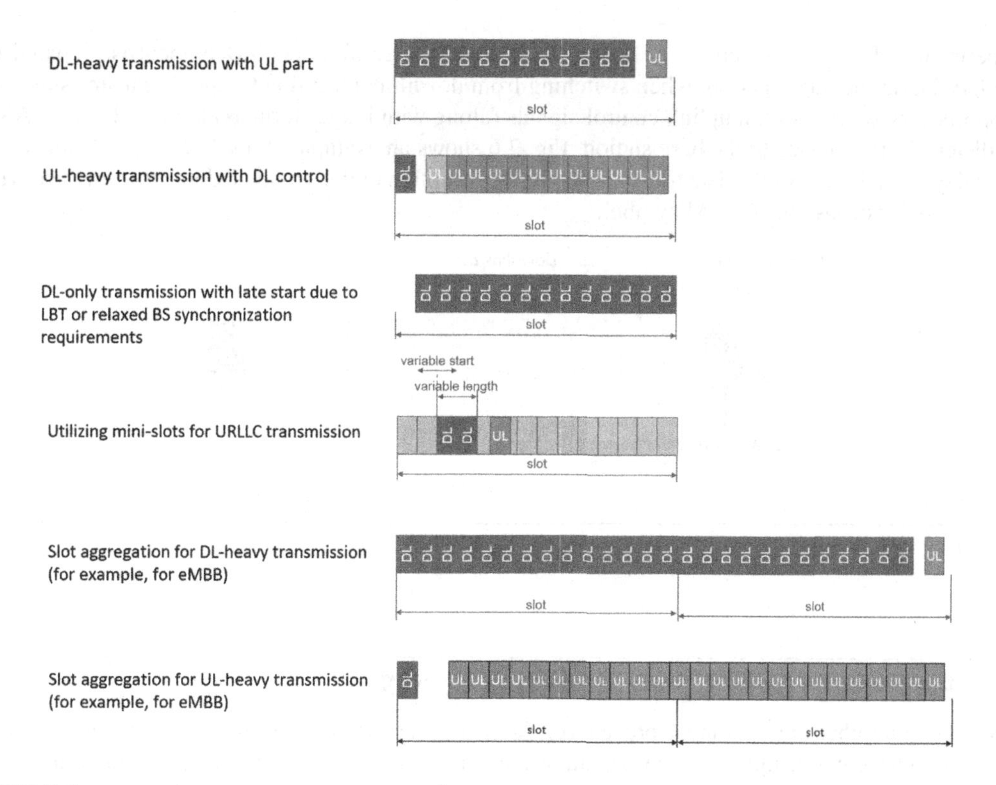

FIGURE 2.5

Examples of NR physical layer frame structure.

to begin immediately without waiting for the start of a slot boundary (ultra-reliable low-latency communications (URLLC), for example). When transmitting in the unlicensed spectrum, it is beneficial to start transmission immediately after the listen-before-talk (LBT) mechanism. When transmitting in the millimeter-wave band, the large amount of bandwidth available implies that the payload supported by a few OFDM symbols is large enough for many of the packets. Fig. 2.5 provides examples of URLLC- and LBT-based transmission in unlicensed spectrum via mini-slots and illustrates that multiple slots can be aggregated for services that do not require extremely low latency (eMBB, for example). Having a longer transmission duration helps to increase coverage or reduce the overhead due to switching (in TDD), transmission of reference signals, and control information.

The same frame structure can be used for FDD, by enabling simultaneous reception and transmission (that is, downlink and uplink can overlap in time). This frame structure is also applicable to device-to-device (D2D) communications. In that case, the downlink slot structure can be used by the device that is initiating (or scheduling) the transmission, and the uplink slot structure can be used by the device responding to the transmission.

To obtain low latency, a slot (or a set of slots in the case of slot aggregation) is front loaded with control signals and reference signals at the beginning of the slot (or set of slots), as illustrated in Fig. 2.6. An NR frame structure also allows for rapid HARQ acknowledgment, in which decoding is performed during the reception of downlink data and the HARQ acknowledgment is prepared by the UE during the guard period, when switching from downlink reception to uplink transmission. NR supports very short duration uplink control signals (along with longer formats) to provide fast HARQ feedbacks from a device to the base station. Fig. 2.6 shows an example of a self-contained slot where the delay from the end of the data transmission to the reception of the acknowledgment from the device is on the order of just one OFDM symbol.

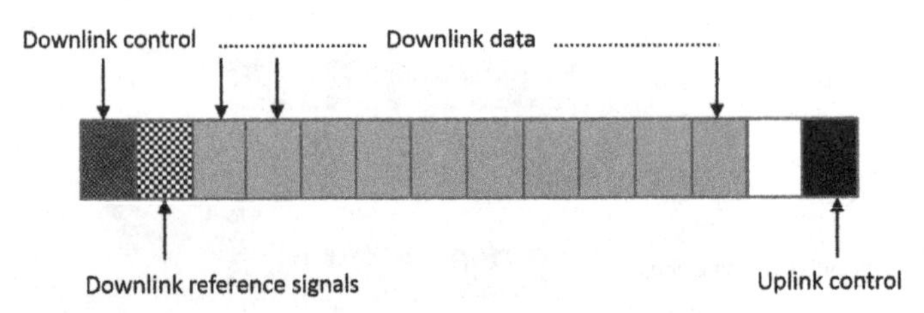

FIGURE 2.6

A self-contained slot.

2.8 PHY PROCEDURES AND MEASUREMENTS

There are a number physical layer procedures defined in the NR specifications, for example for cell search, power control, uplink synchronization and timing control, random access, and beam management, and there are channel-state information related procedures [2,3]. The NR specification also defines physical layer measurements for intra- and inter-frequency handover, inter RAT handover, timing measurements, and measurements for radio resource management [4]. These details (apart from the beam management procedure) are out of scope of this book. We strongly recommend [6] for understanding these details.

2.9 PHYSICAL LAYER CHALLENGES

5G NR is the first cellular technology to operate at millimeter-wave frequencies, support GHz of bandwidths, and utilize a massive number of antennas. These aspects impose a number of challenges for operation of NR physical layer mainly due to less understood radio wave propagation characteristics and hardware impairments (at the base stations and the devices). To enable a high performing NR, it is important to accurately understand the characteristics of radio propagation and hardware impairments. Chapters 3 and 4 address radio wave propagation, channel modeling, and hardware impairment modeling in great detail. In the following, we provide a brief discussion of some of these challenges.

2.9.1 PROPAGATION RELATED CHALLENGES

When going up in frequency the antenna related transmission loss increases as the square of the frequency, for any fixed receive antenna pattern, because the aperture is proportional to the square of the wavelength. Some additional increase in loss due to propagation effects is likely, particularly under NLoS conditions. The challenge here is to utilize advanced multiantenna techniques to steer the signal into favorable directions and to increase the receive antenna aperture. At this point it is not fully clear to what extent the channel conditions may be improved at higher frequencies by means of these antenna techniques. In some scenarios, like LoS, it may even be favorable to go up in frequency. It is clear that the outdoor-to-indoor scenario will be challenging as the penetration loss goes up substantially with frequency. Another effect of using beam-forming and narrow beams is that the channel's dynamic changes will be larger and faster due to sudden blockage of the beam. Furthermore, the directional spread characteristics of the propagation channel is not well known. For channels which are highly dispersive in direction (rich scattering) high gain antennas are not very useful as they will only catch/direct a small fraction of the signal from the transmitter/to the receiver. In this case fully coherent antenna combining techniques are more suitable. However, such techniques are very complex and resource costly when the array antenna sizes are large in terms of the number of elements.

The main challenge for NR is to what extent novel multiantenna techniques may compensate for loss of, or even gain in, performance at higher frequencies comparing with legacy techniques used at lower frequencies. Under LoS conditions it is favorable to use higher frequencies and beam-forming as shown in Chapter 3. Moreover, the propagation losses in NLoS do not increase dramatically with frequency (in the range $0 \log f$ to $6 \log f$) except for outdoor-to-indoor transmissions. However, as early deployments of NR are expected to largely rely on analog beam-forming, the propagation channel is required to be highly directive. There are unfortunately only few highly directionally resolved channel measurements, implying that the possible beam-forming performance at millimeter-wave frequencies is largely unknown.

2.9.2 HARDWARE RELATED CHALLENGES

As regards the demands for increasingly high data rates moving into NR, the need for spectral efficiency and more bandwidth is rapidly increasing. The technology components used to address this increase involve advanced multiantenna techniques such as massive multi user (MU)-MIMO or, moving into the millimeter-wave regime, analog beam-forming. This raises new challenges in terms of efficient radio implementation as both the number of deployed transceivers and their operating frequencies and bandwidths increase.

As one of the important building-blocks in radio transceiver front-ends, the RF power amplifier (PA) continues to play a critical role as a major consumer of power. Operating in dense and highly integrated antenna arrays, the PA may also suffer the effects of mutual coupling unless sufficient isolation is implemented. This introduces another source of distortion caused by load-modulation as the active impedance presented to the amplifier changes over time. Another critical issue in antenna systems using directive transmission is assessing the distortion behavior in terms of how it is distributed over space. With conventional methods, the distortion follows the same pattern as the desired signal, but using MU-MIMO precoding or beam-forming techniques, this is generally not the case.

RF oscillators are another source of imperfections in the analog front-end, which, when moving toward the millimeter-wave bands, becomes a limiting factor. Maintaining a stable oscillation becomes more difficult at very high frequencies as the increasing losses decrease the quality factor of the resonator-tanks and there is a lack of power generation due to fundamental limitations in transistor technology. Thus, as the phase noise increases, coherent transmission may become increasingly difficult. Thus, phase noise needs to be accurately modeled in order to assess its impact and to provide a basis for countermeasures such as tracking. Phase-locked loop (PLL) techniques may be used in order to decrease the narrow-band $1/f$ part of the phase noise, which may have more significant negative effects on OFDM-based transmission, but this comes at the cost of increased white noise levels.

As advanced signal processing hardware and algorithms make it possible to compensate for some of the more prominent impairments, data-converters such analog-to-digital converter (ADC) and digital-to-analog converter (DAC) are crucial components connecting the analog and digital domains. Unfortunately, data-converters do not enjoy the same beneficial scaling as described by Moore's law, as these are mixed signal components and generally require linear transistor technologies. Scaling down the geometries to increase the available processing speed is not necessarily a good option in this case, as it makes the transistors operate more as switches. As shown in the literature, massive MU-MIMO opens up possibilities to decrease the effective resolution. This will, however, inflict an increasing amount of quantization noise which, if left unchecked, may corrupt the signal to a large extent. In order to assess the impact of a coarser quantization, different modeling techniques based on either deterministic behavioral models or stochastic processes may aid us.

Overall, the need for improved mathematical tools capable of assessing the behavior of non-ideal radio components over time/frequency and space is larger than ever. Specifically this is so since these models may serve as a basis for advanced compensation techniques such as digital pre-distortion (DPD) or phase-noise tracking. Current developments in this field have further opened up possibilities of new analysis and understanding of how radio imperfections behave in the context of a large antenna array. Another relevant field of modeling which currently is in its cradle is that of stochastic modeling of radio impairments aimed to aid performance analysis on link- or system-level, as the kind of oversampled passband or baseband data most behavioral models are built upon may not always be available. This modeling framework will be further discussed in Chapter 4.

REFERENCES

[1] 3GPP, TS 38.211 NR Physical channels and modulation, June 2018.
[2] 3GPP, TS 38.213 NR Physical layer procedures for control, June 2018.
[3] 3GPP, TS 38.214 NR Physical layer procedures for data, June 2018.
[4] 3GPP, TS 38.215 NR Physical layer measurements, June 2018.
[5] E. Dahlman, S. Parkvall, J. Sköld, 4G LTE-Advanced Pro and The Road to 5G, Academic Press, 2016.
[6] E. Dahlman, S. Parkvall, J. Sköld, 5G NR: The Next Generation Wireless Access Technology, Academic Press, 2018.

CHAPTER

PROPAGATION & CHANNEL MODELING

3

Radiowave propagation knowledge is crucial for development of wireless communications. The reason is that propagation represents the two most important respects in which mobile and fixed telecommunications differ.

The first is the physical media in which the transmitted electromagnetic waves propagate. In the fixed case the medium is mainly transmission lines, for example optical fibers and coaxial copper cables. The media in the case of mobile communications is in contrast the complete environment in which the transmitter and receiver are embedded. As denser materials both absorb and reflect a substantial fraction of the energy of radio waves, the main medium for propagation is the free space between those materials. Though it is possible to steer the transmitted signal to the desired receiver by utilizing advanced antenna techniques, the propagation environment does not allow for full isolation between links, wherefore signals transmitted from different sources are normally mixed at the receivers, causing substantial interference.

The second respect in which mobile communications are different from the fixed case is that the channel may vary heavily in time and space. Good knowledge of the characteristics of these variations is essential when optimizing transmission techniques as well as when dimensioning and planning networks.

This chapter provides a comprehensive overview of the impact of the radio propagation channel on mobile communications, explaining some key aspects in detail. It is organized as follows. In Section 3.1 the fundamentals of radiowave propagation are explained based on electromagnetic theory. Section 3.2 presents commonly used quantities for the characterization of the propagation channel, whereas corresponding experimental results are exemplified in Section 3.3. Recent 5G channel modeling provided by mainly 3GPP and ITU-R is presented in Section 3.4. Focus is put on understanding the models and the extent to which the modeling is accurate and realistic. Moreover, model components that need further improvements are pointed out. Finally, the chapter is summarized and future work is proposed in Section 3.5.

3.1 PROPAGATION FUNDAMENTALS

This section deals with some fundamental aspects of propagation that are useful for understanding characterization and modeling of the radiowave channel in a mobile communications context. For a comprehensive description of electromagnetic theory the reader is referred to standard textbooks on the topic such as [3], [13]. Propagation concerns the radiowave pathway between the transmit antenna and the receive antenna. Sometimes the antenna is mixed up with propagation effects in propagation modeling, which is a problem which is addressed in Section 3.1.2. Antennas may, however, not

5G Physical Layer. https://doi.org/10.1016/B978-0-12-814578-4.00008-4

be avoided in propagation research where they primarily are used for measuring the electromagnetic fields at specific space points or in specific directions of propagation. This section sheds light on some of the fundamental aspects of radiowave propagation. The first subsection explains what electromagnetic waves are. The second subsection deals with free-space propagation and the third with the basic propagation mechanisms.

3.1.1 ELECTROMAGNETIC WAVES

In the most fundamental respect electromagnetic waves are defined as the set of solutions of Maxwell's equations in space regions free from charges. This set of solutions may be represented by different series expansions. We will here focus on the plane wave expansion, which is very useful for describing propagation in a comprehensible way. Moreover, plane waves are the most common representation used in standard channel models. For plane waves, the electric **E** and magnetic **H** fields are equal and orthogonal to each other and to the Poynting vector,

$$\mathbf{S} = \mathbf{E} \times \mathbf{H}^*, \tag{3.1}$$

which points in the direction of propagation. Defining an orthonormal reference system $(\mathbf{e}_1, \mathbf{e}_2, \mathbf{e}_3)$ the electric field of the plane wave is given by

$$\mathbf{E}(\mathbf{x}, t) = \mathrm{Re}\left[(\mathbf{e}_1 E_1 + \mathbf{e}_2 E_2) e^{i\mathbf{k}\cdot\mathbf{x} - i\omega t}\right] \tag{3.2}$$

where $\mathbf{x}$ is the space coordinate, t is the time, $\mathbf{k} = \mathbf{e}_3 2\pi/\lambda$ where λ is the wavelength, and ω is the angle frequency of the wave. The corresponding magnetic field is given by

$$\mathbf{H} = \mathbf{e}_3 \times \mathbf{E}\sqrt{\frac{\epsilon}{\mu}} \tag{3.3}$$

where ϵ and μ are the permittivity and permeability, respectively. Fig. 3.1 shows three cases of plane waves defined by the following electric fields:

$$\left.\begin{array}{lll} 1) & E_1 = 0, & E_2 = 1 \\ 2) & E_1 = \frac{1}{\sqrt{2}}, & E_2 = \frac{i}{\sqrt{2}} \\ 3) & E_1 = \frac{1}{\sqrt{4}}\left(e^{-i\pi/6} - 1\right), & E_2 = \frac{1}{\sqrt{4}}\left(e^{-i\pi/6} + 1\right) \end{array}\right\} \tag{3.4}$$

where 1) corresponds to linear polarization, 2) to circular polarization and 3) to elliptic polarization. It is clearly seen that the **E** and the **H** fields are orthogonal and equal in strength at each time instance. A plane wave is defined by its six corresponding degrees of freedom: 1) the polarization elliptic axial ratio, 2) the polarization ellipse main axis rotation angle, 3) the field amplitude, 4) the wave phase, 5) the Poynting vector polar angle, and 6) the Poynting vector azimuth angle. These degrees of freedom may also be represented by the electric and the magnetic vector fields at a space point. In this case the electromagnetic wave field is viewed as a sum of plane waves which results in that the magnetic and the electric fields are decorrelated. For any wave vector **k**, however, a plane wave supports only two orthogonal polarization states.

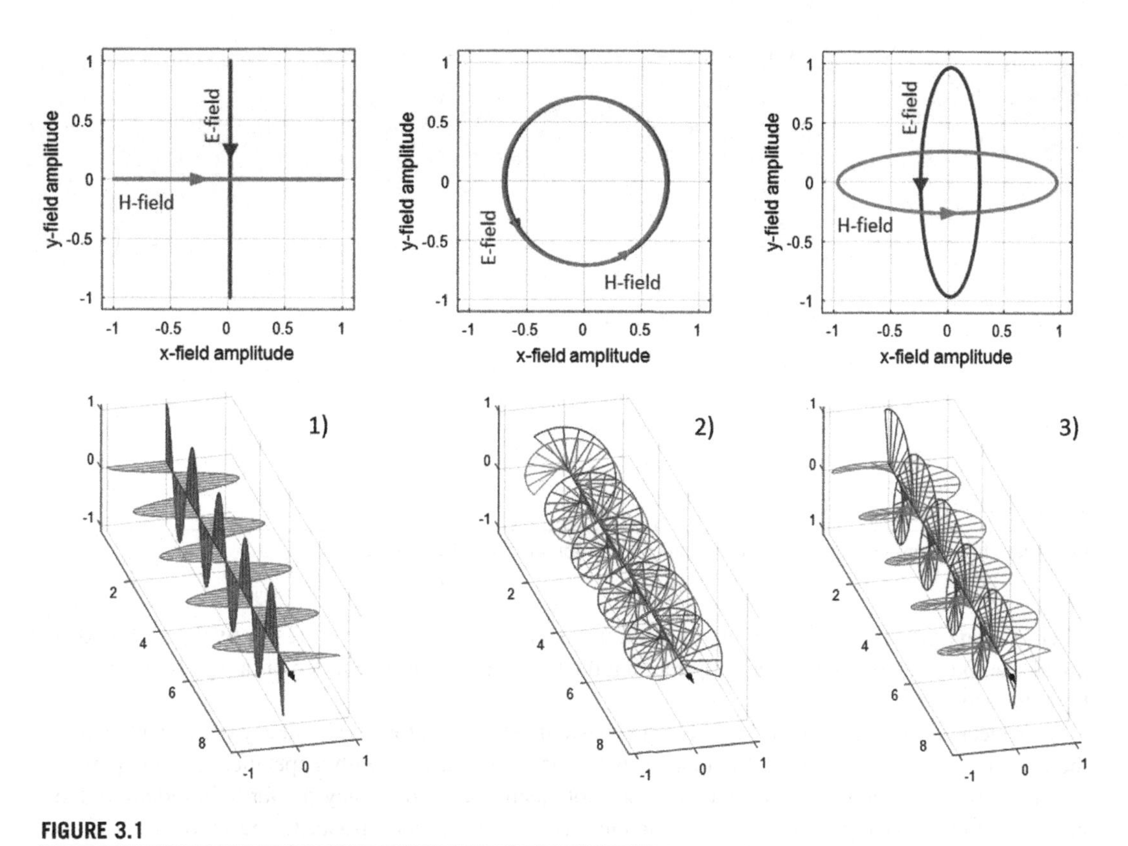

FIGURE 3.1

Plane waves with 1) linear, 2) circular, and 3) elliptic polarizations.

3.1.2 FREE-SPACE PROPAGATION

In order to understand propagation in general it is essential to have full insight into free-space propagation. In principle, propagation should exclude antenna effects. In practice, however, it is not possible to study propagation without the use of antennas. Here we will focus on the case of using some basic antennas being very clear on which effects are due to the antennas and which effects are due to propagation. The isotropically radiating antenna is commonly used as a reference. In free space the transmission loss between two such antennas (shown in Fig. 3.2) is referred to as *free-space basic transmission loss* (by ITU-R [8]) and is in dB units given by

$$L_{\mathrm{bf}} = 20 \log \left(\frac{4\pi d}{\lambda} \right) \tag{3.5}$$

where d is the distance between the antennas and λ is the wavelength. Though the isotropic antenna is commonly used as a reference it has been shown that it is not very practical to realize [16], [33]. There is a corresponding expression for arbitrary antenna patterns,

$$L_{\mathrm{f}} = L_{\mathrm{bf}} - G_t - G_r \tag{3.6}$$

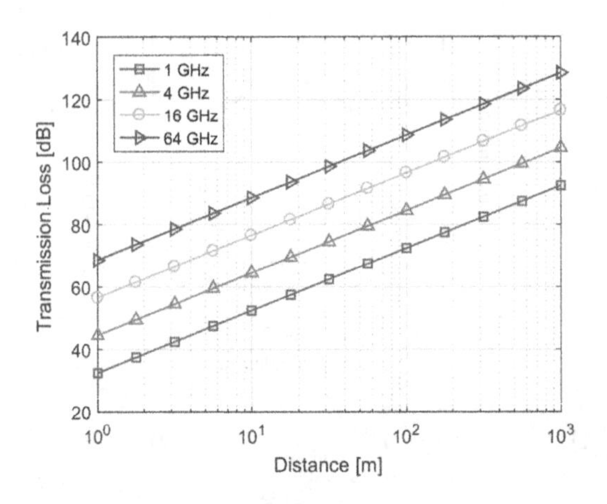

FIGURE 3.2

Free-space basic transmission loss versus distance for different carrier frequencies.

where G_t and G_r are the antenna gains at the transmitter and the receiver, respectively. It should be noted that (3.6) is based on the assumption that the transmit and receive antennas are fully matched in both direction and polarization.

It is clear from Fig. 3.2 that the free-space basic transmission loss is frequency dependent. This is the case, by (3.6), for any type of antenna (dipole, horn, patch, etc.) with a specified antenna pattern. As the shape of a specific type of antenna does not change with frequency the far field radiation does not change either. The size of a specific antenna is, however, proportional to the wavelength λ and thus the aperture (effective antenna area) is proportional to λ^2. As a consequence the received power for free-space transmission using any type of antennas (fixed gain and antenna pattern) depends on the frequency, f, as $-20\log f$ in dB units. This is indeed the result of fixing the antenna gains of (3.6). It should be noted that this frequency dependence is a pure antenna effect. From a propagation perspective there is no frequency dependence when transmitting with a fixed gain antenna as the power flux density at any far field distance is frequency independent in this case.

Instead of fixing the gain of the receive antenna the aperture may be kept constant over frequency. In this way the received signal becomes proportional to the power flux density at the receiver, which is frequency independent when a fixed gain transmit antenna is used. When a fixed antenna aperture is used also in the transmit end of a free-space link, there is actually a reduction of transmission loss with increasing frequency proportional to $-20\log(f)$. This is explained by that the antenna gain of a fixed aperture antenna increases with frequency providing corresponding increased signal strength at the receiver. This gain is, however, valid only in the far field of the antennas. For high frequencies in the millimeter-wave range the far field may be at fairly large distances depending on the sizes of the antenna apertures. In the near field region the transmitted lobe width may be substantially smaller than the receive antenna aperture. In this case the transmission loss may actually be negligible (0 dB). It should, however, be noted that, for physical reasons, there is no possibility of obtaining a transmission gain (i.e., loss < 0 dB) [4]. Moreover, large aperture antennas have corresponding high gains and narrow lobes, which have to be directed towards the antenna in the other end of the link.

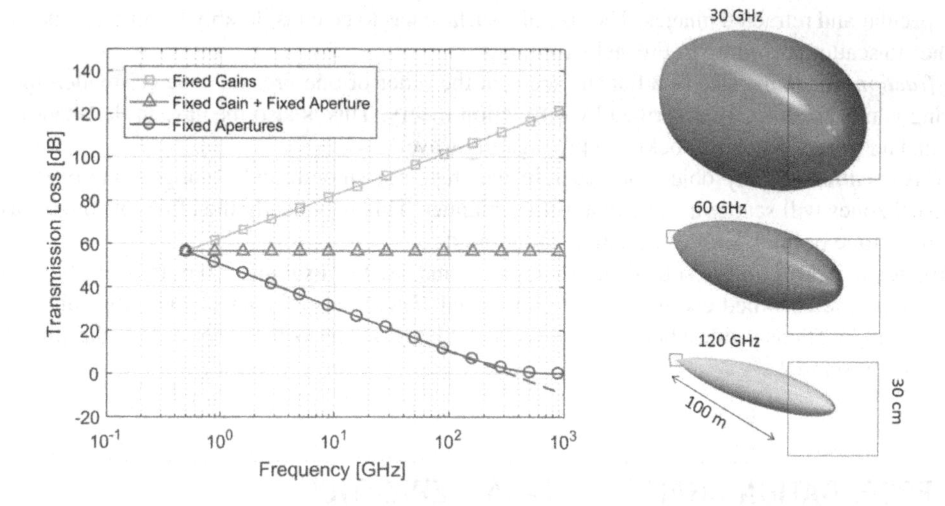

FIGURE 3.3

Free-space transmission loss versus frequency for 1) fixed gains, 2) fixed gain and fixed aperture, and 3) fixed apertures (left hand side), and corresponding lobes for the fixed aperture transmit antenna case (right hand side). The dashed line in the left hand graph shows the loss assuming far field when in the near field region.

All cases described above are illustrated in Fig. 3.3 where a 100-m free-space transmission link is simulated for fixed gain antennas ($G = 5$ dBi), and fixed square antenna aperture (30x30 cm). The fixed aperture gives 5 dBi gain at 500 MHz and 51 dBi gain at 100 GHz. The transition from the far field region to the near field region between 200 GHz and 500 GHz when using fixed aperture antennas in both ends of the link is clearly shown in the figure.

The *free-space basic transmission loss* is commonly referred to as *free-space path loss* in the literature. There are probably historical reasons for this. This use is, however, somewhat problematic as it gives the impression that free-space propagation might be frequency dependent, which is not the case as explained above. For this reason the notion of *path loss* is avoided throughout this book. Instead the notion of *transmission loss* is used when appropriate.

3.1.3 SCATTERING AND ABSORPTION

In a real environment the conditions are rarely similar to free space. Typically there is an abundance of structures and objects in the environment surrounding a transmission link which scatter the propagating electromagnetic waves. In an urban environment these objects are typically man made, like buildings and vehicles. In rural environments vegetation and geographical formations dominate the environment. Scattering can be categorized into a few basic types:

- ***Specular reflection and refraction*** At sufficiently flat and large surfaces electromagnetic waves are scattered fulfilling Fresnel reflection and refraction formulas. In optics this type of scattering results in specular and refracted images. The size of a surface has to be considerably larger than one Fresnel zone[1] to scatter according to Fresnel's formulas.

- ***Diffraction*** When the size of a flat surface is at the order of one Fresnel zone or smaller, the scattering is most accurately described by diffraction theory. This is also the case in the shadow zone behind an object which is blocking a propagating wave.
- ***Diffuse scattering*** Any object or structure which is not smooth and homogeneous over several Fresnel zones will scatter a wave in a diffuse manner. This is typically the effect of rough surfaces such as stone or brick walls, vegetation and ground.
- ***Absorption*** Absorption is substantial in some scenarios. For example when transmitting through vegetation the absorbed energy may be substantial. This is also the case when transmitting from outdoor to locations inside a building. Moreover, at large distances the absorption in the atmosphere may be substantial.

3.2 PROPAGATION CHANNEL CHARACTERIZATION

When developing radio transmission techniques it is crucial to know well the characteristics of the radio propagation channel. Under conditions close to free space, like for microwave links, the main challenge is to determine the time-varying transmission loss of the direct path. In this case the loss is mainly varying due to rain and consequently the corresponding probability of rain intensity is of interest. For cellular communications, however, the channel is characterized by the scattering environment and the mobility of terminals and some of the scattering objects which are not stationary. As a consequence, the received signal may vary considerably over time and space. Moreover, in cellular scenarios the propagation is affected by multipath propagation, i.e., the received signal is a mix of a multitude of different waves scattered around in the environment arriving with different directions and delays. To account for the multipath, the channel is commonly modeled as a set of discrete (plane) waves which combine at the receive antenna. Each discrete wave is determined by its pathway from the transmit antenna, via scattering (except for the direct path), to the receive antenna. In typical mobile communications scenarios the number of paths needed to characterize the channel may be very large up to tens of thousands. For the plane wave expansion the channel H_{mn} between transmit antenna element n and receive antenna element m is mathematically characterized by

$$H_{mn} = \sum_{l=1}^{N} \mathbf{g}_m^{\text{rx}}\left(-\mathbf{k}_l^{\text{rx}}\right)^{\text{T}} \cdot \mathbf{A}_l \cdot \mathbf{g}_n^{\text{tx}}\left(\mathbf{k}_l^{\text{tx}}\right) \cdot \exp\left[i\left(\mathbf{k}_l^{\text{tx}} \cdot \mathbf{r}_n^{\text{tx}} - \mathbf{k}_l^{\text{rx}} \cdot \mathbf{r}_m^{\text{rx}} + \omega\tau_l + \omega_{\text{D}_l} t\right)\right] \tag{3.7}$$

where $\mathbf{A}_l$ is the complex polarimetric amplitude matrix of the lth of totally N plane waves, $\mathbf{g}_m^{\text{rx}}\left(-\mathbf{k}_l^{\text{rx}}\right)$ and $\mathbf{g}_n^{\text{tx}}\left(\mathbf{k}_l^{\text{tx}}\right)$ are the complex polarimetric antenna pattern vectors for the corresponding wave vectors $\mathbf{k}_l^{\text{rx}}$ and $\mathbf{k}_l^{\text{tx}}$, $\mathbf{r}_n^{\text{tx}}$ and $\mathbf{r}_m^{\text{rx}}$ are the position vectors of the receive and transmit antenna elements relative to corresponding antenna reference points, ω is the angular frequency ($\omega = 4\pi f$), τ_l is the wave propagation delay between transmit and receive antenna reference points, ω_{D_l} is the Doppler angular

[1]If the pathway via any point at the border of a reflective surface is more than one wavelength longer than the specularly reflected path, the reflecting surface is larger than one Fresnel zone.

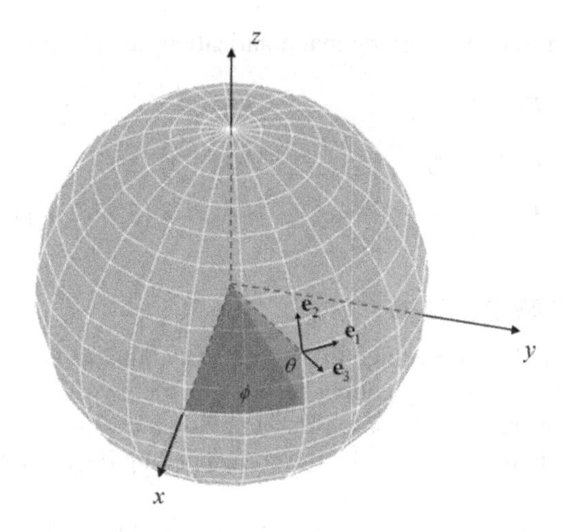

FIGURE 3.4

Local antenna reference system. The coordinate system for polarization is given by $\mathbf{e}_1 = \hat{\boldsymbol{\phi}}$, $\mathbf{e}_2 = \hat{\boldsymbol{\theta}}$ and $\mathbf{e}_3 = \hat{\mathbf{r}}$.

frequency and t is the time. The antenna patterns are defined based on local reference systems. For this purpose fixed local coordinate systems (x, y, z, or in polar coordinates, r, ϕ, θ) are used. Based on the fixed reference systems additional direction related reference systems are used for specifying the corresponding polarimetric pattern components in the different directions. The axes of these direction relative reference systems are defined by corresponding unit vectors pointing in the directions of increasing azimuth, $\mathbf{e}_1 = \hat{\boldsymbol{\phi}}$, and elevation, $\mathbf{e}_2 = \hat{\boldsymbol{\theta}}$, angles and increasing radius distance, $\mathbf{e}_3 = \hat{\mathbf{r}}$, as shown in Fig. 3.4. For free-space propagation and use of unit gain antennas the polarimetric amplitude is given by

$$\mathbf{A} = \frac{\lambda}{4\pi d} \begin{bmatrix} -1 & 0 \\ 0 & 1 \end{bmatrix} \tag{3.8}$$

where the minus sign is due to that the receiver coordinate system is rotated 180 degrees relative to the transmitter coordinate system. With both transmit and receive antennas having any of the three polarization cases of Eq. (3.4) the channel gain is given by

$$\left| \left(\mathbf{g}^{\text{rx}}\right)^{\text{T}} \cdot \mathbf{A} \cdot \mathbf{g}^{\text{tx}} \right| = \frac{\lambda}{4\pi d} \tag{3.9}$$

which, as expected, corresponds to the freespace-loss L_{bf} given by Eq. (3.5). When using receive antenna polarizations which are orthogonal to the transmit antenna polarizations, we have

$$\left| \left(\mathbf{g}^{\text{rx}}\right)^{\text{T}} \cdot \mathbf{A} \cdot \mathbf{g}^{\text{tx}} \right| = 0. \tag{3.10}$$

With regard to (3.4) the corresponding orthogonal polarizations are given by

$$\left.\begin{array}{lll} 1) & E_1 = 1, & E_2 = 0 \\ 2) & E_1 = -\frac{1}{\sqrt{2}}, & E_2 = \frac{i}{\sqrt{2}} \\ 3) & E_1 = \frac{1}{\sqrt{4}}\left(e^{-i\pi/6}+1\right), & E_2 = \frac{1}{\sqrt{4}}\left(e^{-i\pi/6}-1\right) \end{array}\right\}. \tag{3.11}$$

3.2.1 FREQUENCY-DELAY DOMAIN

The channel response of Eq. (3.7) is specified in the frequency domain, i.e., it is a function of the radio frequency f. There is a complementary equivalent way to specify the channel in the delay domain providing the corresponding channel response of an impulse, $h(\tau)$, as a function of the delay τ. The relation between these two domains is given by the corresponding Fourier transforms,

$$\left.\begin{array}{lll} h(\tau) & = & \int_{-\infty}^{\infty} H(f)\exp(i2\pi f\tau)\,df \\ H(f) & = & \int_{-\infty}^{\infty} h(\tau)\exp(-i2\pi f\tau)\,d\tau \end{array}\right\}. \tag{3.12}$$

For radio data transmissions, modulation of a continuous wave around a specific carrier frequency f_0 is utilized. The speed of the modulation and corresponding bitrate is proportional to the bandwith B used. For a channel, the band-limiting filter characteristics determine the shape of the corresponding impulse response. This may be exemplified by means of a channel having a single multipath component with a delay of τ_1 and a uniform bandpass filter

$$\left.\begin{array}{lll} h(\tau) & = & \frac{B}{2}\text{sinc}\left[\pi B(\tau-\tau_1)\right]\cdot\left[\exp(i2\pi f_0 t)+\exp(-i2\pi f_0 t)\right] \\ H(f) & = & \begin{cases} 0 & \text{if } |f| > f_0+\frac{B}{2} \text{ and } |f| < f_0-\frac{B}{2} \\ \frac{1}{2} & \text{if } |f| < f_0+\frac{B}{2} \text{ and } |f| > f_0-\frac{B}{2} \end{cases} \end{array}\right\}. \tag{3.13}$$

The corresponding channel responses for $f_0 = 2$ GHz, $B = 200$ MHz, and $\tau_1 = 10$ ns are shown in Fig. 3.5. In the passband the fast oscillations of $h(\tau)$, due to the carrier frequency f_0, are evident. Moreover, both positive and negative frequencies of $H(f)$ are needed for providing the real channel response. For convenience the channel is commonly described in the baseband meaning that the frequency is translated to zero mean, i.e., $f' = f - f_0$

$$\left.\begin{array}{lll} h(\tau) & = & B\text{sinc}\left[\pi B(\tau-\tau_1)\right] \\ H(f') & = & \begin{cases} 0 & \text{if } |f'| > \frac{B}{2} \\ 1 & \text{if } |f'| < \frac{B}{2} \end{cases} \end{array}\right\}. \tag{3.14}$$

In the following the baseband representation of the channel will be used, if the passband representation is not specifically indicated, with f representing the frequency in the baseband.

The uniform frequency filter in Eq. (3.14) results in substantial ringings, or side-lobes, around the main impulse in the delay domain. This can be mitigated by choosing a different frequency filter such

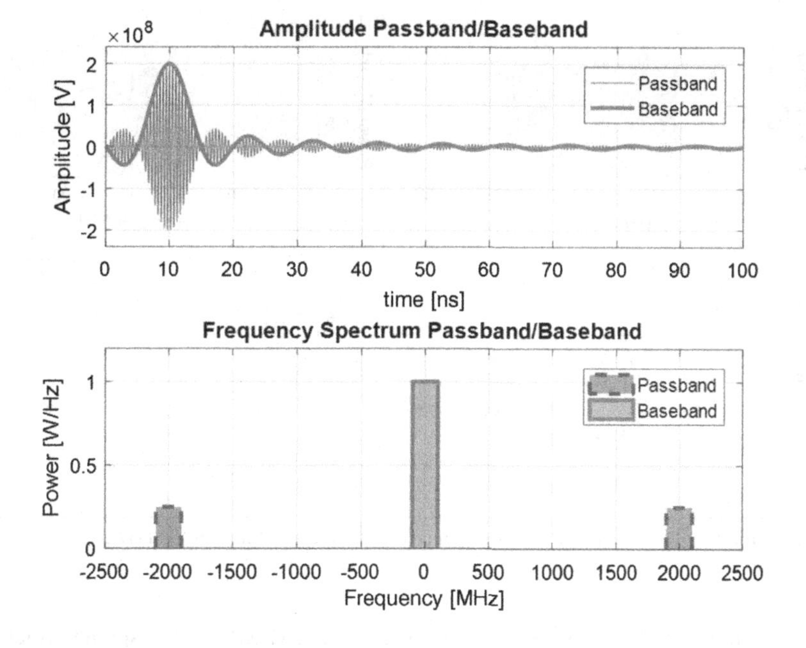

FIGURE 3.5

Example of a channel in passband and baseband with carrier frequency at 2000 MHz using 200 MHz bandwidth. The power of the baseband channel is twice the power of the passband channel to facilitate the comparison in the delay domain.

as the commonly used Hann filter

$$\left.\begin{aligned} h(\tau) &= B\operatorname{sinc}[\pi B(\tau-\tau_1)]+ \\ &\quad \tfrac{B}{2}\operatorname{sinc}[\pi B(\tau-\tau_1)+\pi]+\tfrac{B}{2}\operatorname{sinc}[\pi B(\tau-\tau_1)-\pi] \\ H(f) &= \begin{cases} 0 & \text{if } |f|>\frac{B}{2} \\ \cos\left(\frac{\pi}{B}f\right) & \text{if } |f|<\frac{B}{2} \end{cases} \end{aligned}\right\}. \tag{3.15}$$

In Fig. 3.6, the channel responses of a uniform frequency filter and a Hann filter are shown. The side-lobes for the Hann filter are clearly smaller than for the uniform filter.

When modeling a channel in frequency domain a sum of different waves having different delays and amplitudes, as a result of scattering in the environment, is used. We will illustrate this by the typical exponentially decaying channel,

$$\left.\begin{aligned} H(f) &= \textstyle\sum_{l=1}^{N} a_l \exp(i2\pi f\tau_l) \\ a_l &= \exp\left(-\frac{\tau_l}{\sigma_\tau}+i\phi_l\right) \end{aligned}\right\} \tag{3.16}$$

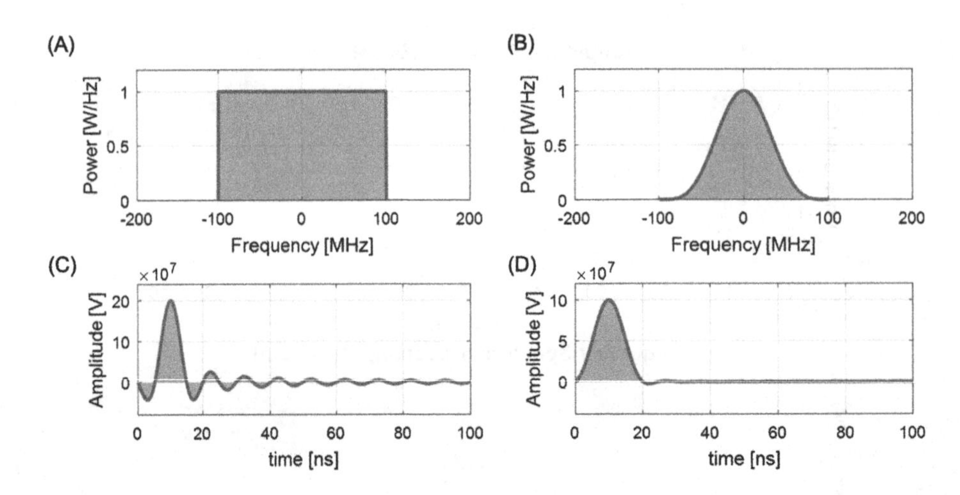

FIGURE 3.6

Channel responses for a uniform frequency filter (A) and (C), and a Hann filter (B) and (D).

where ϕ_l is a random phase and σ_τ is the root mean square (RMS) delay spread which is a useful measure defined by

$$\left.\begin{aligned} \mu_\tau &= \frac{\sum_{l=1}^{N} \tau_l |a_l|^2}{\sum_{l=1}^{N} |a_l|^2} \\ \sigma_\tau &= \sqrt{\frac{\sum_{l=1}^{N} (\mu_\tau - \tau_l)^2 |a_l|^2}{\sum_{l=1}^{N} |a_l|^2}} \end{aligned}\right\}. \tag{3.17}$$

For delay spreads substantially larger than the one over the bandwidth, i.e., $\sigma_\tau > \frac{1}{\mathrm{B}}$ the channel becomes frequency selective, meaning that the signal is fading substantially over this band. In Fig. 3.7 the corresponding channels are shown for $\sigma_\tau = 20$ ns and $\sigma_\tau = 100$ ns. Two cases of bandwidths are shown, $B = 10$ MHz and $B = 100$ MHz, applying Hann filtering in the delay domain to suppress the side-lobes. As the delay domain resolution is proportional to the used bandwidth, more multipath components are resolved at the higher bandwidth. An important channel property is that the channel is frequency flat over a bandwidth which is less than that needed to resolve multipath components as shown in Fig. 3.7D. A measure commonly used to characterize the frequency selectivity is the coherence bandwidth defined as the bandwidth over which the channel is correlated at some level, typically over 0.9. For the channels in Fig. 3.7 the corresponding coherence bandwidths are 1 MHz for $\sigma_\tau = 100$ ns, and 5 MHz for $\sigma_\tau = 20$ ns.

3.2.2 DOPPLER-TIME DOMAIN

The channel characteristics in Doppler-time domain are fully analogous with frequency-delay domain. In this case the channel variations in time t are specified by the Doppler frequencies f_{D} of the corre-

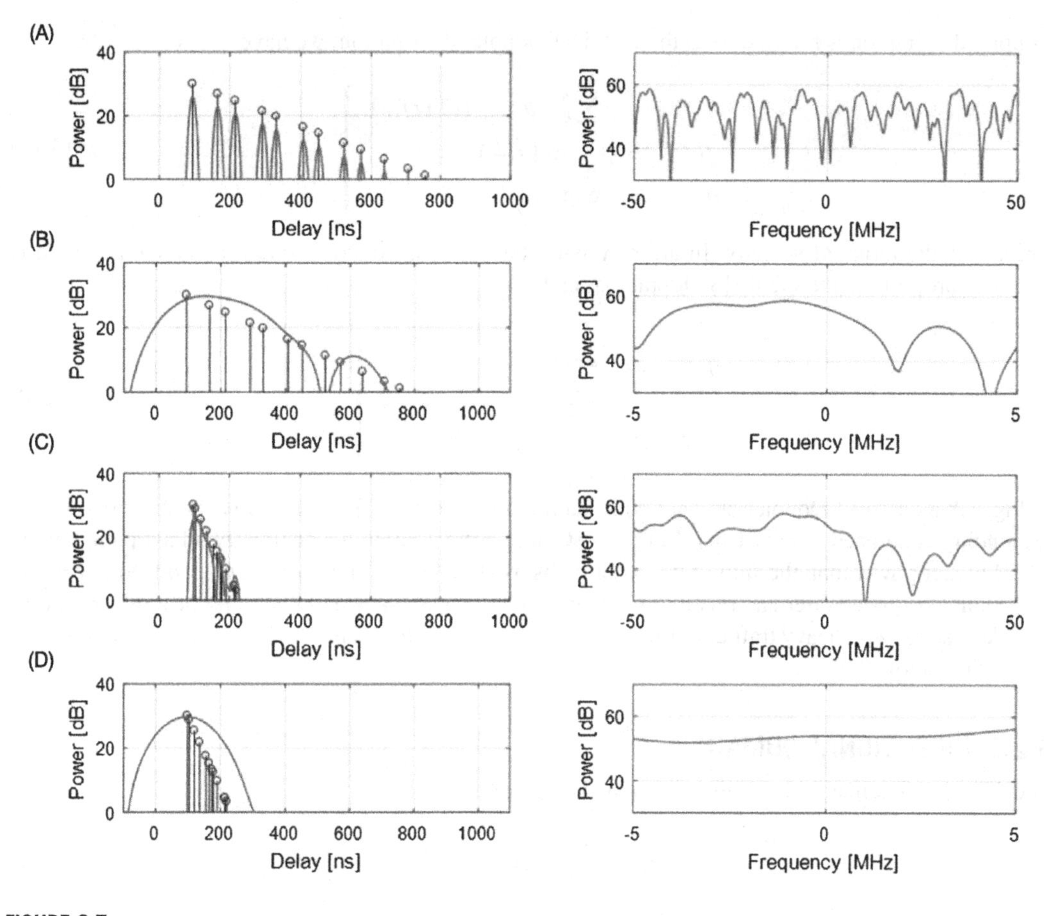

FIGURE 3.7

Channel responses in delay and frequency domains for $\sigma_\tau = 100$ ns, (A) and (B), and $\sigma_\tau = 20$ ns, (C) and (D), for $B = 100$ MHz, (A) and (C), and $B = 10$ MHz, (B) and (D).

sponding multipath components,

$$\left.\begin{aligned} h(t) &= \int_{-\infty}^{\infty} H(f_\mathrm{D}) \exp(i2\pi f_\mathrm{D} t)\, df_\mathrm{D}, \\ H(f_\mathrm{D}) &= \int_{-\infty}^{\infty} h(t) \exp(-i2\pi f_\mathrm{D} t)\, dt \end{aligned}\right\}. \tag{3.18}$$

When modeling a channel in the time domain a sum of different waves having different Doppler frequencies is used. The main reason for having different Doppler frequencies is that either end, or both ends, of a radio link are moving causing a Doppler up-shift (down-shift) in frequency when the antenna is moving towards (away from) a wave. Another reason is that significant scatterers in the environment, like vehicles and trees, might be moving. We will illustrate this by a channel having a uniform directional distribution of waves in the horizontal plane around a moving terminal. The corresponding

Doppler distribution is referred to as the classical Doppler distribution; we have

$$\left.\begin{array}{lcl} h(t) & = & \sum_{l=1}^{N} a_l \exp\left(i 2\pi t f_{\mathrm{D}_l}\right) \\ f_{\mathrm{D}_l} & = & \frac{v}{\lambda} \cos\left(\frac{2\pi l}{N}\right) \\ a_l & = & \exp\left(i\phi_l\right) \end{array}\right\} \tag{3.19}$$

where v is the terminal velocity. In analogy with the frequency-delay domain the coherence time is inversely proportional to the RMS Doppler spread σ_{f_D},

$$\left.\begin{array}{lcl} \mu_{f_\mathrm{D}} & = & \frac{\sum_{l=1}^{N} f_{\mathrm{D}_l} |a_l|^2}{\sum_{l=1}^{N} |a_l|^2} \\ \sigma_{f_\mathrm{D}} & = & \sqrt{\frac{\sum_{l=1}^{N} \left(\mu_{f_\mathrm{D}} - f_{\mathrm{D}_l}\right)^2 |a_l|^2}{\sum_{l=1}^{N} |a_l|^2}} \end{array}\right\}. \tag{3.20}$$

In Fig. 3.8 a classical Doppler channel with maximum frequency $f_{\mathrm{D}_{\max}} = 100$ Hz is shown. The corresponding coherence time is 1 ms. In Fig. 3.8C and D the case with a stationary added path, having 10 dB higher power than the sum of the other paths, is also shown. This case corresponds to a scenario with stationary transmitter and receiver in an environment with substantial amount of moving scatterers, like a street with heavy traffic. For this case the coherence time is infinite, as the correlation in time never goes below 0.9.

3.2.3 DIRECTIONAL DOMAIN

The directional domain relates directly to the Doppler domain by

$$f_{\mathrm{D}_l} = \frac{\mathbf{v} \cdot \mathbf{k}_l}{2\pi} = \frac{v}{\lambda} \cos\left(\theta_l\right) \tag{3.21}$$

where θ_l is the angle between the velocity $\mathbf{v}$ and the wave vector $\mathbf{k}_l$. The direction information may be obtained from Doppler information in terms of the $\mathbf{k}$ vector,

$$\left.\begin{array}{lcl} k_{x,y,z} & = & 2\pi \frac{f_{\mathrm{D}x,y,z}}{v_{x,y,z}} \\ u_{x,y,z} & = & \lambda \frac{f_{\mathrm{D}x,y,z}}{v_{x,y,z}} \end{array}\right\} \tag{3.22}$$

where $\mathbf{u}$ is the direction unit vector. In practice, the corresponding directional spectrum is obtained by performing a Fourier transformation of three dimensional spatial channel samples as described in Section 3.3.2.1.

The directional spread is commonly characterized by means of the angular spread in azimuth and elevation (or polar) angles. Angle spreads are problematic, however, as they are both cyclic and non-Euclidian. To overcome the problem that angles are cyclic the cut in angle range may be placed at the point that minimizes the spread. The other problem—that angles are non-Euclidian variables—is more serious. When characterizing the channel by azimuth and elevation spreads the channel is not invariant under rotation of the coordinate system.

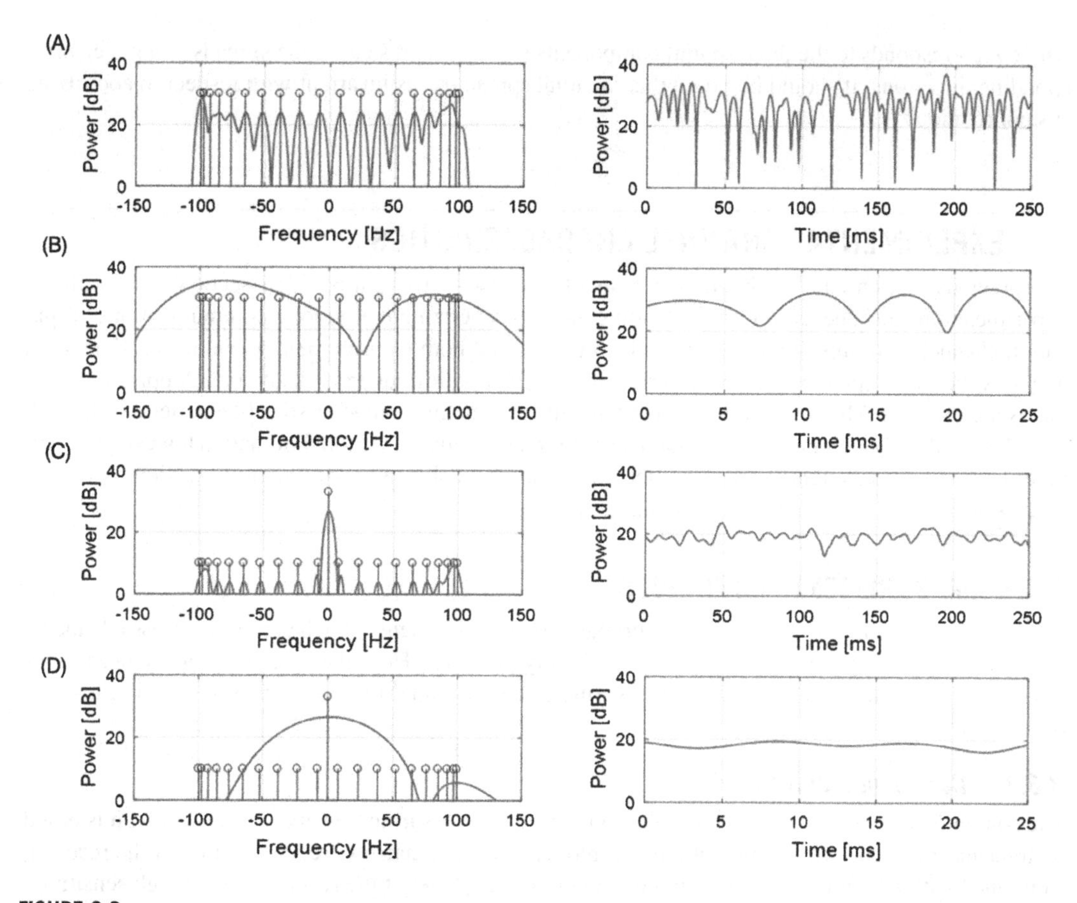

FIGURE 3.8

Channel responses in Doppler and time domains for classical Doppler with $\sigma_{f_D} = 70$ Hz, (A) and (B), and for classical Doppler with a stationary added path with $\sigma_{f_D} = 20$ Hz, (C) and (D), for $T = 250$ ms, (A) and (C), and $T = 25$ ms, (B) and (D).

An alternative directional spread definition, proposed in [19], does not suffer from these problems. It is based on a Doppler spread in three dimensions normalized so that it is equal to the angular spread when the spread is small. This directional spread σ_{dir} is defined by

$$\left.\begin{aligned} \mu_{u_n} &= \frac{\sum_{l=1}^{N} u_{n,l}|a_l|^2}{\sum_{l=1}^{N}|a_l|^2} \\ \sigma_{\text{dir}_n} &= \frac{180}{\pi}\sqrt{\frac{\sum_{l=1}^{N}\left(\mu_{u_n}-u_{n,l}\right)^2|a_l|^2}{\sum_{l=1}^{N}|a_l|^2}} \\ \sigma_{\text{dir}} &= \sqrt{\sigma_{\text{dir}_x}^2+\sigma_{\text{dir}_y}^2+\sigma_{\text{dir}_z}^2} \end{aligned}\right\} \tag{3.23}$$

where n corresponds to the three spatial components (x, y, z) representing the spreads along the corresponding directions. It should be noted that the total spread σ_{dir} is invariant with respect to coordinate system rotations.

3.3 EXPERIMENTAL CHANNEL CHARACTERISTICS

The previous section laid the theoretical basis of radiowave propagation. This section deals with the experimental results needed for understanding and characterizing real mobile communications propagation channels. For this purpose one needs to understand both measurement techniques and analysis methods. For a complete characterization of the channel both frequency-delay and Doppler-time domains are required. Moreover, to understand and utilize antenna characteristics like pattern, lobewidth and MIMO, the channel needs to be characterized in the spatial/directional domain. However, this domain is actually equivalent to the Doppler-time domain as shown in the previous section. All these aspects are addressed in some depth in this section.

3.3.1 MEASUREMENT TECHNIQUES

There are several different techniques for propagation measurements which are more or less advanced. The corresponding hardware designs may be highly complex. Here the most commonly used types of equipment and techniques used for measuring a radio channel response and transmission loss are briefly described.

3.3.1.1 Continuous Wave

The continuous wave (CW) method is commonly used for measuring transmission loss only. It is based on transmitting a sine wave signal at a fixed frequency using a narrow frequency filter at the receiver. Combined with a high transmit power and a low noise amplifier at the receiver a very high sensitivity may be obtained. The hardware solutions are typically relatively compact and simple, enabling extensive sampling of the signal strength over large areas in a fast and convenient way. One drawback is, however, that the multipath of the channel is not resolved resulting in substantial spatial fading.

Fig. 3.9 shows a measured signal CW signal at 5.1 GHz from a measurement route in a street microcell under non line-of-sight (NLoS) conditions. In order to reduce the fading due to the multipath effect a sliding average over 1.7 m has been applied.

3.3.1.2 Vector Network Analyzer

In contrast to CW measurements, a vector network analyzer (VNA) allows maximum possible measurement bandwidths. The basic principle is to perform a frequency swept sampling of the channel over a predefined bandwidth. For providing corresponding channel responses in the delay domain, it is convenient to use discrete Fourier transformation (DFT) methods. As the VNA measurement principle is based on measuring fully coherent ratios between the transmitted and received signal both the transmitter and the receiver antennas have to be connected to the VNA with RF cables. The advantage is that the received signal is fully synchronized with the transmitted signal, which enables absolute delay measurements and coherent averaging over long times for the suppression of noise. A substantial drawback is that the mobility is limited by the RF cabling. Moreover, a single frequency sweep may

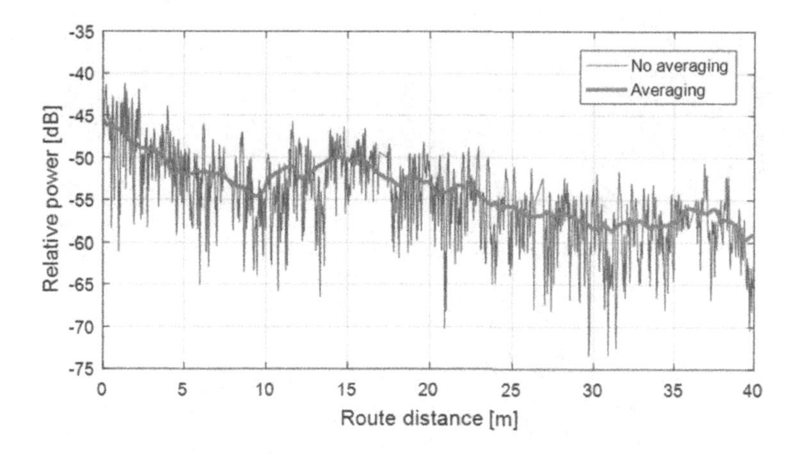

FIGURE 3.9

Continuous wave measurement at 5.1 GHz in a street microcell under NLoS condition.

take several seconds depending on the SNR at the receiver location which requires that both ends of the link are stationary. Another drawback is that the range is limited due to signal attenuation by the RF cables, which may be several dB per meter depending on the radio frequency. The range may be extended considerably to many kilometers, however, by replacing the RF cables by optical fibers using RF-to-opto and opto-to-RF converters. In Fig. 3.10 the measured channel response at 58.7 GHz using 2 GHz bandwidth is shown for a NLoS microcellular street scenario [22]. In this specific measurement an optical fiber was used to extend the range to more than 100 m. The large bandwidth results in a large fraction of the rich multipath of this channel being resolved.

3.3.1.3 Correlation-Based Channel Sounding

The main advantage with correlation-based channel sounders is that they are both mobile and wideband. To achieve this, a dedicated sounding signal is transmitted periodically and then correlated as a function of delay at the receiver. Typically OFDM or pseudo-random sequences are used together with a sliding correlator in delay. This technique allows both mobile and wideband measurements. The trade-off is between channel sampling frequency and noise suppression. Moreover, as the transmitted signal is modulated in both phase and amplitude, there are limitations due to amplifier non-linearities. More details of this type of equipment are presented in [32] and [15].

3.3.1.4 Directional Characteristics

The directional characteristics of the propagation channel are of particular interest when going up in carrier frequency into the millimeter-wave range. At these frequencies the use of omnidirectional antennas limits the possible range due to the substantial increase of transmission loss as a result of the decreasing antenna apertures. For this reason beam-forming techniques are required for focusing the transmission and reception in propagation directions which minimize the loss. There are basically two methods used for this purpose in propagation measurements.

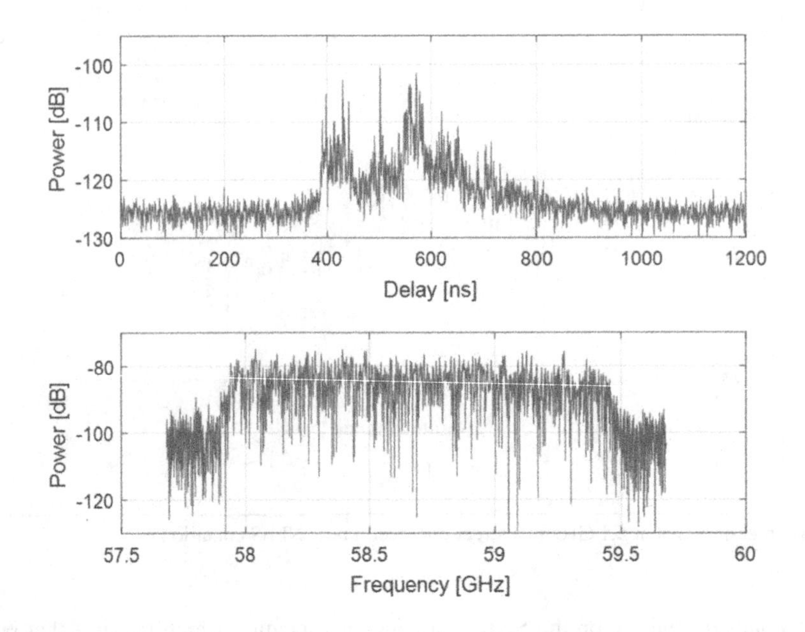

FIGURE 3.10

VNA channel measurement at 58.7 GHz and 2 GHz BW in a street microcell under NLoS condition. The baseband power delay profile is shown together with the channel response in the passband.

The first method is to use physical directive antennas, such as horn and parabolic reflector antennas, which are rotated in elevation and azimuth to scan the space angle. This method is not sensitive to time-varying channel conditions and is therefore suitable for CW and correlation-based channel sounding.

The second method is to use the so-called virtual antenna method. This method utilizes space sampling of the propagation channel where a single physical antenna is moved to different space positions by means of a robotic antenna positioning system. The directional characteristics of the measured channel are then determined off-line by means of array antenna techniques. The advantage of this method is that very high resolution and suppression of side-lobes may be achieved. The drawback is that the spatial sampling means long measurement times, up to many hours for large arrays, and therefore also it requires phase locked transmitter and receiver and stationary channel conditions. This method is suitable for VNA-based channel sounding.

3.3.2 ANALYSIS METHODS

Acquiring good quality measurement data requires considerable effort and skill. The raw measurement data is, however, of little use without thorough and accurate analysis. Providing reliable and accurate analyzed results takes even more effort than performing the actual measurements. In this section some commonly used analysis methods and their corresponding advantages and disadvantages are described. Moreover, requirements for providing comparability between different measurements and/or frequency ranges are provided.

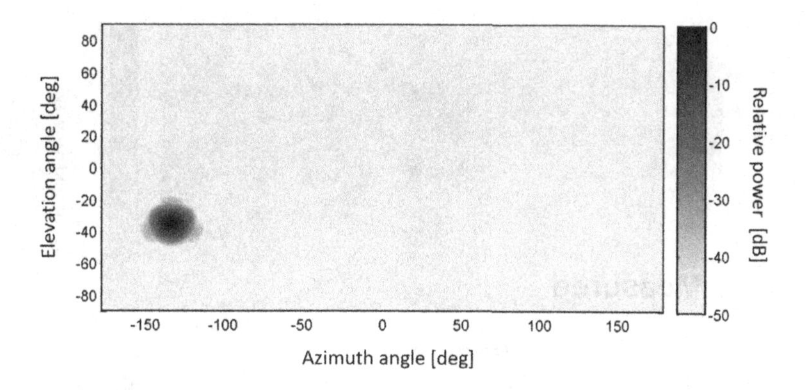

FIGURE 3.11

Simulated response of a plane wave impinging from $-135°$ in azimuth and $-35°$ in elevation for a cubic virtual array.

3.3.2.1 Spectral Analysis

The spectral methods are based on direct analysis of the measured signal and corresponding power distributions using Fourier techniques for transformations between frequency and delay, and between Doppler and time domains, as described in Sections 3.2.2 and 3.2.1. For directional analysis, direct angular channel sampling using directive physical antennas or virtual antennas may be used.

For a stationary channel, the spatial samples are analogous to time samples of a moving terminal. When the channel is sampled in three spatial dimensions it is possible to determine the corresponding power spectrum for all three components of the wave vector $\mathbf{k}$.

One example of this method using a cubic virtual antenna of $25^3 = 625$ samples is shown in Fig. 3.11. The space samples are transformed to the $\mathbf{k}$ domain by DFT using a Hann filter over all three space dimensions for reducing the side-lobes. The corresponding directional spectrum is obtained by filtering out values from the $\mathbf{k}$ domain cube with a fixed radius $\sqrt{k_x^2 + k_y^2 + k_z^2} = |\mathbf{k}|$. The performance of this method is impressive, providing side-lobe suppression of more than 50 dB. This is in contrast to the use of physical antennas for which the side-lobe level typically is suppressed by less than 30 dB. To bring the measured channel to the discrete format of (3.7) each peak of the multidimensional spectrally measured channel data is identified as a multipath component with corresponding phase and amplitude [21]. An example for an indoor NLoS scenario is shown in Fig. 3.12, where the directly measured channel is shown together with corresponding synthesized channel using 400 estimated multipath components. It is clear that the modeled channel agrees well with the directly measured channel.

3.3.2.2 Superresolution Methods

Assuming that the discrete plane wave model (3.7) is valid, the theoretically possible accuracy is limited only by the signal-to-noise ratio. There are many superresolution methods reported in the literature, of which the most popular ones are based on maximization of the likelihood P. The corresponding log-

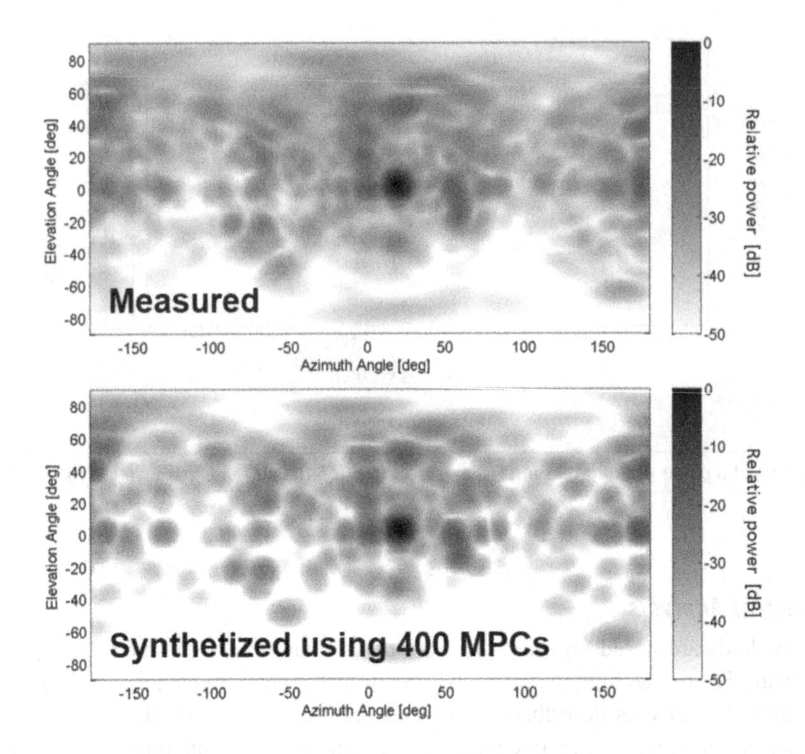

FIGURE 3.12

Measured indoor LoS channel at 60 GHz and corresponding synthesized channel using 400 MPCs for a cubic virtual array.

likelihood function to minimize is given by

$$-\log P = N_{\mathrm{rx}} N_{\mathrm{tx}} N_{\mathrm{f}} \log(\pi\sigma^2) + \frac{1}{\sigma^2} \sum_{m=1}^{N_{\mathrm{rx}}} \sum_{n=1}^{N_{\mathrm{tx}}} \sum_{k=1}^{N_{\mathrm{f}}} \left| \tilde{H}_{\mathrm{mnk}} - H_{\mathrm{mnk}} \right|^2 \tag{3.24}$$

where m and n are the space samples in the receive and transmit ends, respectively, k is the index over frequency, $\tilde{H}_{\mathrm{mnk}}$ and H_{mnk} are the modeled and measured channel responses, respectively, and σ^2 represents the power of the noise which is assumed to follow a zero mean complex Gaussian distribution over the channel samples. In order to find the most probable set of plane waves, which would mimic the measured channel best, the log-likelihood function is minimized with respect to the model parameters.

A free search over all model parameters and measurement samples is practically not possible due to the huge computational effort. One common method for reducing this effort is SAGE [5], which is based on maximizing the likelihood for one parameter at a time and iterate until the minimum is found. The problem with this method is that plane waves, which are closely spaced in angle or delay, are strongly correlated. As a consequence the increase in computational effort is very large and therefore the convergence is very slow. This problem has been addressed by means of gradient methods, like

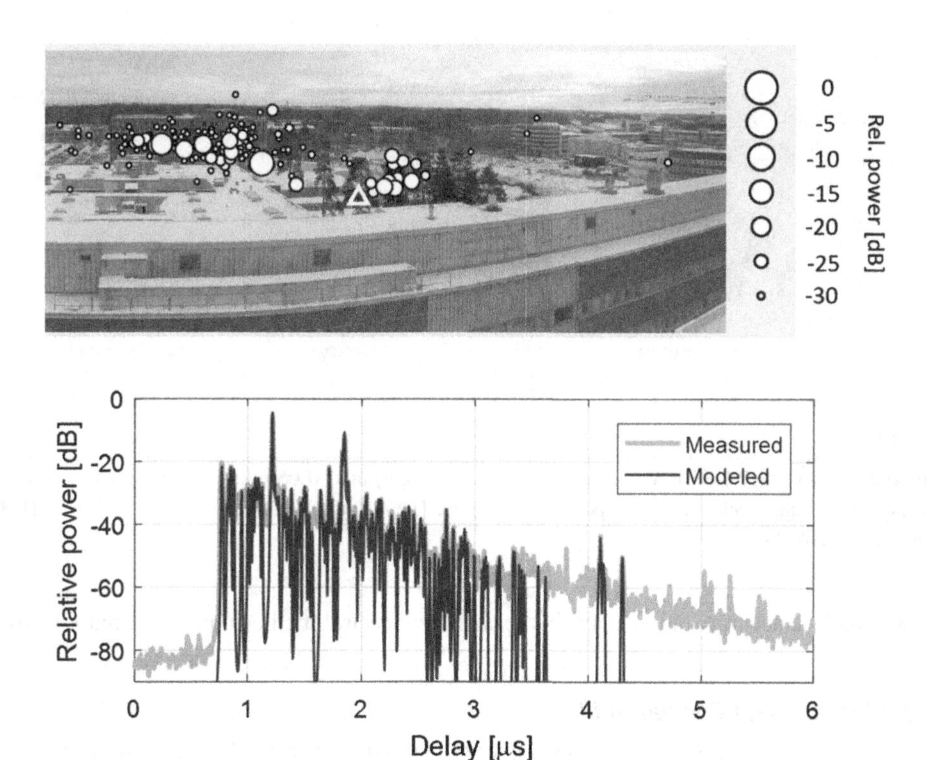

FIGURE 3.13

The lower graph shows power delay profiles from a NLoS urban macrocell scenario at 5 GHz. Both the directly measured profile (upper light) and the profile based on the super-resolved estimates (lower dark) are shown. The upper graph shows the directions of the estimated plane waves as viewed at the BS location where the triangle indicates the direction of the mobile station.

RIMAX [34], [30], which utilize differentiation around local likelihood maxima to achieve fast convergence of the correlated parameters. The problem with this method is, however, that some correlation or coupling between most of the multipath components remains, meaning that the likelihood needs to be maximized simultaneously for all waves. A method to decouple the multipath components is proposed in [17]. As the method also provides reduction of the data size and parameter space, many orders of magnitude better computational efficiency is obtained. An example of the application of this method is shown in Fig. 3.13 which is from an urban macrocell scenario at 5.1 GHz where a virtual planar array of 10x25 elements with 2-cm spacing was used at the BS. It is clear that the super-resolved estimates account for the main part of the received power. Moreover, the main directions of the waves, as viewed from the BS, show that the most significant scatterers are trees and exterior building walls.

Though superresolution methods may provide a high accuracy of significant discrete multipath components, the method suffers from incapability of estimating the diffuse or dense multipath component of the channel. As this component may be substantial, superresolution methods commonly

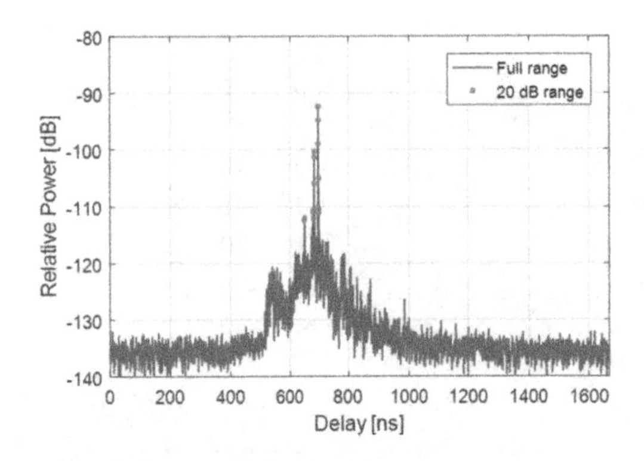

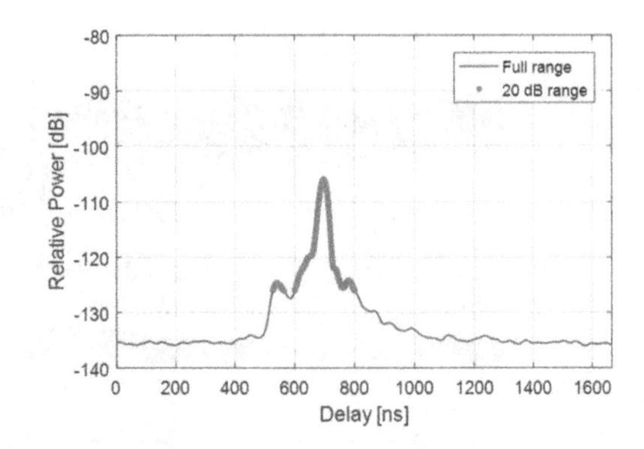

FIGURE 3.14

Power delay profiles using 2 GHz BW (left) and 80 MHz BW (right) in a NLoS microcellular scenario. Corresponding 20-dB dynamic ranges below the main peak are indicated. The resulting delay spreads are 7 ns for 2 GHz BW and 28 ns for 80 MHz BW.

give substantially inaccurate output. For this reason, the spectral methods are preferred for providing reliable output.

3.3.2.3 Measurement Comparability

The amount of propagation measurement campaigns performed worldwide is indeed extensive. Many of the corresponding results are compared and summarized in statistical analyses performed by the research community. In those analyses it is, however, important to be thorough and to fulfill the requirements for ensuring that the measurements are comparable. For this purpose a list of critical requirements which need to be fulfilled, for different campaigns and frequency bands, has been identified [28]:

- Equal measurement bandwidth (providing equal delay resolution)
- Comparable antenna pattern, either physical or synthesized
- Equal dynamic power range in the respective domain of analysis (e.g., delay, angle)
- Same environment and same antenna locations (for comparing different frequency bands)

It has e.g. been found that the requirement of equal bandwidths is critical for avoiding a fictitious decrease of delay spread with increasing frequency. The basic problem is that substantially larger measurement bandwidths are available at higher frequencies in the millimeter-wave range, wherefore those measurements typically are performed using substantially larger bandwidths. The need for equalizing the bandwidths used in the analysis is illustrated in Fig. 3.14 where a 20-dB dynamic range below the main peak has been used for determining the RMS delay spread. The impact of the different used bandwidths is considerable, giving delay spreads of 7 ns and 28 ns for 2 GHz and 80 MHz bandwidths, respectively. The reason why the equalization of bandwidth is required is that any strong discrete multipath component that is resolved has a peak power proportional to the used bandwidth, while the non-resolved components stay at a constant level irrespective of the bandwidth used. As a fixed dynamic power range below the main peak typically is used in the analysis, the difference of using different bandwidths is substantial.

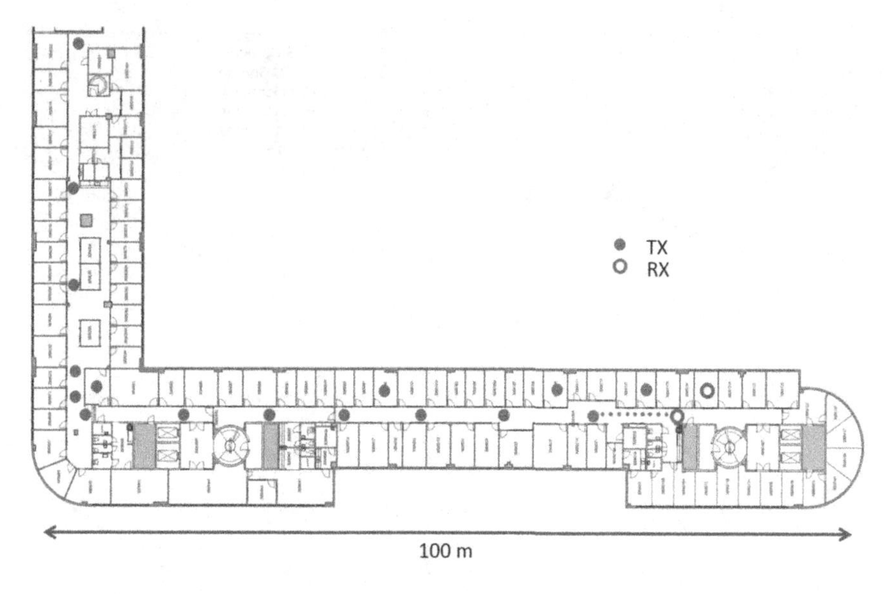

FIGURE 3.15

Floorplan of indoor office measurement scenario.

3.3.3 TRANSMISSION LOSS MEASUREMENTS

The loss in received signal strength due to propagation effects is one of the most basic and one of the most important characteristics of the radio propagation channel. In this section the current empirical understanding of transmission loss is presented and illustrated with a number of measurement examples from real cellular scenarios with focus on the frequency dependency in the range 1–100 GHz. Most of the measurements have been performed using vertical omni dipole antennas having very similar patterns for all frequencies. Additionally, vertical patch antennas or open waveguides have been used in outdoor-to-indoor measurements at the outdoor transmitter location. As measurements performed around 60 GHz are subject to oxygen absorption of about 1.5 dB/100 m, this loss has been compensated for in those measurements. The reason for this is the desire to facilitate the modeling by making possible smooth interpolation/extrapolation over the full frequency range and, depending on the need, adding the oxygen absorption to the baseline model. Moreover, all measurement data is provided in terms of the loss in excess of free-space loss in order to avoid any impact of antenna frequency dependence and to focus on pure propagation effects. For this purpose, all measurement data is carefully calibrated by line-of-sight (LoS) short range (0.1–1.0 m) measurements.

3.3.3.1 Indoor Office Scenario

This measurement example is from an indoor office environment. The basic layout shown in Fig. 3.15 is a corridor with office rooms along both sides. At the end of the corridor there is a 90 degrees turn. The receive (Rx) antenna is placed at two locations, one in the corridor and one inside an adjacent office room. The transmit antenna is placed at different locations both in the corridor and inside office

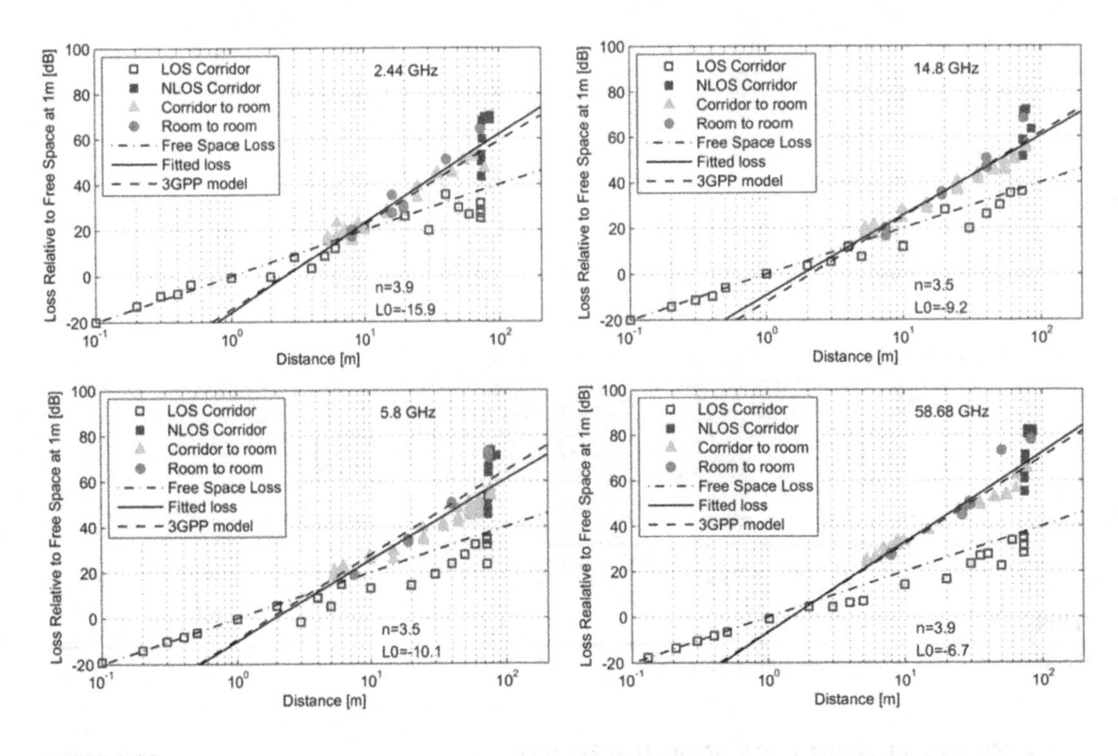

FIGURE 3.16

Transmission loss, in excess free-space loss, versus distance for 2.44, 5.8, 14.8, and 58.68 GHz measured in the indoor office environment.

rooms. The exterior walls of the building are made of brick and the interior walls of plasterboard and glass.

Fig. 3.16 summarizes the main results of the transmission loss analysis. The loss in dB units, L, relative to free-space power at 1-m distance is plotted for the frequencies, 2.44 GHz, 5.8 GHz. 14.8 GHz, and 58.68 GHz, for the different LoS and NLoS scenarios. A two parameter exponent model,

$$L = 10n\log(d) + L_0, \tag{3.25}$$

has been fitted to the measured loss, in dB units, where d is the distance between transmitter and receiver in meters. The corresponding model by 3GPP [2] accounts also for the frequency dependence,

$$L_{3GPP} = 38.3\log(d) - 15.1 + 4.9\log f \tag{3.26}$$

where f is the carrier frequency in GHz. The two last terms correspond to the frequency dependent term L_0 in (3.25). It is clear from the figure that the measurement data agrees very well with the 3GPP model. Moreover, the frequency trend is clear as the propagation loss increases about 5 dB per decade in addition to the free-space loss.

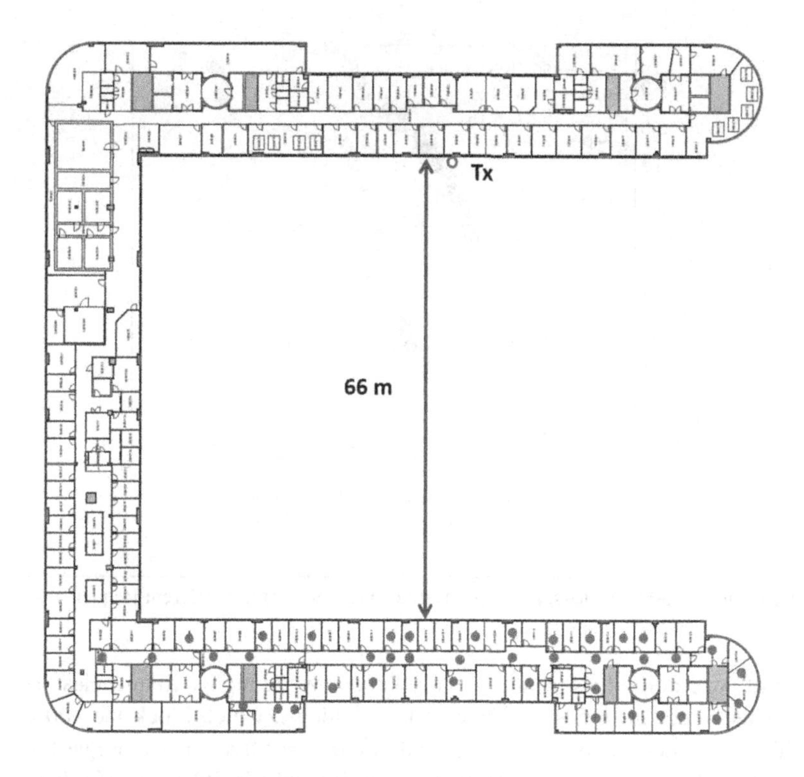

FIGURE 3.17

Outdoor-to-indoor measurement scenario. The indoor Rx locations are marked with filled circles and the Tx location is marked with an open circle.

3.3.3.2 Outdoor-to-Indoor Scenario

An outdoor-to-indoor multifrequency measurement campaign has been performed in an eight floors office building in an urban environment as depicted in Fig. 3.17. The transmitter is located in an open window at the top floor of the building and the received signal is measured at two slightly shifted (30 cm) positions at 40 indoor locations across the inner yard on the same floor. At the top floor of the building the exterior wall is covered with metal. The windows are, however, pure glass without metalization.

Between 2.44 GHz and 14.8 GHz the building penetration loss ranges from around 0 dB up to 30 dB (Fig. 3.18). The lower end of penetration loss around 0 dB is similar for all frequencies, while the highest losses around 45 dB occur only at 58.68 GHz. The minimum loss, due to penetration of the exterior wall/window only, is in the range 0–5 dB with the highest values for 5.8 GHz and 58.68 GHz. This non-monotonic dependence on frequency may be explained assuming that the three layers of glass, in the window frames, cause constructive or destructive interference, as an effect of multiple reflections, resulting in periodic varying attenuation as a function of frequency. Subsequent measurements show that the window loss is about 2, 10, 0, and 6 dB at 2.44, 5.8, 14.8, and 58.68 GHz, respectively, which confirms this effect and explains the measured minimum penetration loss. Moreover, it is clear

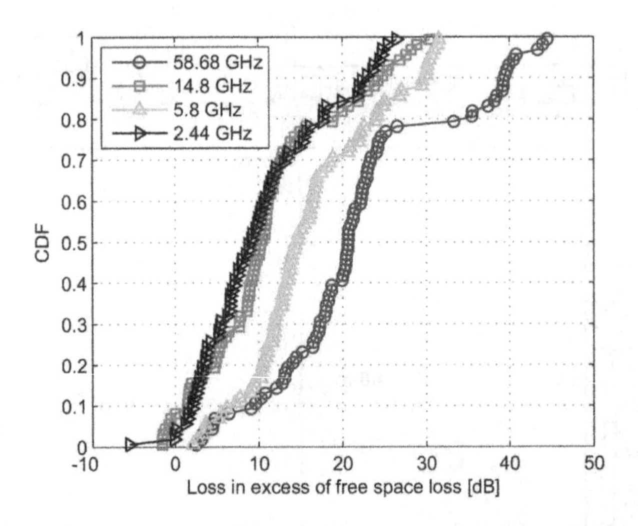

FIGURE 3.18

CDFs of measured outdoor-to-indoor loss, in excess of free-space loss, for the different frequencies.

that the spread of penetration loss is substantially larger for the higher frequencies. This may partly be the result of the venetian blinds, in some of the windows, which block the vertically polarized waves at the higher frequencies but are transparent at the lower frequencies. Regarding median loss, the measurement results agree very well with the corresponding 3GPP model [2]. It should be noted that the 3GPP model does not account for any non-monotonic frequency effects. Moreover, there is an increasing spread of the measured loss with increasing frequency, which also is not accounted for in the 3GPP model. Further details on outdoor to indoor propagation modeling are given in the channel modeling section, Section 3.4.1.1.

3.3.3.3 Outdoor Street Scenario

Outdoor street measurements have been made in an urban area consisting of mainly modern office building blocks of about 100 m length and 25 m height (Fig. 3.20). The measurements were performed in both LoS and NLoS in a street canyon, of about 20 m width, with both transmitter and receiver antennas located about 1.5 m above ground. In Fig. 3.19 the excess loss is shown for all frequencies. In LoS a multipath gain of up to 5 dB (relative to free space) is observed, which is similar at all frequencies. This gain is due to additional paths from reflections off the ground and exterior walls. In the NLoS region behind the corner of the building a substantial increase in the excess loss is observed. This loss is substantially lower than what is expected by knife edge diffraction at the corner, as indicated in Fig. 3.19. Further, the frequency dependence is much weaker than what is expected from diffraction. This result suggests that the dominating propagation mechanisms in NLoS must be different from diffraction, e.g. specular and/or diffuse scattering by objects or rough exterior walls. Moreover, it is clear that the oxygen compensation at 60 GHz is substantial, up to 4 dB, for the NLoS data, which is more than what is expected from the link distance only. This is, however, explained by the lengths of significant reflected propagation paths being substantially larger than the link distance. The measured

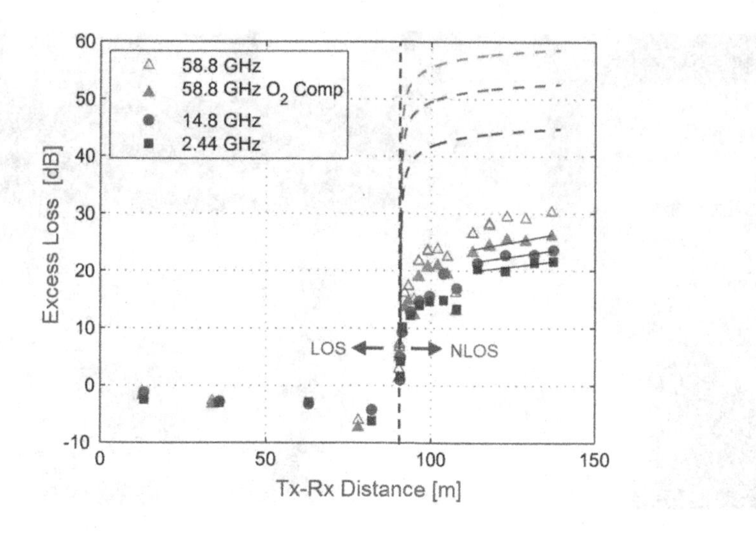

FIGURE 3.19

Measured loss versus distance for RX2 for the scenario shown in Fig. 3.20. The corresponding loss by knife edge diffraction is marked with dashed lines.

frequency dependence of the excess loss (about 3 log f [dB]) is clearly less than what is expected by knife edge diffraction (about 10 log f [dB]). It is, however, somewhat higher than that of the 3GPP channel model (1.3 log f [dB]) reported in [2].

In order to get some further insight into the propagation mechanisms, visual ray tracing has been performed for two measurement locations at 60 GHz as shown in Fig. 3.20. The first transmitter location (TX1) is in NLoS but very close to LoS. The first arriving path is attenuated by diffraction. It is possible to reconstruct the pathway of the strongest path assuming one specular reflection off an exterior building wall along the street. The second transmitter location (TX2) is substantially further down the street into the NLoS region. At the delay corresponding to the path diffracted around the corner, no signal above the noise floor is observed. The first cluster of weak paths is observed at propagation distances substantially longer than the diffraction path length. This cluster is likely to be caused by scatterers and/or rough surfaces in the area of the street corner. The strongest peak stands out having around 20 dB higher power level than the rest of the power delay profile. A plausible corresponding pathway (matching the propagation length of the measured peak) is possible to reconstruct assuming four specular reflections off exterior building walls. This shows that specular paths may be important even far into the NLoS region. However, for most of the NLoS locations such pronounced peaks were not observed.

3.3.3.4 Outdoor Urban Over Rooftop Scenario

To measure transmission loss for an urban outdoor over rooftop propagation scenario (macrocell scenario) is very challenging at higher frequencies, particularly in the millimeter-wave range as the transmission loss increases substantially when using practical omni antennas. This problem may be mitigated somewhat by using directive antennas at the BS location above the roof tops. However, at the UE locations on the ground, substantial angle spread is expected both in elevation and azimuth due to the UE being embedded in the urban clutter. To solve the problem by using high transmit power is

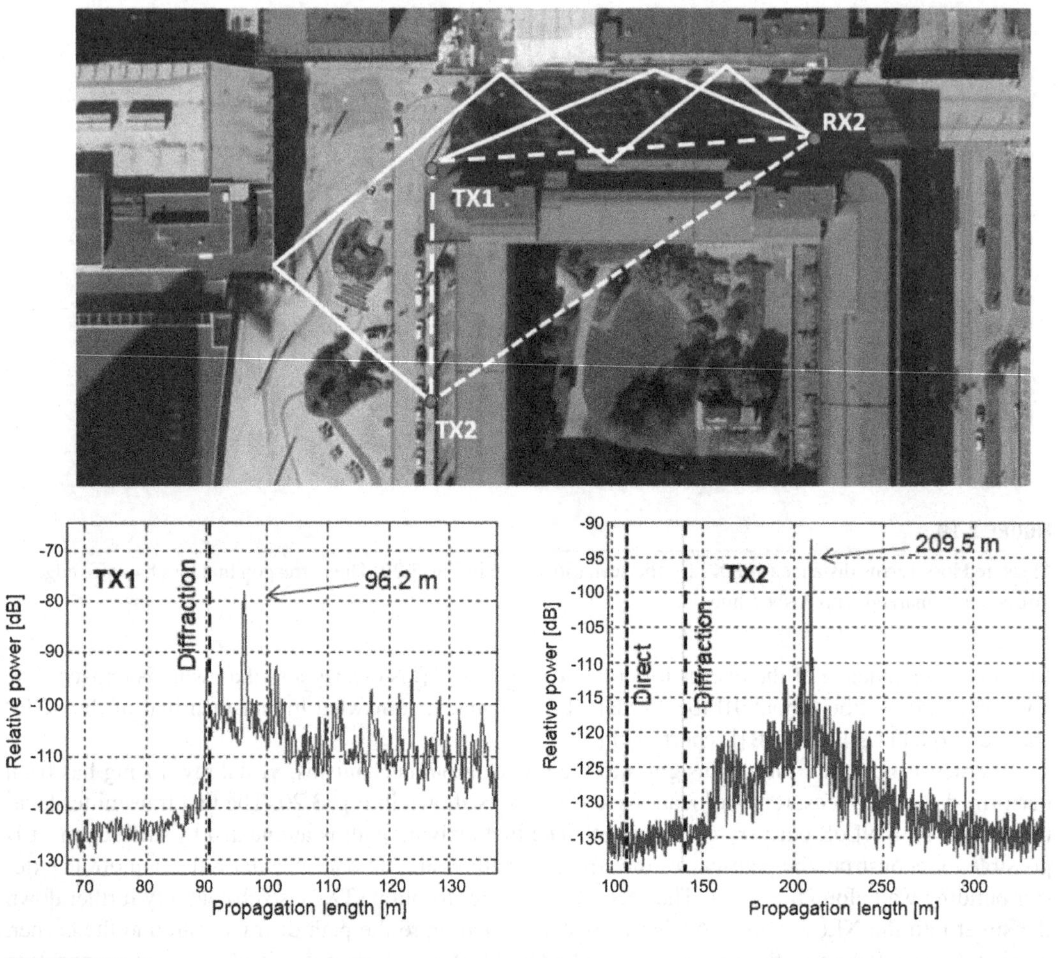

FIGURE 3.20

Power delay profiles in NLoS locations TX1 and TX2 at 60 GHz and corresponding reconstructed pathways.

practically not possible at millimeter-wave frequencies. A more practical solution is to use very high performance LNAs. Moreover, for CW measurements a very high suppression of noise is possible by means of Doppler filtering. This technique has been used in two independent multifrequency measurement campaigns in Aalborg [27] and Tokyo [23]. In the Aalborg campaign no frequency dependency was observed. This may, however, be the effect of a limited dynamic range at the highest frequency 28 GHz. In the Tokyo campaign the measurement sensitivity was better at the higher frequencies, allowing measurements up to 67 GHz.

The 3GPP modeling of urban macrocell transmission loss [2] is largely based on the Aalborg measurement campaign and is, relative to free-space, given by

$$L_{\text{UMa}} = 19.08\log(d) - 18.9 \text{ [dB]} \tag{3.27}$$

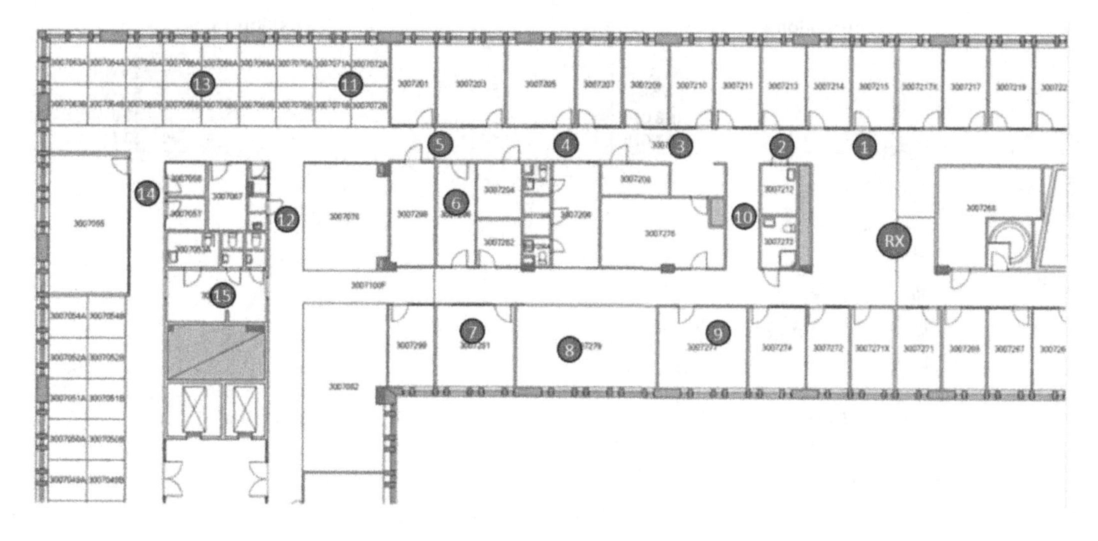

FIGURE 3.21

Indoor channel impulse response measurement scenario.

where d is the distance in meters and f the frequency in GHz. The corresponding model provided by ITU-R in [12] is based on the Tokyo measurements and is given by

$$L_{\mathrm{UMa}} = 23.9\log(d) - 38.7 + 3.0\log f \text{ [dB]}. \tag{3.28}$$

Except that the ITU-R model is frequency dependent and the 3GPP model is not, they are similar. The empirical basis for whether the loss for urban macrocell scenarios in general is frequency dependent or not is a subject for further investigation.

3.3.4 DELAY DOMAIN MEASUREMENTS

The delay domain is important for providing the characterization of the frequency selectivity of the channel as described in Section 3.2.1. Moreover, it is critical for optimizing transmission waveforms with respect to delay spread (see Chapter 6). As 3GPP has chosen OFDM for NR, cyclic prefix length optimization is directly related to the delay spread of the channel. In this section measured delay domain properties for a wide frequency range and important propagation scenarios are presented. The general frequency trends and comparisons with the 3GPP channel model are saved for Section 3.3.4.4.

3.3.4.1 Indoor Office

A multifrequency measurement campaign has been performed in an indoor office scenario as shown in Fig. 3.21. The receiver was placed at a fixed location and the transmitter at 15 different locations mainly in NLoS. All requirements for measurement comparability over different frequencies (provided in Section 3.3.2.3) are fulfilled. The channel is measured at 2.4, 5.8, 14.8, and 58.7 GHz. In Fig. 3.22 corresponding power delay profiles of two exemplary Tx locations are shown together with the average

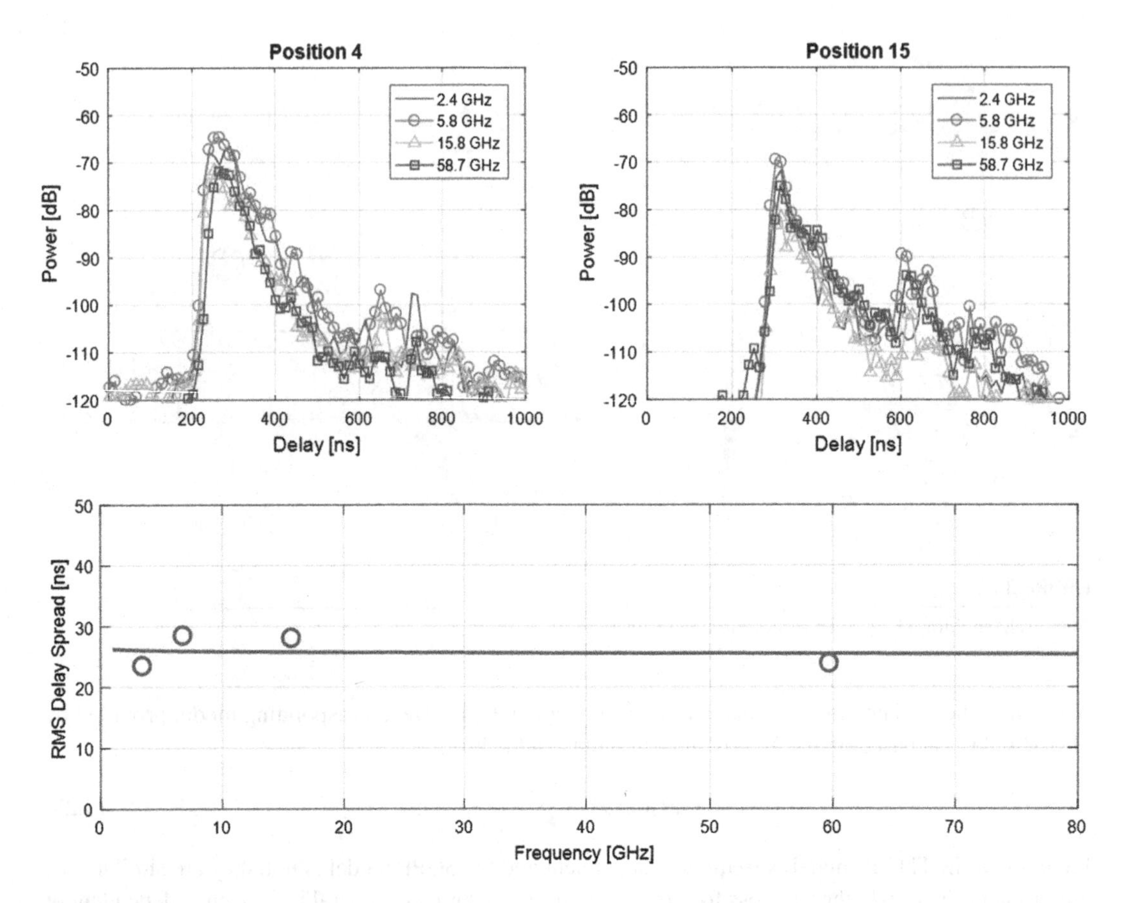

FIGURE 3.22

Power delay profiles for TX positions 4 and 14 (upper graphs) and average delay spread as a function of frequency together with fitted 3GPP type of model (lower graph).

RMS delay spread, σ_{DS}, as a function of frequency. No obvious frequency trend is observed, neither in power delay profiles nor in the average delay spread values. The independence is in fact well inside the 95% confidence range when fitting the 3GPP type of the model

$$\sigma_{DS} = 10^{\beta}(1+f)^{\alpha} \tag{3.29}$$

where α and β are model parameters and f is the carrier frequency in GHz. The fitted values for the curve shown in Fig. 3.22 are $\alpha = -0.01 \pm 0.05$ and $\beta = -7.58$.

3.3.4.2 Outdoor-to-Indoor

The outdoor-to-indoor measurements described in Section 3.3.3.2 have been further analyzed to determine delay spread and the corresponding frequency dependency. It is here important to point out that the dynamic range of the impulse responses in many locations is limited, down to below 10 dB.

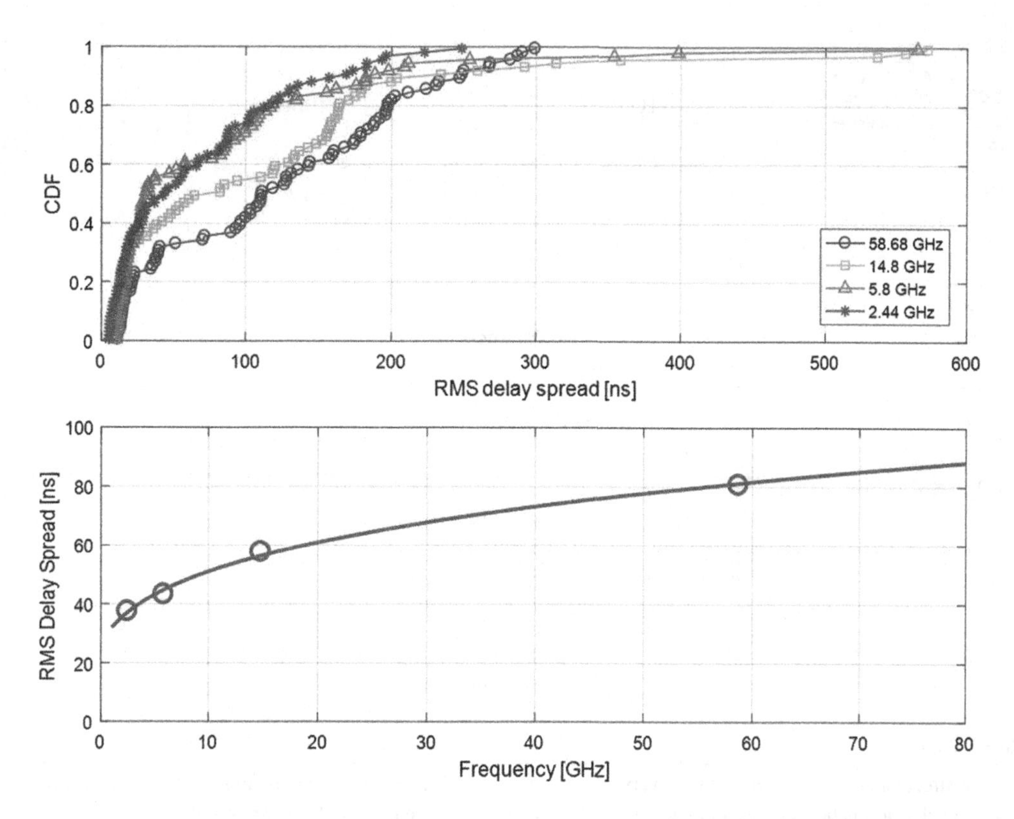

FIGURE 3.23

Outdoor-to-indoor CDFs of RMS delay spread for the different frequencies (upper graphs) and average delay spread as a function of frequency together with fitted 3GPP type of model (lower graph).

This means that the absolute delay spread sometimes may be underestimated. However, as the same dynamic range was used for all frequencies at each location no bias in the frequency dependency is introduced. The trend seems to be that the delay spread goes up with frequency as shown in Fig. 3.23. It is likely that the longer delays are caused by multiple reflections over the inner yard. Strong specular reflections are expected due to windows and/or metal tiles covering the exterior walls. However, these tiles and/or windows are smaller than one Fresnel zone at the lower frequencies, resulting in that the corresponding reflections are non-specular and therefore attenuated at the lower frequencies. It should be noted that this observed frequency trend is not general, but specific to the geometry of the scenario, as pointed out later in this section.

3.3.4.3 Outdoor Street Canyon Scenario

The outdoor street measurements described in Section 3.3.3.3 have also been analyzed to determine the corresponding delay spread characteristics with focus on the NLoS region. In this region the trend is that the delay spread seems to be independent of frequency, as shown in Fig. 3.24. The fitted model shows indeed that there is no significant frequency dependence.

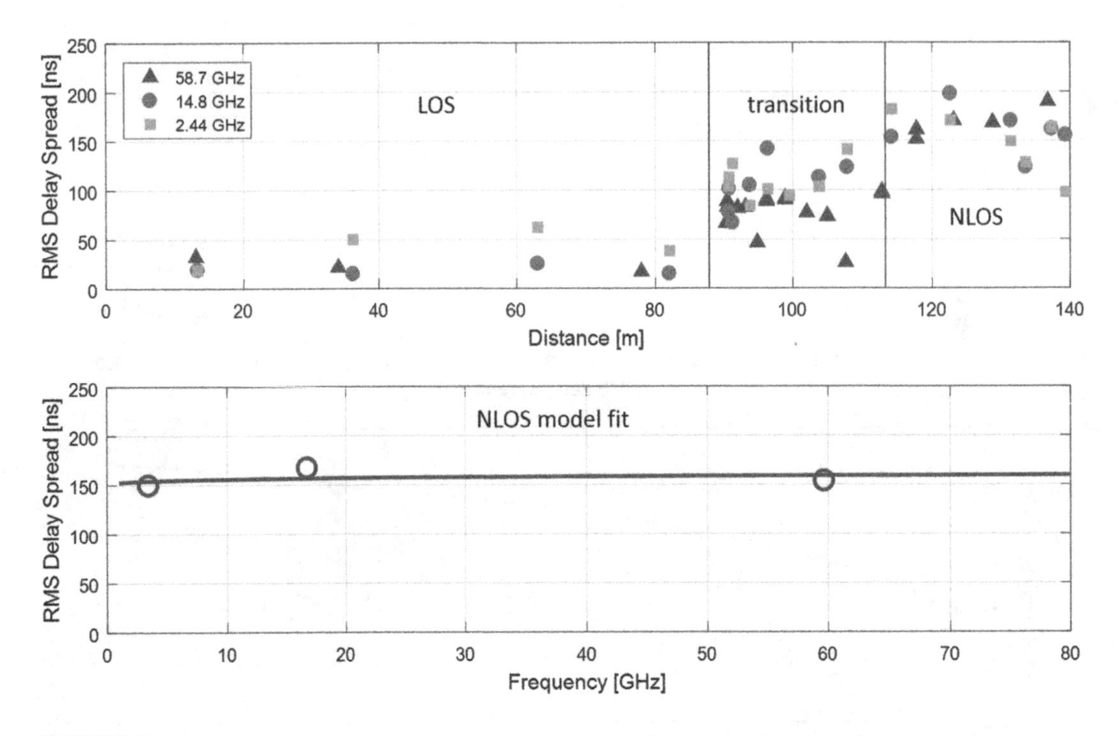

FIGURE 3.24

Outdoor street canyon RMS delay spread versus distance for the different frequencies (upper graph) and average delay spread as a function of frequency together with fitted 3GPP type of model (lower graph).

3.3.4.4 General Frequency Trend in Delay Domain

The experimental delay domain results presented so far indicate, except for the outdoor-to-indoor scenario, that there is no clear frequency trend. Previous results by e.g. 3GPP [2] indicate that the delay spread generally decreases when the frequency increases. However, when developing the 3GPP model the requirements for comparability between different frequency bands (provided in Section 3.3.2.3) were not thoroughly fulfilled, wherefore the corresponding results might be questioned. The EU funded project mmMAGIC [25], [26] has undertaken extensive channel measurements in which the requirements for comparability between different frequency bands were carefully seen to be fulfilled. Corresponding model parameters for five scenarios have been determined by statistically combining 15 independent measurement campaigns by six organizations. In Fig. 3.25 the 3GPP type of model (3.29) fitted to the mmMAGIC measurement data as well as the corresponding 3GPP model values are shown. There is a clear discrepancy between the two model fits, where the 3GPP model parameters show a strong general decrease of delay spread as a function of frequency, which is absent in the mmMAGIC data. Only street canyon LoS and indoor office in LoS show a slightly decreasing trend, within the 95% confidence range, for the mmMAGIC data.

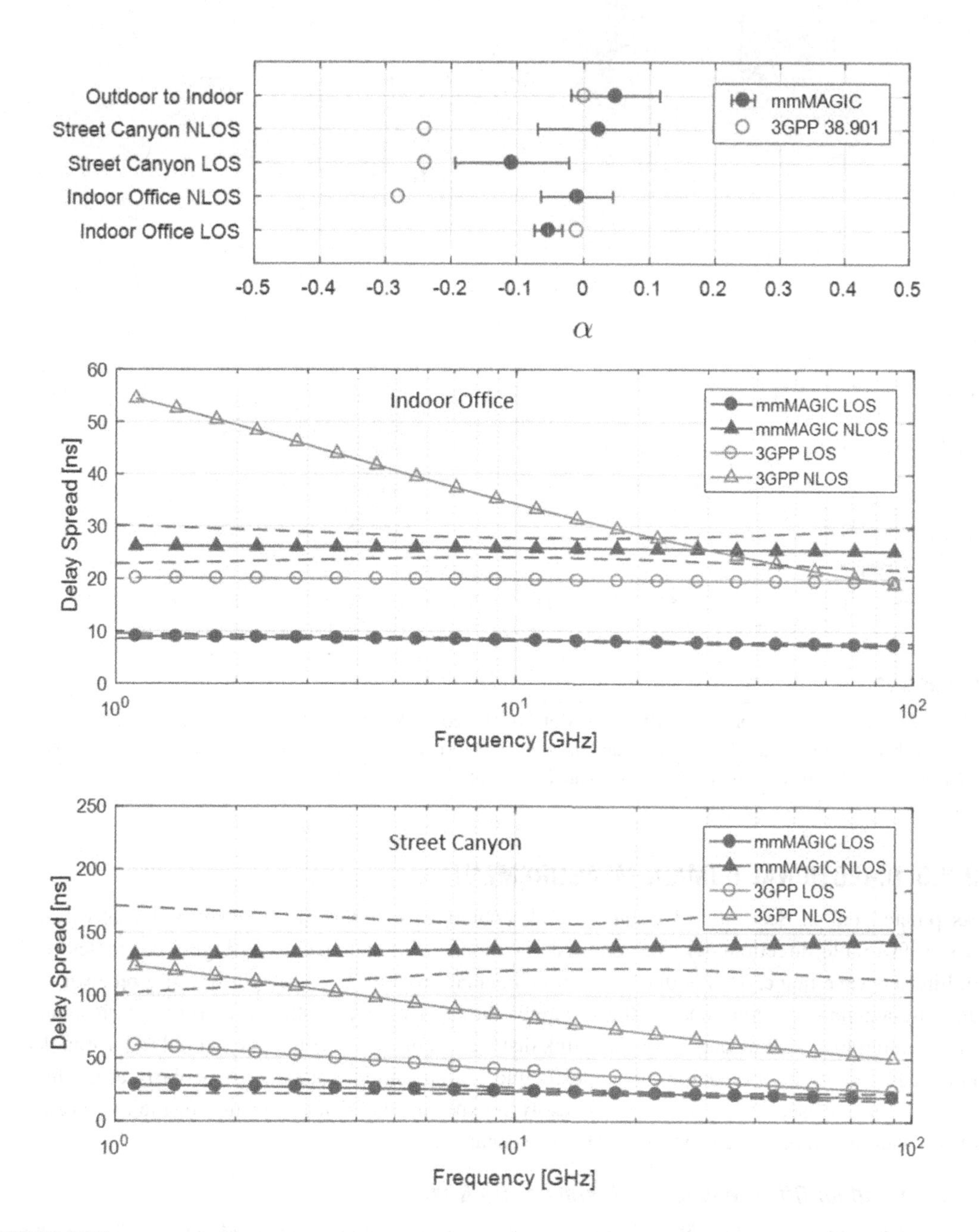

FIGURE 3.25

Delay spread model (3.29) fitted to mmMAGIC channel measurements and corresponding 3GPP model. The dashed lines indicate 95% the confidence limits for the mmMAGIC fits.

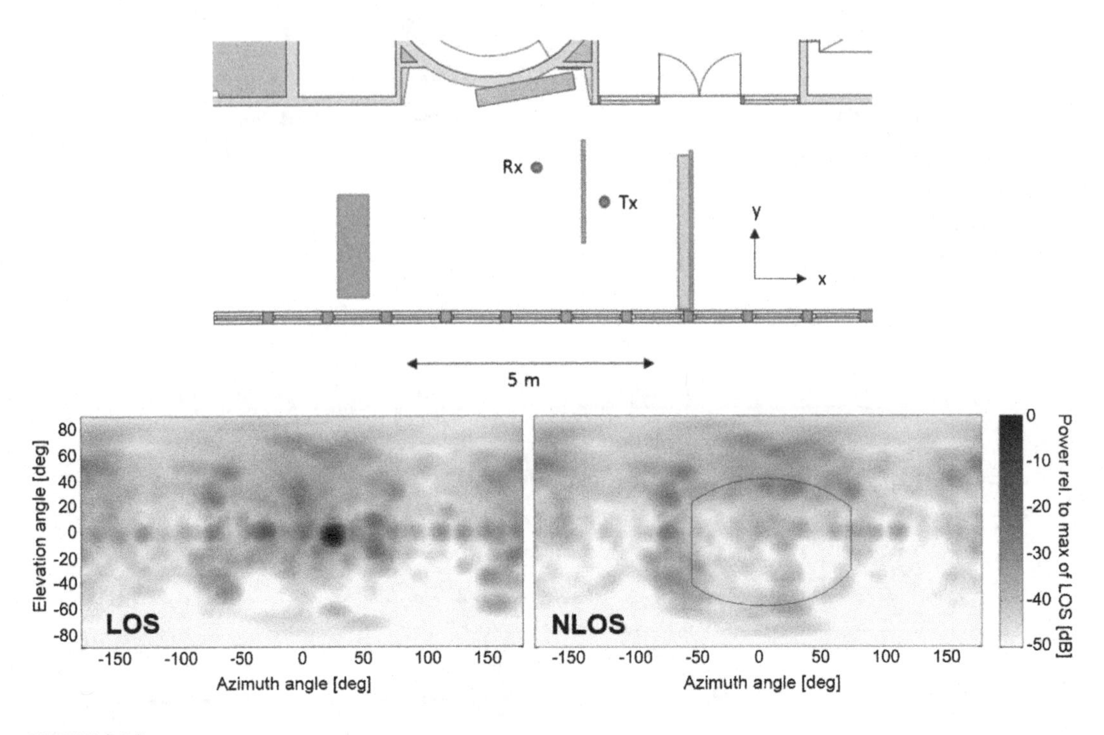

FIGURE 3.26

Measurement scenario (upper graph) where the whiteboard used for NLoS measurements is marked with a vertical line between the Rx and Tx antennas, and directional power spectra (lower graphs) for the LoS and NLoS measurements. The whiteboard contour is indicated with a black line in the NLoS graph.

3.3.5 DIRECTIONAL DOMAIN MEASUREMENTS

As pointed out in Sections 3.1.2 and 3.3.1.4 techniques to point the transmit and receive antennas in favorable directions will be required for mobile communications at high frequencies in the millimeter-wave range. This is due to the fundamental property that the aperture of a receiver using an omni-antenna is proportional to the square of the carrier wavelength and therefore corresponding transmission losses allow only very short link distances. For this reason it is essential to have in depth knowledge about the directional properties of radio propagation channels, particularly at the higher frequencies. This section presents highly resolved experimental characterization of the radio channel over a wide frequency range for some selected scenarios.

3.3.5.1 Indoor Office Wideband Results at 60 GHz

Channel measurement data at 58.7 GHz from an indoor office scenario, depicted in Fig. 3.26, have been analyzed using the spectral analysis method described in Section 3.3.2.1. For this purpose the channel has been sampled using a vertical dipole (2 dBi gain) antenna in both the transmit and the receive end of the link. The spatial samples were obtained by means of a 3D antenna positioning robot providing a

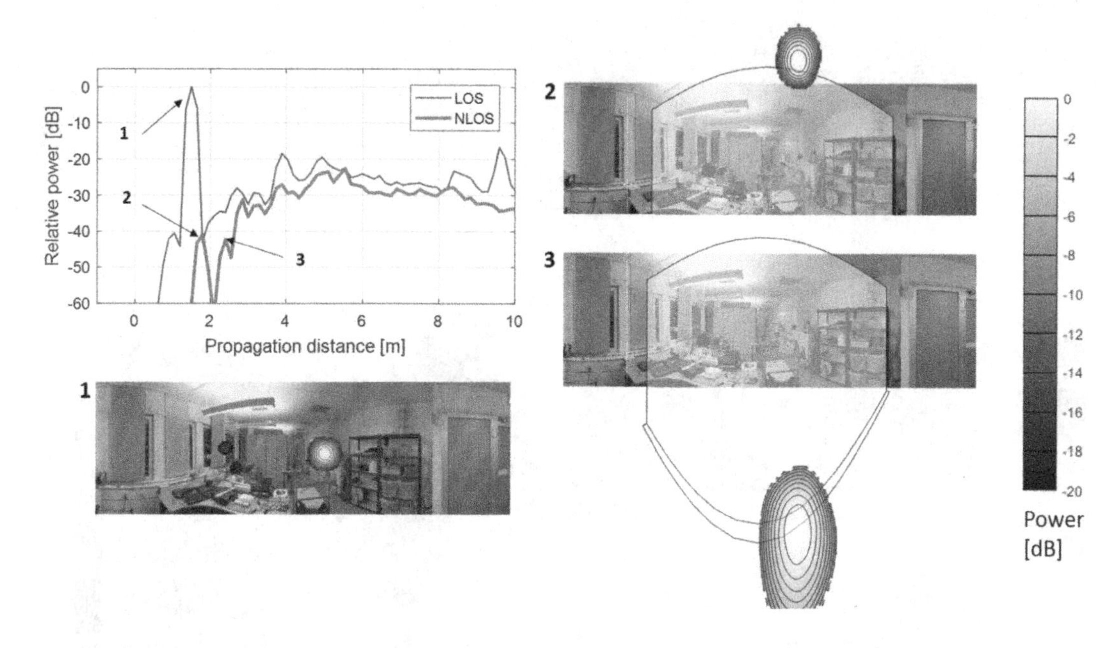

FIGURE 3.27

Power delay profiles for both LoS and NLoS (upper left graph) together with directional distributions of the first arriving paths on top of panoramic photographs.

virtual antenna array the size of 25x25x25 = 15625 elements using a spatial sampling distance of 0.4 wavelengths.

The measurements were performed in an indoor office environment at 1.5 m Tx–Rx distance under both LoS and NLoS conditions in the 57.68–59.68 GHz band. For the NLoS measurement a 2 m x 1.2 m large metal whiteboard was placed between the Tx and Rx antennas. The locations of Tx and Rx were the same in both the LoS and the NLoS measurements. In Fig. 3.26 the full directional spectra for both the LoS and the NLoS measurements are shown. Due to the short distance the LoS measurement is clearly dominated by the direct path. In contrast to LoS, the NLoS measurement is spatially much richer, having around ten strong paths in different directions. However, except for the direct path, and a few other strong paths blocked by the whiteboard, the two measurements show very similar directional characteristics. The channel seems to be composed of some distinct directions on top of a more smooth (diffuse) background. There seems to be a rich distribution of diffuse paths in all directions, except for those directions which correspond to the parts of the floor which are empty (no furniture).

In Fig. 3.27 corresponding power delay profiles for the LoS and the NLoS cases are shown for the first arriving paths. The first path is clearly dominant in the LoS case whereas the later reflected paths dominate in the NLoS case. There is, however, an early path also in the NLoS case which is suppressed by 40 dB relative to the LoS path due to diffraction over the upper edge of the whiteboard. There is also a path arriving somewhat later which is diffracted at the lower edge of the whiteboard. Both diffracted paths are shown on top of a panoramic photograph in Fig. 3.27.

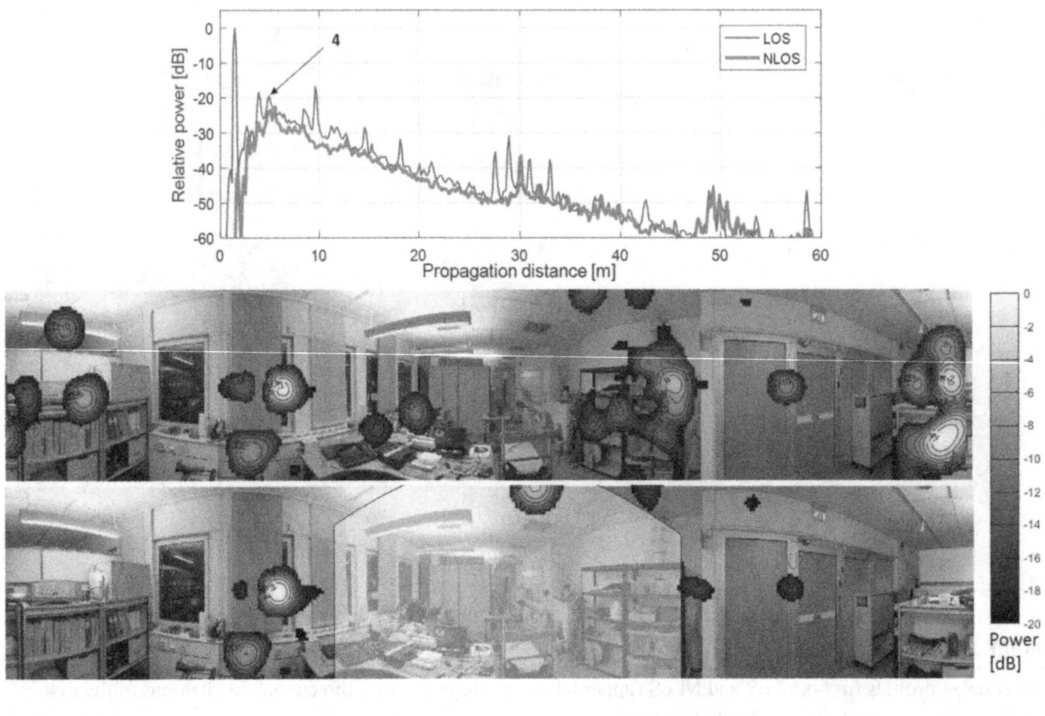

FIGURE 3.28

Power delay profiles for both LoS and NLoS (upper graph) together with directional distributions of the strongest path in NLoS on top of panoramic photographs.

When analyzing the strong peak number 4 of the power delay profile, shown in Fig. 3.28, it is clear that there are some significant scatterers such as objects on the table and the bookshelves. There are also a couple of strong reflections off a window and a wall segment. Comparing the LoS and the NLoS cases it is clear that some of the high power directions are absent in the NLoS graph due to the corresponding pathways being blocked by the whiteboard. An important conclusion that can be drawn is that simple ray tracing, based on the geometry of the room, would not succeed to model the rich scattering caused by furniture and other objects in the room.

3.3.5.2 Indoor Office Multifrequency Results

The measurements of previous section have been extended, adding two lower frequencies at 5.8 GHz and 14.8 GHz. In order to provide comparable characteristics over all measured frequencies the requirements of Section 3.3.2.3 are fulfilled in both measurements and analysis. For this purpose the measurement bandwidths have been equalized in the analysis, meaning that the 14.8 GHz and the 58.7 GHz measurement data are reduced to the bandwidth of the 5.8 GHz measurement data of 150 MHz. This equalization is important to avoid the effect that specular spikes, of the power delay profile, are amplified at higher frequencies due to the much larger bandwidth being available. Furthermore, the measurements at 58.7 GHz are affected by attenuation due to oxygen absorption. In order to provide re-

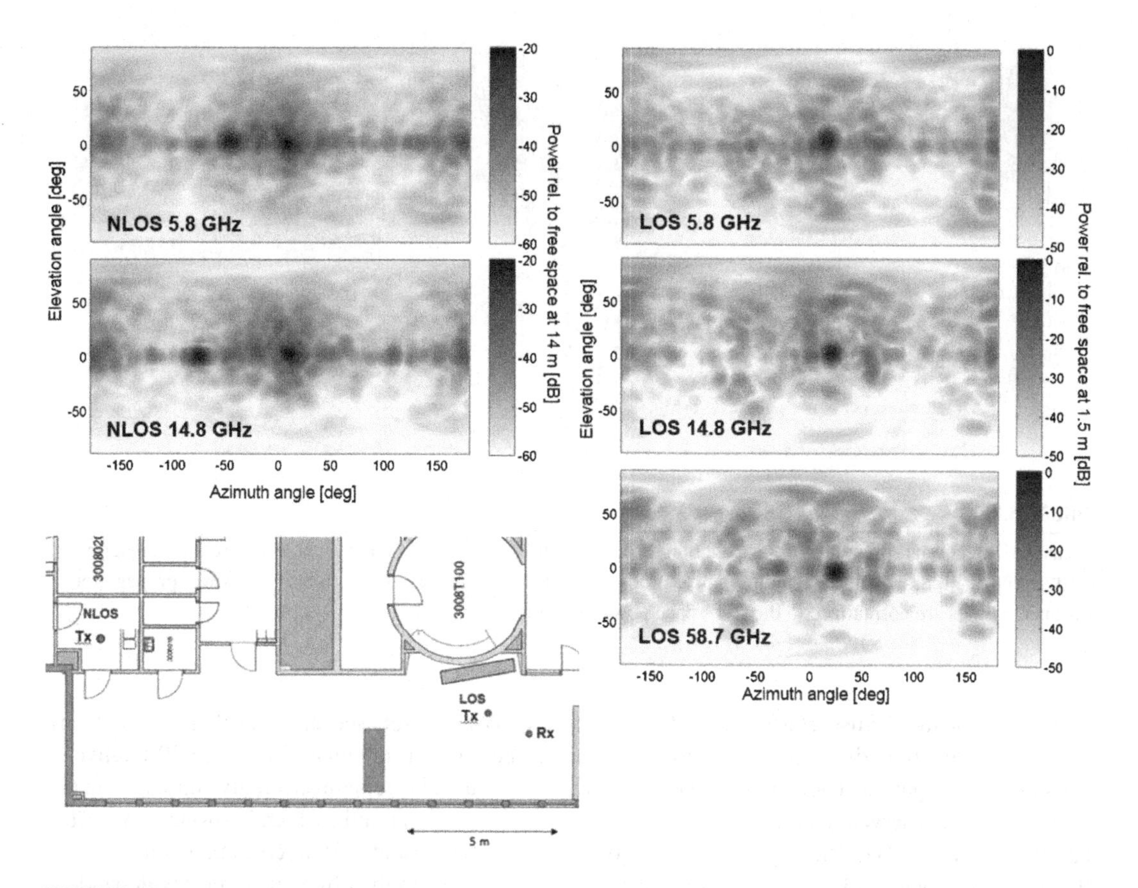

FIGURE 3.29

Measurement scenario (lower left graph), and directional power spectra for the NLoS (upper and middle left graphs) and LoS (right graphs) measurements for the different frequencies.

sults suitable for frequency consistent channel modeling and interpolation, this attenuation is removed in the analysis by compensating the power delay profiles with 1.5 dB per 100 m propagation distance at this frequency. It should be noted that though the link distance itself does not motivate this compensation, scattered paths may have substantially longer propagation distances, as seen in Fig. 3.31.

The LoS scenario is the same as in the previous section. In the NLoS measurements, the Rx antenna was placed in a small kitchen at the end of the office space (see Fig. 3.29). The distance between the Tx and the Rx antennas in the NLoS case was 14 m. It should be noted that the NLoS scenario was measured only at 5.8 GHz and 14.8 GHz due to the suspension of access to the indoor environment in the middle of the campaign. The directional power spectra are strikingly similar for all frequencies. One small frequency dependent difference is observed in the LoS graphs where a band of higher signal power around zero degree elevation angle is most pronounced at 5.8 GHz and least pronounced at 14.8 GHz. This difference is explained by the reflectivity of the windows which was found to be substantially higher at 5.8 GHz than at the other frequencies. The NLoS graphs differ from the LoS

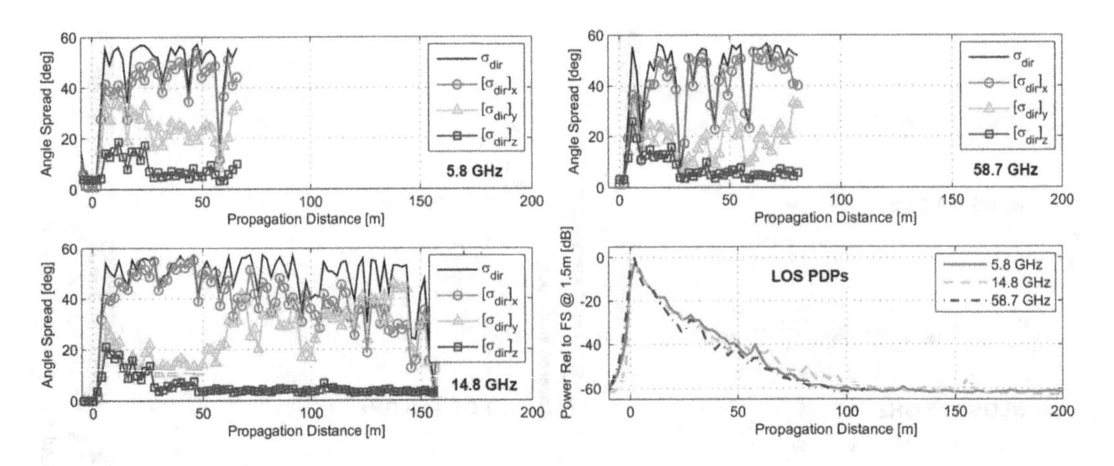

FIGURE 3.30

Measured directional spread versus propagation distance for the LoS scenario and the three different frequencies (left and upper right) and corresponding power propagation distance profiles (lower right). The power propagation distance profiles are normalized to the free-space power at 1.5 m.

graphs in that the diffuse cluster around the main peak is more focused and that there are not one but a couple of strong directions. Furthermore, the received power is attenuated about 20 dB relative to free-space propagation. The graphs of the two measured frequencies remain strikingly similar, as in the LoS case. One observed difference is the peak at −50 degrees in azimuth, which is strong at 5.8 GHz and weak at 14.8 GHz. The opposite effect is observed for the peak at −75 degrees in azimuth where the power is strong at 14.8 GHz and weak at 5.8 GHz. This is again the effect of frequency dependent window attenuation/reflection.

Directional spreads, σ_{dir}, according to the rotation invariant definition of Section 3.2.3 have been determined for the different scenarios and frequencies. In Fig. 3.30 the directional spreads for the LoS scenario versus propagation distance together with corresponding PDPs are shown. Again it is striking how similar the profiles for the different frequencies are. The directional spread is shown only for the part of the power propagation distance profiles for which there is sufficient signal above the noise floor. Basically the same characteristics are observed at all frequencies. For the LoS spike the spread is small, around 5 degrees. At other delays the spread is typically saturated at 57.3 degrees, which by definition is the maximum possible spread, as shown in Section 3.2.3. At a few delays where strong reflections occur the directional spread goes down. Another striking observation is that the elevation spread decays fast to very small values. For longer delays the directional spread in the x-dimension, which is the longest dimension of the room, dominates. Except for short delays the directional spreads in the different dimensions seem to be proportional to the corresponding lengths of the room. A likely explanation is that the power decays faster for smaller room dimensions, due to more frequent interactions with the corresponding walls, floor and ceiling, resulting in a smaller directional spread in those dimensions.

In Fig. 3.31 the corresponding graphs are shown for the NLoS scenario. The characteristics are very similar to those of the LoS scenario. One of the main observations is that there is a strong echo at about

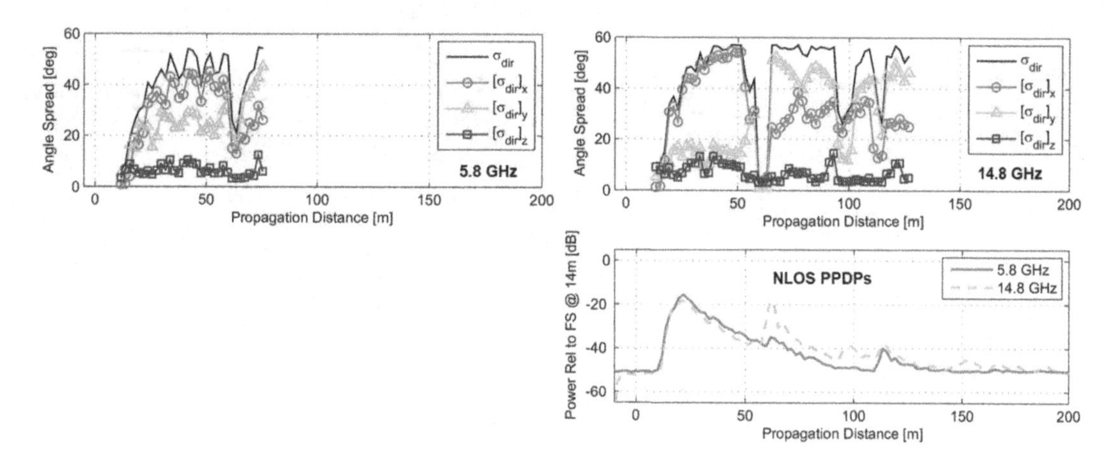

FIGURE 3.31

Measured directional spread versus propagation distance for the NLoS scenario and the three different frequencies (left and upper right) and corresponding power propagation distance profiles (lower right). The power propagation distance profiles are normalized to the free-space power at 14 m.

60 m propagation distance at 14.8 GHz, which is not observed at 5.8 GHz. The reason for the difference is that the windows reflect/attenuate differently at different frequencies. This is an effect of the three layers of glass (non-metalized) in the windows of the room. Due to multiple reflections between these layers different frequencies are attenuated differently when transmitted through the windows. Dedicated window attenuation measurements, performed in conjunction with the channel measurements, show that the window attenuation at 14.8 GHz is negligible, while it is around 10 dB at 5.8 GHz. The assumption that the strong echo at 14.8 GHz is due to a pathway going out through a window reflected off a neighboring building and in again through another window explains exactly the power difference of about 20 dB (relative to 5.8 GHz) which is twice the window attenuation at 5.8 GHz. It also explains why the directional spread in the y-direction increases at delays when strong radio waves enter the room from outside, by reflection off the adjacent building, as the power of waves propagating along the y-dimension of the room then is increased.

The total directional spreads, corresponding to summing the power of all delays (using angular distributions shown in Fig. 3.29), are shown in Fig. 3.32. Due to the shorter LoS link distance at 58.7 GHz (1.5 m) the relatively stronger LoS peak introduces a corresponding bias in directional spread (smaller spread) as compared with 5.8 and 14.8 GHz. In order to remove this bias the LoS peak has been decreased by 2.5 dB at 58.7 GHz in the analysis. No obvious frequency trend is observed. The characteristics are very similar for all frequencies where the elevation spread is small around 10 degrees and the directional spreads in x- and y-dimensions are substantially larger, between 20 and 40 degrees. Moreover, the directional spread in the y-dimension is substantially larger for the NLoS scenario due to the pathway which goes out of the building and in again, after reflecting off the adjacent building at 14.8 GHz and due to a strong window reflection at 5.8 GHz.

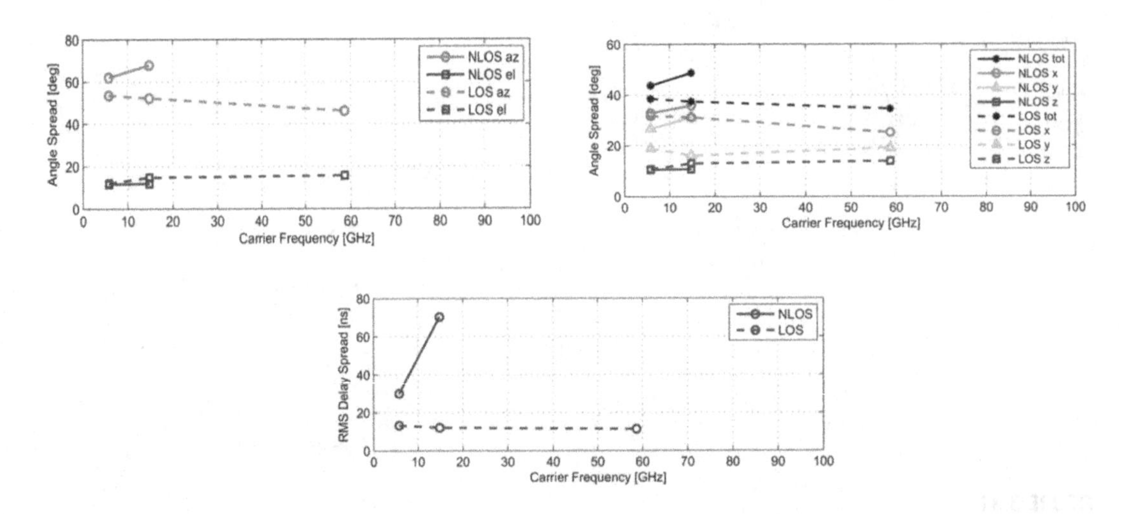

FIGURE 3.32

Total directional spreads (upper two) and delay spreads (lower) versus carrier frequency for the different measurement scenarios. The upper left plot show ordinary r.m.s. azimuth and elevation spreads, while the upper right show directional spreads according to the new improved definition according to the definition of Section 3.2.3.

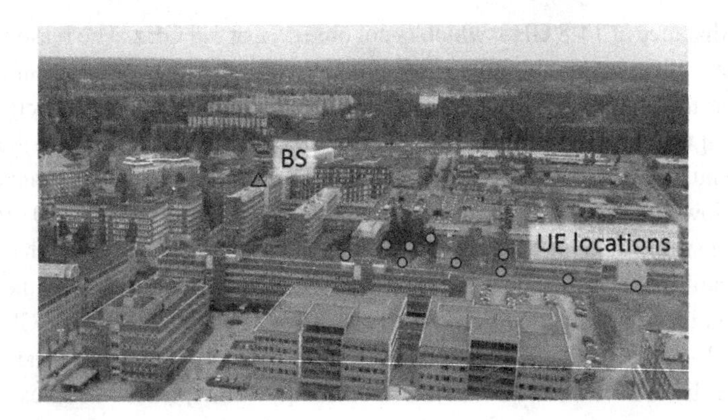

FIGURE 3.33

Photograph of measurement scenario.

3.3.5.3 Urban Macrocell Outdoor Results at 5 GHz

A measurement campaign in an urban outdoor macrocell scenario, depicted in Fig. 3.33, has been performed at 5.25 GHz using 200 MHz bandwidth [18]. At the BS a directional patch antenna with 7 dBi gain (90° beamwidth) and vertical polarization was used. A virtual planar array of 10x25 elements ($N_{\text{horizontal}} \times N_{\text{vertical}}$), with 2 cm (0.35λ) spacing was formed by means of a robotic antenna positioning system providing a spatial accuracy better than 0.1 mm. In the user equipment (UE) end an

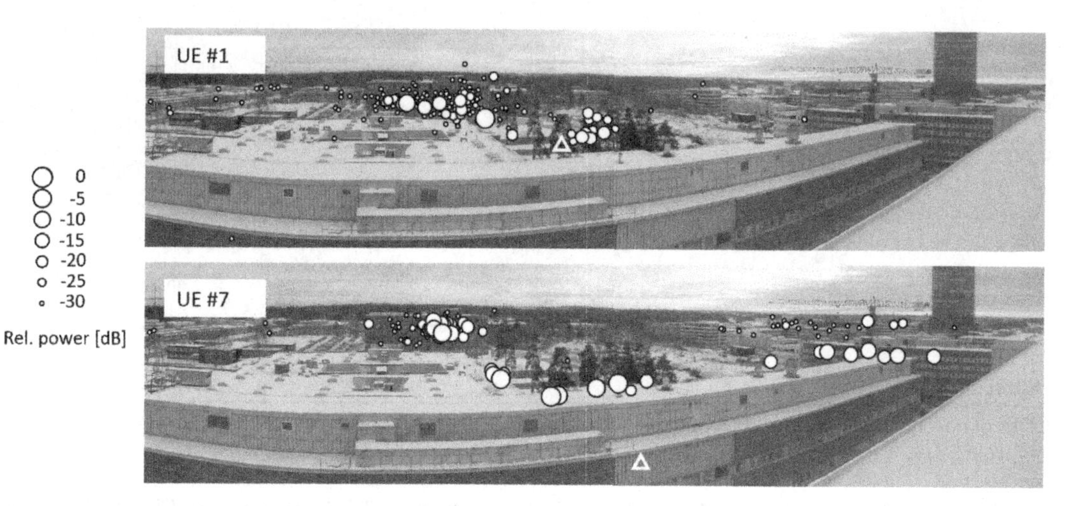

FIGURE 3.34

Estimated paths superimposed on a panoramic photograph taken from the BS location at the roof. The directions of the UE locations are indicated with triangles. It should be noted that the UEs are under NLoS conditions so that they are not visible from the BS.

ordinary vertical dipole antenna was used. The measurement data analysis is based on the superresolution method described in Section 3.3.2.2 and Ref. [17].

The total measurement time for a single UE location was around 7 min, since the antenna positioning system took about 1 s for each change of antenna position. Moreover, as it was a bit windy during the measurements, the movement of some significant trees affected the measurement data. At specific delays and directions of arrival and departure the corresponding radio waves are distorted due to time variations created by the moving trees. The result of these time variations is that the power of waves scattered in those trees will largely appear as randomly distributed in angle, i.e., as noise. The remaining coherent power scattered off the trees is reduced correspondingly, meaning that the significance of the trees is underestimated.

For each UE location $N = 500$ waves have been estimated. As described in [18] the estimates are initialized by finding peaks in the angle domain above the noise floor. Moreover, the requirement that the standard deviation of the errors of the estimates is less than 40 degrees in angle and 20 m in propagation distance is set. The corresponding result of a power delay profile is shown in Fig. 3.13 of Section 3.3.2.2 demonstrating that the estimates account for most of the power of the measured channel. To ensure a high reliability of the presented results the analysis is based on paths having estimation errors with standard deviations less than 2 degrees in elevation and 4 degrees in azimuth. The corresponding estimated plane waves superimposed on a panoramic photograph are shown in Fig. 3.34. It is clear that the dominant paths are diffracted over the roof tops and/or reflected off neighboring buildings. It seems that the main propagation mechanisms are reflections off neighboring buildings which are under LoS conditions from both the BS and the UE. At some UE locations (e.g. UE location 7) diffraction over the rooftops seems to be important.

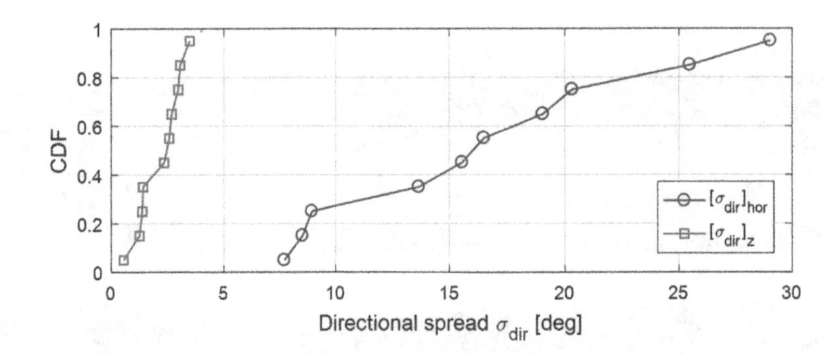

FIGURE 3.35

CDFs of directional spreads in z-direction and horizontal (x/y) plane.

The directional spreads according to the definition in Section 3.2.3 are shown in Fig. 3.35. In this case the spread in the z-direction is very similar to ordinary elevation spread and the spread in the horizontal plane is very similar to ordinary azimuth spread. The spread in z-direction (elevation) is very small, in the range 1–4 degrees, whereas the spread in the horizontal plane (azimuth) is between 7 and 30 degrees.

3.4 CHANNEL MODELING

This section deals with commonly used channel models specified in standardization bodies like 3GPP, ITU-R, and also international research projects and networks like METIS and mmMAGIC. The focus is put on the recent models with particular emphasis on 5G specific features. All these models are geometric stochastic channel models (except the METIS map based model) for which each channel realization is Monte Carlo generated according to each parameter's probability distribution. To achieve continuity in space and time the random distributions are auto-correlated in those dimensions. The provisioning of the full correlation matrix between all channel parameters and all autocorrelation functions is a huge challenge. Moreover, to provide values within proper confidence ranges requires a massive amount of measurement data. Even if all needed correlations are provided it is a considerable computational task to perform all autocorrelations needed for providing spatial consistency. The alternative approach employed in the METIS map based model is based on simplified geometry ray tracing. In this model, all correlations and spatial consistency are provided automatically as a result of the model being based fully on deterministic and physical principles. It accounts for all important scattering mechanisms, i.e., specular reflections, diffraction, scattering by rough surfaces and objects, and outdoor-to-indoor penetration. The drawback of this model is its corresponding implementation complexity.

The complexity of channel models has been increasing with each new generation of mobile communications. This is a result of the fact that each new generation has been providing better performance by utilizing more of the physical propagation channel's degrees of freedom. In the second generation (GSM) the complexity was relatively low as narrow bandwidths and low degree of spatial diversity were utilized. Through the third and the fourth generations both bandwidths and spatial diversity techniques were extended by e.g. introducing soft handover and MIMO transmission.

The main focus of the section is on the recent model for 5G developed by 3GPP [2] and ITU-R [9]. This model provides a successful extension of transmission loss modeling into the mm-Wave range up to 100 GHz. Moreover, extended bandwidth (up to 2 GHz) and highly resolved directional characterization of paths is provided. Though some of the very highly resolved characteristics are not fully validated the model is likely to be accurate enough for the early 5G networks which do not utilize the largest bandwidths and narrowest antenna beams.

3.4.1 5G STOCHASTIC CHANNEL MODELS

In this section the 3GPP channel model for the development of 5G in the frequency range 0.5–100 GHz [2], which also has been adopted by ITU-R for IMT-2020 [9], is described. The model is largely based on previous generation modeling [1], [6] and is of the geometry-based stochastic type of channel model. The mmMAGIC channel model [26] is largely harmonized with the 3GPP/ITU-R model with some improvements and additions. The channel H_{mn} is defined by a set of plane waves according to (3.7) for which the polarimetric amplitudes $\mathbf{A}_l$, directions $\mathbf{k}_l^{\mathrm{tx}}$ and $\mathbf{k}_l^{\mathrm{rx}}$, delays τ_l, and Doppler frequencies ω_{D_l} are empirically characterized with the corresponding stochastic distributions. Only the line-of-sight component is deterministically defined by the geometry.

3.4.1.1 Transmission Loss Modeling

The transmission loss modeling is based on assuming isotropic antennas in both ends of the link using the basic transmission loss definition of ITU-R Recommendation P.341 [8]. Any dependence on specific antenna patterns is modeled separately by combining the antenna pattern with corresponding multipath distributions as described in the next section. The transmission loss (referred to as path-loss in the ITU-R and 3GPP specifications) properties of the models are summarized in Table 3.1. These models are largely validated by the measurements presented in Section 3.3.3 for the indoor office and street canyon scenarios. The losses for the line-of-sight cases are close to the free-space loss up to a break point, after which the loss is proportional to $40\log(d)$. Under non-line-of-sight conditions, there is no frequency dependence of the loss relative to free space for the urban macro (UMa) and the rural macro (RMa) scenarios. However, for the indoor and street canyon scenarios there is a small increase of the loss relative to free space with increasing frequency.

Further, it is assumed that there is a log-normal stochastic location variation of the loss around the distance dependent functions. The corresponding standard deviations range between 4 and 8 dB. These variations are correlated in space by the exponential autocorrelation function

$$R = \exp\left(-\frac{|\Delta d|}{d_{\mathrm{cor}}}\right) \tag{3.30}$$

where Δd is the distance between two points in space and d_{cor} is a constant.

The building penetration loss, L_{O2I}, is modeled by

$$L_{\mathrm{O2I}} = 5 - 10\log\left(\alpha 10^{\frac{-L_{\mathrm{glass}}}{10}} + \beta 10^{\frac{-L_{\mathrm{concrete}}}{10}}\right) + \gamma d_{\mathrm{2D\text{-}in}} \tag{3.31}$$

Table 3.1 3GPP transmission loss models (with permission from 3GPP)

Scenario	LOS/NLOS	Pathloss [dB], f_c is in GHz and d is in meters, see note 6	Shadow fading std [dB]	Applicability range, antenna height default values
RMa	LOS	$PL_{RMa\text{-}LOS} = \begin{cases} PL_1 & 10\text{m} \le d_{2D} \le d_{BP} \\ PL_2 & d_{BP} \le d_{2D} \le 10\text{km} \end{cases}$, see note 5 $PL_1 = 20\log_{10}(40\pi d_{3D} f_c/3) + \min(0.03h^{1.72},10)\log_{10}(d_{3D}) - \min(0.044h^{1.72},14.77) + 0.002\log_{10}(h)d_{3D}$ $PL_2 = PL_1(d_{BP}) + 40\log_{10}(d_{3D}/d_{BP})$	$\sigma_{SF} = 4$ $\sigma_{SF} = 6$	$h_{BS} = 35\text{m}$ $h_{UT} = 1.5\text{m}$ $W = 20\text{m}$ $h = 5\text{m}$ h = avg. building height W = avg. street width The applicability ranges: $5\text{m} \le h \le 50\text{m}$ $5\text{m} \le W \le 50\text{m}$ $10\text{m} \le h_{BS} \le 150\text{m}$ $1\text{m} \le h_{UT} \le 10\text{m}$
	NLOS	$PL_{RMa\text{-}NLOS} = \max(PL_{RMa\text{-}LOS}, PL'_{RMa\text{-}NLOS})$ for $10\text{m} \le d_{2D} \le 5\text{km}$ $PL'_{RMa\text{-}NLOS} = 161.04 - 7.1\log_{10}(W) + 7.5\log_{10}(h) - (24.37 - 3.7(h/h_{BS})^2)\log_{10}(h_{BS}) + (43.42 - 3.1\log_{10}(h_{BS}))(\log_{10}(d_{3D}) - 3) + 20\log_{10}(f_c) - (3.2(\log_{10}(11.75h_{UT}))^2 - 4.97)$	$\sigma_{SF} = 8$	
UMa	LOS	$PL_{UMa\text{-}LOS} = \begin{cases} PL_1 & 10\text{m} \le d_{2D} \le d'_{BP} \\ PL_2 & d'_{BP} \le d_{2D} \le 5\text{km} \end{cases}$, see note 1 $PL_1 = 28.0 + 22\log_{10}(d_{3D}) + 20\log_{10}(f_c)$ $PL_2 = 28.0 + 40\log_{10}(d_{3D}) + 20\log_{10}(f_c) - 9\log_{10}((d'_{BP})^2 + (h_{BS} - h_{UT})^2)$	$\sigma_{SF} = 4$	$1.5\text{m} \le h_{UT} \le 22.5\text{m}$ $h_{BS} = 25\text{m}$
	NLOS	$PL_{UMa\text{-}NLOS} = \max(PL_{UMa\text{-}LOS}, PL'_{UMa\text{-}NLOS})$ for $10\text{m} \le d_{2D} \le 5\text{km}$ $PL'_{UMa\text{-}NLOS} = 13.54 + 39.08\log_{10}(d_{3D}) + 20\log_{10}(f_c) - 0.6(h_{UT} - 1.5)$	$\sigma_{SF} = 6$	$1.5\text{m} \le h_{UT} \le 22.5\text{m}$ $h_{BS} = 25\text{m}$ Explanations: see note 3
		Optional $\text{PL} = 32.4 + 20\log_{10}(f_c) + 30\log_{10}(d_{3D})$	$\sigma_{SF} = 7.8$	
UMi - Street Canyon	LOS	$PL_{UMi\text{-}LOS} = \begin{cases} PL_1 & 10\text{m} \le d_{2D} \le d'_{BP} \\ PL_2 & d'_{BP} \le d_{2D} \le 5\text{km} \end{cases}$, see note 1 $PL_1 = 32.4 + 21\log_{10}(d_{3D}) + 20\log_{10}(f_c)$ $PL_2 = 32.4 + 40\log_{10}(d_{3D}) + 20\log_{10}(f_c) - 9.5\log_{10}((d'_{BP})^2 + (h_{BS} - h_{UT})^2)$	$\sigma_{SF} = 4$	$1.5\text{m} \le h_{UT} \le 22.5\text{m}$ $h_{BS} = 10\text{m}$
	NLOS	$PL_{UMi\text{-}NLOS} = \max(PL_{UMi\text{-}LOS}, PL'_{UMi\text{-}NLOS})$ for $10\text{m} \le d_{2D} \le 5\text{km}$ $PL'_{UMi\text{-}NLOS} = 35.3\log_{10}(d_{3D}) + 22.4 + 21.3\log_{10}(f_c) - 0.3(h_{UT} - 1.5)$	$\sigma_{SF} = 7.82$	$1.5\text{m} \le h_{UT} \le 22.5\text{m}$ $h_{BS} = 10\text{m}$ Explanations: see note 4
		Optional $\text{PL} = 32.4 + 20\log_{10}(f_c) + 31.9\log_{10}(d_{3D})$	$\sigma_{SF} = 8.2$	

(continued on next page)

Table 3.1 *(continued)*

Scenario		Path loss	Shadow fading	Applicability range
InH - Office	LOS	$PL_{\text{InH-LOS}} = 32.4 + 17.3\log_{10}(d_{3D}) + 20\log_{10}(f_c)$	$\sigma_{SF} = 3$	$1\text{m} \le d_{3D} \le 150\text{m}$
	NLOS	$PL_{\text{InH-NLOS}} = \max(PL_{\text{InH-LOS}}, PL'_{\text{InH-NLOS}})$ $PL'_{\text{InH-NLOS}} = 38.3\log_{10}(d_{3D}) + 17.30 + 24.9\log_{10}(f_c)$	$\sigma_{SF} = 8.03$	$1\text{m} \le d_{3D} \le 150\text{m}$
		Optional $PL'_{\text{InH-NLOS}} = 32.4 + 20\log_{10}(f_c) + 31.9\log_{10}(d_{3D})$	$\sigma_{SF} = 8.29$	$1\text{m} \le d_{3D} \le 150\text{m}$

Note 1: Breakpoint distance $d'_{BP} = 4\, h'_{BS}\, h'_{UT}\, f_c/c$, where f_c is the centre frequency in Hz, $c = 3.0\times10^8$ m/s is the propagation velocity in free space, and h'_{BS} and h'_{UT} are the effective antenna heights at the BS and the UT, respectively. The effective antenna heights h'_{BS} and h'_{UT} are computed as follows: $h'_{BS} = h_{BS} - h_E$, $h'_{UT} = h_{UT} - h_E$, where h_{BS} and h_{UT} are the actual antenna heights, and h_E is the effective environment height. For UMi $h_E = 1.0$m. For UMa h_E=1m with a probability equal to $1/(1+C(d_{2D}, h_{UT}))$ and chosen from a discrete uniform distribution uniform(12,15,…,(h_{UT}-1.5)) otherwise. With $C(d_{2D}, h_{UT})$ given by

$$C(d_{2D}, h_{UT}) = \begin{cases} 0 & , h_{UT} < 13\text{m} \\ \left(\frac{h_{UT} - 13}{10}\right)^{1.5} g(d_{2D}) & , 13\text{m} \le h_{UT} \le 23\text{m} \end{cases},$$

where

$$g(d_{2D}) = \begin{cases} 0 & , d_{2D} \le 18\text{m} \\ \frac{5}{4}\left(\frac{d_{2D}}{100}\right)^3 \exp\left(\frac{-d_{2D}}{150}\right) & , 18\text{m} < d_{2D} \end{cases}.$$

Note that h_E depends on d_{2D} and h_{UT} and thus needs to be independently determined for every link between BS sites and UTs. A BS site may be a single BS or multiple co-located BSs.

Note 2: The applicable frequency range of the PL formula in this table is $0.5 < f_c < f_H$ GHz, where $f_H = 30$ GHz for RMa and $f_H = 100$ GHz for all the other scenarios. It is noted that RMa pathloss model for >7 GHz is validated based on a single measurement campaign conducted at 24 GHz.

Note 3: UMa NLOS pathloss is from TR36.873 with simplified format and $PL_{\text{UMa-LOS}}$ = Pathloss of UMa LOS outdoor scenario.

Note 4: $PL_{\text{UMi-LOS}}$ = Pathloss of UMi-Street Canyon LOS outdoor scenario.

Note 5: Break point distance $d_{BP} = 2\pi\, h_{BS}\, h_{UT}\, f_c/c$, where f_c is the centre frequency in Hz, $c = 3.0 \times 10^8$ m/s is the propagation velocity in free space, and h_{BS} and h_{UT} are the antenna heights at the BS and the UT, respectively.

Note 6: f_c denotes the center frequency normalized by 1GHz, all distance related values are normalized by 1m, unless it is stated otherwise.

where α and β are, respectively, the fractions of glass and concrete area of exterior building walls, L_{glass} and L_{concrete} are the corresponding losses of the multipane windows and concrete, γ is the loss per meter horizontal penetration depth $d_{\text{2D-in}}$ into the building. In the 3GPP and ITU models $\gamma = 0.5$ dB/m. A more thorough analysis by mmMAGIC shows that γ is uniformly distributed between 0.5 and 1.5 dB/m. It has been found that buildings may be grouped in a high-loss and a low-loss category. Buildings in the high-loss category are referred to as "thermal efficient," as the corresponding construction materials and methods result in considerable penetration losses. Particularly the thin metal coatings used for blocking heat radiation through windows also attenuates radio waves in the order of 20–30 dB. The low loss category corresponds to traditional buildings for which non-coated windows are used. The loss caused by the exterior wall construction material, L_m, is given by

$$L_m = a + bf. \tag{3.32}$$

The corresponding material parameters are given in Table 3.2.

It is assumed that there is a log-normal location variability of the penetration loss due to internal irregularities of the buildings such as furniture, interior walls elevator shafts, etc. The corresponding standard deviations are $\sigma_{\text{low}} = 4.4$ dB and $\sigma_{\text{high}} = 6.5$ dB for the low-loss and the high-loss categories, respectively. The mmMAGIC model has refined the variability by introducing a frequency dependence

Table 3.2 3GPP penetration loss material parameters

Material	a [dB]	b [dB/GHz]
Standard multipane glass	2.0	0.2
IRR glass	23.0	0.3
Concrete	5.0	4.0

of the corresponding standard deviation,

$$\sigma = 4 + \mathrm{k}_\sigma f \tag{3.33}$$

where k_σ has been estimated to be 0.08 dB/GHz for the low loss category.

In Fig. 3.36 the building penetration loss of the 3GPP model as a function of frequency is shown for the two categories of buildings (traditional and thermal efficient) for 5%, 50%, and 95% levels of the location variability distributions. For comparison the corresponding losses of the ITU-R Recommendation P.2109 for building entry loss are shown. The ITU-R model is empirical and is based on extensive measurement data whereas the 3GPP model is based on simplified physical principles. At the 50% level the two models agree well for frequencies below 50 GHz. However, the frequency trend is substantially stronger for the 3GPP model. This may be explained by the fact that the values of Table 3.2 correspond to a summed glass layer thickness of 24 mm for traditional and 36 mm for thermal efficient windows [7]. This is about three times the thicknesses found in real buildings. It is also clear that the ITU-R model [10] reflects that the spread increases with frequency, which the 3GPP mode does not.

In the 3GPP and ITU-R IMT2020 models the dependence of the propagation angle of incidence relative to the exterior wall is accounted for by the addition of a constant of 5 dB in (3.31). As a consequence the spread of the loss distributions is reduced. This is clearly observed in Fig. 3.36 where the 5% and 95% levels are shifted substantially more for the model of the ITU-R Recommendation P.2109 [10]. As this Recommendation was developed to support spectrum sharing studies between e.g. IMT and satellites the dependence of the elevation angle is accounted for by

$$L(\theta) = L(\theta = 0) + \alpha_\theta \theta \tag{3.34}$$

where θ is the elevation angle of the path relative to the exterior wall, and α_θ is a constant which is estimated to be about 20 dB/90 degrees.

3.4.1.2 Multipath Directional and Delay Modeling

The distributions of amplitudes, delays and directions of multipath components are generated based on first and second order moments of closed form stochastic distributions. Furthermore, the distributions are specified at two levels: the intercluster level and the intra-cluster level. This means that at the higher level the stochastic distributions of clusters of multipath components are generated. At the lower level the corresponding distribution within clusters are generated. The motivation for this two level clustering is historical and based on observations of power delay profiles [31]. It may, however, be questioned if this topology really can be justified when comparing with highly resolved empirical data.

In the delay domain, both the probability and the power of clusters follow distributions which are exponentially decaying with a log-normal shadowing added. In angle domains, i.e., elevation and

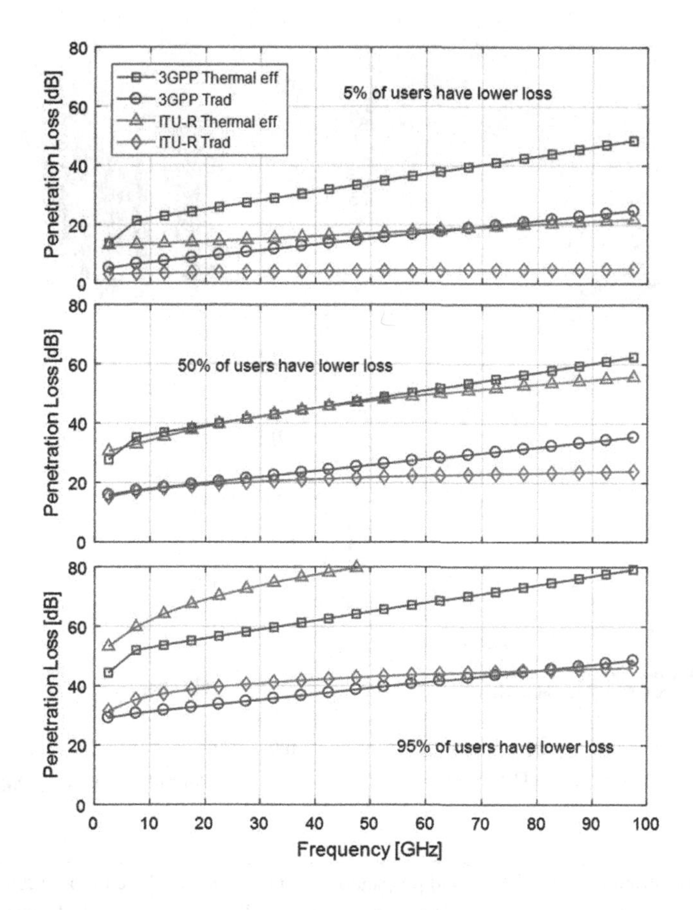

FIGURE 3.36

Building penetration loss versus frequency for the microcell scenario and the 3GPP and ITU-R P.2109 models at 5%, 50%, and 95% probability levels.

azimuth, the distributions are wrapped Gaussian functions of the cluster power. This is somewhat problematic, as the probability of having many clusters in the same direction becomes low which is in contrast with empirical data. In the baseline model each cluster consists of 20 multipath components. These are at fixed delays cluster-wise except for the two strongest clusters which are subdivided into three sublevels of fixed delays. All 20 subpaths have fixed power within each cluster and have a tabulated distribution to provide a Laplacian power distribution in angle. In Fig. 3.37 multipath component distributions in azimuth and propagation distance are shown for the indoor office scenario (same environment as in Section 3.3.4.1) in NLoS at 60 GHz. Both very high resolution measurement data [14], [26] and a corresponding realization of the 3GPP model are shown. The measurements were performed in an office environment using a 50 cm wide and 12.5 cm high planar array and 2 GHz bandwidth providing extreme resolution in direction. It is clear that the measured distributions do not show the effect

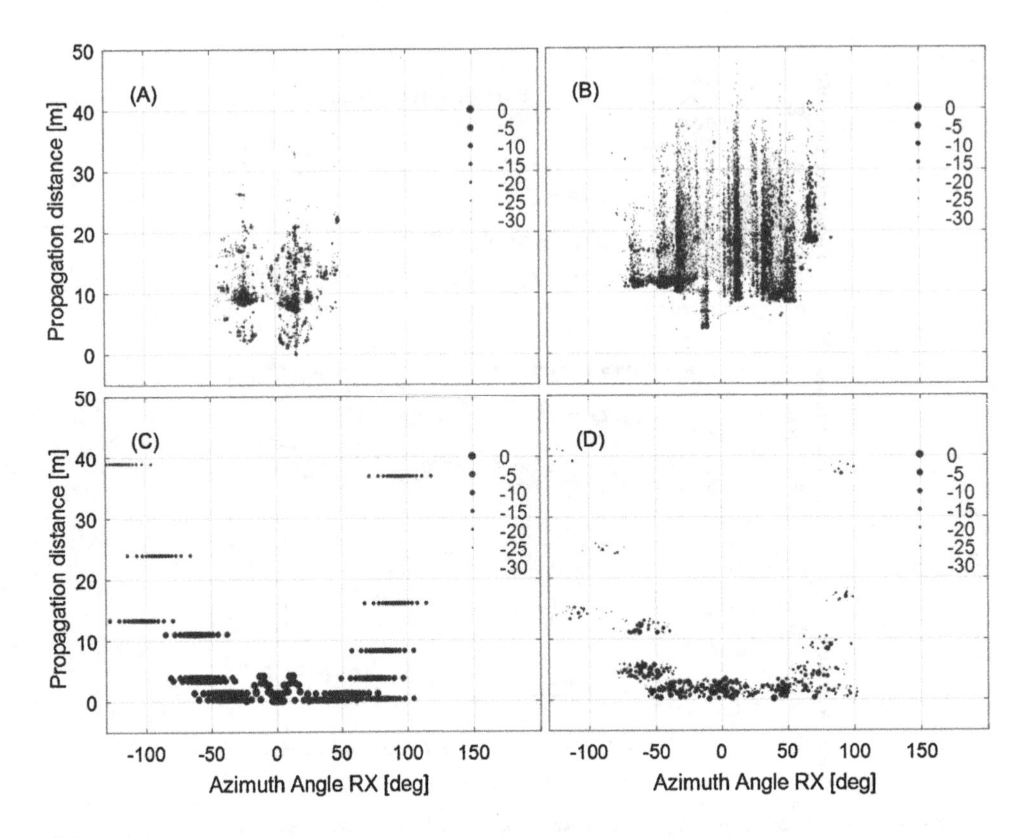

FIGURE 3.37

Multipath component distributions in azimuth and propagation distance for the office indoor scenario at 60 GHz under NLoS conditions. Measurement data are shown in (A) for a medium obstructed NLoS scenario, in (B) for a heavily obstructed NLoS scenario, and a random realization of the 3GPP model for the base line case in (C); and for the high resolution option in (D).

that clusters at longer delays are shifted from the main direction. This model property might lead to a non-realistic decrease of delay spread when narrow-beam antennas are used.

It is clear that the baseline model which uses 20 subpaths per cluster at fixed delays provides a synthetic output which does not agree well with the highly resolved measurements. This effect is even more pronounced when looking at the power ordered distributions of the multipath components, as shown in Fig. 3.38. In the measurements the power of the MPCs decay substantially with the power ordered number. At MPC number 20 the measured power is between 7 dB and 15 dB below the maximum power whereas there is no corresponding decay for the baseline 3GPP model. This is problematic to the extent that large arrays or very narrow beams are used as each MPC then may be resolved. The corresponding spatial multiplexing performance, of e.g. interference suppression or MIMO capacity, would then be unrealistically good as shown in the channel modeling chapter of [29] and [20]. For this reason, 3GPP provides an optional way to model the multipath components to be used in the case

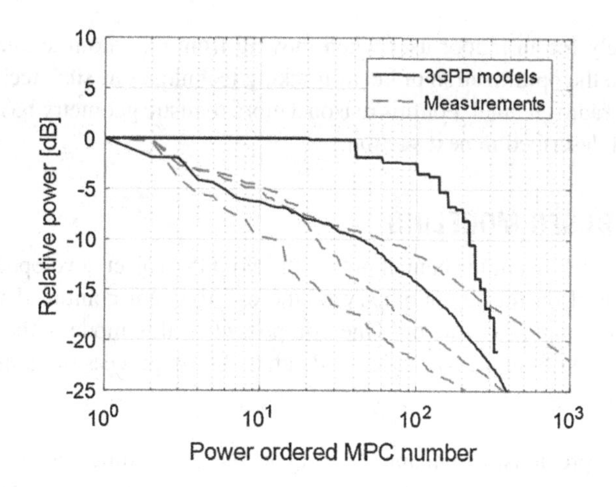

FIGURE 3.38

Power ordered multipath component distributions for the office indoor scenario at 60 GHz under NLoS conditions for the 3GPP baseline model and the high resolution option and corresponding distributions for three measurement locations.

of very large antenna arrays and/or large bandwidths. The distributions of the clusters are kept the same as for the baseline model. Within each cluster, however, a considerably larger number of MPCs is allowed. Corresponding distributions are uniform in angle and delay. The components are, however, power weighted using a Laplacian function in angle and an exponential function in delay. The result of this optional method is shown in Figs. 3.37 and 3.38. It is clear that the distribution of MPC powers is much more realistic with the high resolution option. Also the angle and delay distributions within clusters appear to be much more realistic.

The 3GPP model shows a strong frequency dependent decrease of both directional and delay spreads for most of the scenarios. This frequency trend is not confirmed by measurements presented in Sections 3.3.4 and 3.3.5. The explanation for the observed trend might be that all of the requirements for ensuring comparability between different frequency bands provided in Section 3.3.2.3 were not fulfilled in the measurements which the 3GPP model was based on. As pointed out in Section 3.3.4.4 a very thorough analysis performed by mmMAGIC showed no or very weak frequency trends of the delay spread.

3.4.1.3 Spatial Consistency

In order to provide realistic model output in the case of mobility, i.e., when the UE is moving or in the case of MU-MIMO, a spatial distribution of channel realizations is correlated using function (3.30). In the case of the baseline model only the intercluster parameters are correlated. However, for the high resolution option also the intra-cluster parameters are correlated. The corresponding correlation distances range from 10 to 50 m depending on the parameter and scenario. This method will indeed make the channel variations continuous as the UE moves. There is, however, no support for ensuring that changes are realistic in e.g. Doppler and death and birth processes. For e.g. an outdoor user, the channel conditions may be quite stationary until the user moves around a corner in a street crossing.

The same effect is likely for an indoor user when moving from one room to another. This may have some critical impact on the optimization of beam-tracking techniques as such techniques will be based on the dynamics of the radio channel. For this reason a more realistic geometry based option is provided by the blockage model described in next section.

3.4.2 GEOMETRY-BASED MODELING

In order to provide realistic dynamic channel output the METIS project developed an alternative channel model which is based on the 3D geometry of the environment combined with electromagnetic material properties and simple ray tracing. One component of this model—the blockage model—is particularly useful for describing the dynamics of death and birth process of paths in a realistic way.

3.4.2.1 Blockage

As has been pointed out previously, high antenna gain and corresponding narrow beams are needed for compensating for decreasing antenna apertures when going up in frequency. As a consequence, any moving object that at some time instant blocks the main beam will result in a dramatic reduction of the received signal. For this purpose METIS developed a model for blockage [24], which later was adopted as an additional feature of the 3GPP model. This model is based on 3D diffraction by a rectangular screen where the signal of each MPC is attenuated based on the geometry of the corresponding path. The model is based on standard closed form mathematical expressions, making it simple and computationally efficient.

The METIS blockage model was later considerably improved by mmMAGIC by accounting for the phase differences of the pathways over the four edges providing accurate output for nearly all geometries. This is in contrast with the standard Fresnel approximation which provides an accurate output only for the case when Tx and Rx are at large distances and perpendicularly oriented relative to the screen. Due to its general validity and good accuracy, the mmMAGIC model has been adopted by ITU-R in Recommendation P.526 [11].

In Fig. 3.39 the outputs of both models are shown for a 4x4 m screen at 4 GHz. It is clear that the mmMAGIC model provides a highly accurate output, as it follows the exact Kirchoff integral solution almost perfectly. The output of the METIS blockage model follows closely the peaks of the mmMAGIC model, meaning that it underestimates the loss somewhat. It is, however, likely that the average signal strength of the mmMAGIC model is within 3 dB of the METIS model. Given that the METIS model is substantially less complex, this model might be preferred in many cases.

3.5 SUMMARY AND FUTURE WORK

In the past decade there have been extensive measurement and modeling efforts for understanding and characterization of propagation for development and optimization of 5G mobile communications. One of the major challenges has been to provide understanding of how the propagation characteristics might change when going from legacy carrier frequencies around 2 GHz up to frequencies higher than 80 GHz. Even if many of the propagation characteristics are fairly similar at those higher frequencies, the antenna sizes scale with wavelength. Consequently, the aperture of any type of antenna is proportional to the square of the wavelength, meaning that the received power scales with $-20\log f$ [dB]

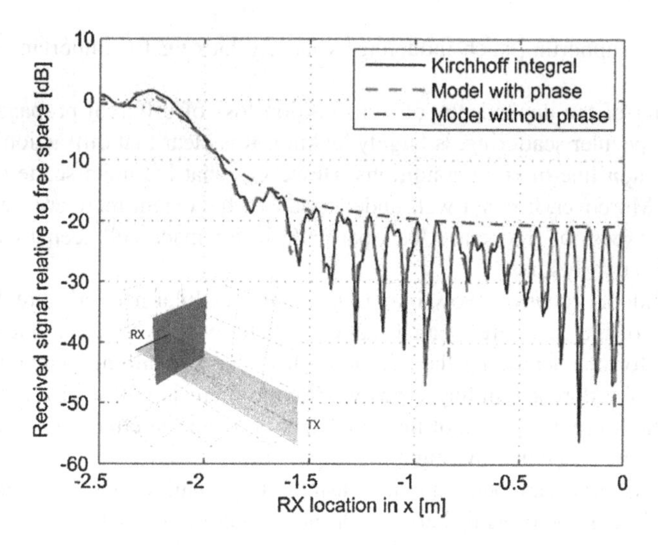

FIGURE 3.39

METIS (without phase) and mmMAGIC (with phase) blockage models at 4 GHz for a 4x4 m screen with Rx 1 m behind the screen and Tx 100 m away in front of the screen. The signal for the RX moving from the center behind the screen to LoS is shown.

relative to the power of the impinging waves. The corresponding increase in loss is more than 30 dB when going from 2 GHz to 80 GHz. However, by utilizing the available area using array antennas or other beam-forming techniques this increase in transmission loss may be compensated for and even turned to a gain when the frequency goes up as shown in Section 3.1.2. In order to optimize advanced antenna transmission techniques utilizing the directional domain, it is crucial that corresponding channel modeling is realistic. As shown in this chapter the knowledge in this area has indeed made progress in the past years. This knowledge is largely implemented in 5G propagation modeling, provided 3GPP and ITU-R, ensuring successful development and optimization of initial 5G cellular communications systems. As higher frequencies and more narrow beams are utilized in later releases of 5G systems higher requirements are put on channel modeling accuracy. For this reason a number of areas in which propagation model improvements would be valuable in order to ensure successful development and optimization of mobile communications systems in the longer term have been identified and summarized as follows:

- Current modeling of highly resolved channel characteristics, in both the delay and the direction domains, is largely arbitrary. Comparing with corresponding measurement data it is clear that there are substantial differences in the structure of multipath distributions and their clustering. Particularly, joint angular and polarimetric distributions for both ends of the link are poorly characterized for outdoor and outdoor-to-indoor scenarios.
- The importance to have accurate knowledge of dynamic variations of the channel including transmission loss is increasingly important with increasing carrier frequency, to make possible optimization of methods to quickly adapt beam-forming to multipath component death and birth processes.

Experimental data supporting such modeling is largely lacking for important scenarios and frequency ranges.

- The understanding of the dependence of transmission loss of different propagation mechanisms, particularly non-specular scattering, is largely lacking. It is clear that diffraction is not very significant under most non-line-of-sight conditions. However, what the main scattering is caused by is not well known. Moreover, it is not well understood to what extent there are frequency dependent trends of loss in excess of free-space loss, particularly for macrocell scenarios. Here, the height dependence is also poorly known.
- The outdoor-to-indoor scenario is very important as most cellular networks are deployed outdoors, while about 80% of the users are located indoors. Quite extensive measurement campaigns have been undertaken for characterizing the additional loss due to building penetration. As there is a vast distribution of different building types worldwide there is still a substantial need for more measurement data. Moreover, angle of incidence and polarization effects are poorly characterized, motivating further experimental investigations.
- New propagation scenarios are becoming increasingly important for 5G. For example machine type communications in e.g. factories or between drones or between drones and ground have recently drawn attention. These scenarios are not well characterized and need more empirical data.

These are only examples of identified areas for future work. Many additional areas, which are not easily foreseen today, are likely to be important in the future. It is, however, a fact that profound knowledge of propagation in general is increasingly important, as more of the degrees of freedom of electromagnetic propagation are utilized in mobile communications, and that all empirical data provided as a basis for modeling will be highly valuable also in the future.

REFERENCES

[1] 3GPP, Study on 3D channel model for LTE, Technical Report 36.873, 2017.
[2] 3GPP, Study on channel model for frequencies from 0.5 to 100 GHz, Technical Report 38.901, 2018.
[3] L. Barclay, Propagation of Radiowaves, 2, Iet, 2003.
[4] G.v. Borgiotti, Maximum power transfer between two planar apertures in the Fresnel zone, IEEE Transactions on Antennas and Propagation 14 (2) (1966) 158–163.
[5] B.H. Fleury, M. Tschudin, R. Heddergott, D. Dahlhaus, K.I. Pedersen, Channel parameter estimation in mobile radio environments using the sage algorithm, IEEE Journal on Selected Areas in Communications 17 (3) (1999) 434–450.
[6] ITU-R, Guidelines for evaluation of radio interface technologies for IMT-Advanced, Report M.2135, 2009.
[7] ITU-R, Effects of building materials and structures on radiowave propagation above about 100 MHz, Recommendation ITU-R P.2040, 2015.
[8] ITU-R, The concept of transmission loss for radio links, Recommendation ITU-R P.341-6, 2016.
[9] ITU-R, Guidelines for evaluation of radio interface technologies for IMT-2020, Report M.2412, 2017.
[10] ITU-R, Prediction of building entry loss, Recommendation ITU-R P.2109, 2017.
[11] ITU-R, Propagation by diffraction, Recommendation ITU-R P.526, 2018.
[12] ITU-R, Propagation data and prediction methods for the planning of short-range outdoor radiocommunication systems and radio local area networks in the frequency range 300 MHz to 100 GHz, Recommendation ITU-R P.1411, 2018.
[13] J.D. Jackson, Electrodynamics, Wiley Online Library, 1975.
[14] N. Jaldén, J. Medbo, H. Asplund, N. Tompson, D. Sundman, Indoor high-resolution channel characterization, in: Antennas and Propagation (ISAP), 2016 International Symposium on, 2016, pp. 618–619.
[15] V.M. Kolmonen, P. Almers, J. Salmi, J. Koivunen, K. Haneda, A. Richter, F. Tufvesson, A.F. Molisch, P. Vainikainen, A dynamic dual-link wideband mimo channel sounder for 5.3 GHz, IEEE Transactions on Instrumentation and Measurement 59 (4) (2010) 873–883.

[16] H. Matzner, M. Milgrom, S. Shtrikman, A study of finite size power isotropic radiators, in: Electrical and Electronics Engineers in Israel, 1995, Eighteenth Convention of, 1995, pp. 1–4.
[17] J. Medbo, F. Harrysson, Efficiency and accuracy enhanced super resolved channel estimation, in: Antennas and Propagation (EUCAP), 2012 6th European Conference on, 2012, pp. 16–20.
[18] J. Medbo, H. Asplund, J.E. Berg, N. Jalden, Directional channel characteristics in elevation and azimuth at an urban macrocell base station, in: Antennas and Propagation (EUCAP), 2012 6th European Conference on, 2012, pp. 428–432.
[19] J. Medbo, H. Asplund, J.E. Berg, 60 GHz channel directional characterization using extreme size virtual antenna array, in: Personal, Indoor, and Mobile Radio Communications (PIMRC), 2015 IEEE 26th Annual International Symposium on, 2015, pp. 176–180.
[20] J. Medbo, P. Kyosti, K. Kusume, L. Raschkowski, K. Haneda, T. Jamsa, V. Nurmela, A. Roivainen, J. Meinila, Radio propagation modeling for 5G mobile and wireless communications, IEEE Communications Magazine 54 (6) (2016) 144–151.
[21] J. Medbo, N. Seifi, H. Asplund, Frequency dependency of measured highly resolved directional propagation channel characteristics, in: European Wireless 2016; 22nd European Wireless Conference; Proceedings of, 2016, pp. 1–6.
[22] J. Medbo, D. Sundman, H. Asplund, N. Jaldén, S. Dwivedi, Wireless urban propagation measurements at 2.44, 5.8, 14.8 & 58.68 GHz, in: General Assembly and Scientific Symposium of the International Union of Radio Science (URSI GASS), 2017 XXXIInd, 2017, pp. 1–4.
[23] J. Medbo, C. Larsson, B.E. Olsson, F.S. Chaves, H.C. Nguyen, I. Rodriguez, T.B. Sorensen, I.Z. Kovács, P. Mogensen, K.T. Lee, J. Woo, M. Sasaki, W. Yamada, The development of the ITU-R terrestrial clutter loss model, in: Antennas and Propagation (EUCAP), 2018 12th European Conference on, 2018.
[24] METIS, METIS channel models, Deliverable D1.4, 2015.
[25] mmMagic, Measurement campaigns and initial channel model for the frequency range 6–100 GHz, Deliverable D2.1, 2016.
[26] mmMagic, Measurement results and final mmMAGIC channel models, Deliverable D2.2, 2017.
[27] H.C. Nguyen, I. Rodriguez, T.B. Sorensen, L.L. Sanchez, I. Kovacs, P. Mogensen, An empirical study of urban macro propagation at 10, 18 and 28 GHz, in: Vehicular Technology Conference (VTC Spring), 2016 IEEE 83rd, 2016, pp. 1–5.
[28] S.L.H. Nguyen, J. Medbo, M. Peter, A. Karttunen, K. Haneda, A. Bamba, R. D'Errico, N. Iqbal, C. Diakhate, J.M. Conrat, On the frequency dependency of radio channel's delay spread: analyses and findings from mmMAGIC multi-frequency channel sounding, in: Antennas and Propagation (EUCAP), 2018 12th European Conference on, 2018.
[29] A. Osseiran, J.F. Monserrat, P. Marsch, 5G Mobile and Wireless Communications Technology, Cambridge University Press, 2016.
[30] A. Richter, Estimation of radio channel parameters: Models and algorithms, Ph.D. thesis, Technischen Universität Ilmenau, 2005.
[31] A.A. Saleh, R. Valenzuela, A statistical model for indoor multipath propagation, IEEE Journal on Selected Areas in Communications 5 (2) (1987) 128–137.
[32] S. Salous, Multi-band multi-antenna chirp channel sounder for frequencies above 6 GHz, in: Antennas and Propagation (EuCAP), 2016 10th European Conference on, 2016, pp. 1–4.
[33] W. Scott, K.S. Hoo, A theorem on the polarization of null-free antennas, IEEE Transactions on Antennas and Propagation 14 (5) (1966) 587–590.
[34] R. Thomä, M. Landmann, A. Richter, Rimax-a maximum likelihood framework for parameter estimation in multidimensional channel sounding, in: Proceedings of the International Symposium on Antennas and Propagation (ISAP'04), 2004, pp. 53–56.

CHAPTER

MATHEMATICAL MODELING OF HARDWARE IMPAIRMENTS

4

As Gordon Moore accurately predicted more than 50 years ago [19], the number of transistors in a dense integrated circuit doubles approximately every two years. In retrospect we have seen the industry follow Moore's observation and it is clear that this development has paved the way for high speed digital communication systems with highly integrated radio transceivers. This development has led to a movement of the border between analog and digital domains in which we today see radio transceivers which are becoming predominantly digital in design.

Through the development of high speed, real-time processing, techniques for compensating imperfections in analog hardware are today commonly used extensively due to stringent requirements imposed by the physical layer design. At the basis for any effective compensation algorithm lies a behavioral model mimicking the behavior of an imperfect analog component. Algorithms such as DPD for linearization of power amplifiers and phase-noise tracking all rely on accurate behavioral modeling in order to perform accurate mitigation.

One aspect that underlines the importance of radio hardware impairment modeling is the constant increase of throughput which is enabled by either using large channel bandwidths at millimeter-wave frequencies or by increasing the spectral efficiency by a denser modulation format and spatial multiplexing. In both of these directions, new challenges face the radio hardware design and the importance of accurate impairment models grows. One particular technology component important for 5G is that of multiantenna techniques such as massive MU-MIMO for spatially multiplex users at sub-6 GHz, along with different methods for analog or hybrid beam-forming techniques aimed toward higher frequencies in the millimeter-wave frequency bands. In both of these areas, impairment models may not only serve as a basis for advanced compensation mechanisms, but also provide valuable insights in how the impairments behave in the spatial domain.

Impairment modeling has also in recent years gained interest from researchers working in the field of communication systems design. Until now, however, may of the models used have been oversimplified for analytical convenience, [3]. However, under some circumstances these simplified models, which are quite often just additive and independent noise, may provide misguiding insights as they do not accurately represent the statistical properties of the impairments, [17].

Conventional behavioral models are however lacking in this aspect as they require time-domain signals, often oversampled. Since knowing the statistical properties of the distortion suffices for conventional link-level analysis which often relies on asymptotics, alternative approaches may be considered in which the impairments are modeled as a correlated noise process. As will be shown in this chapter, the properties of this noise process may be determined using parameterizations of the corresponding behavioral model and knowledge of the transmit covariance.

On a final note—the underlying mathematics behind most common behavioral models used to capture effects of hardware impairments may be more than a century old but, as we will illustrate in

5G Physical Layer. https://doi.org/10.1016/B978-0-12-814578-4.00009-6

this chapter, the modeling work is still evolving in order to support the design of future radio access technologies and advanced antenna systems.

In this chapter, we will overview and discuss the current state-of-the-art in behavioral modeling of some important radio hardware subsystems. We will start with the power amplifier and the Volterra series commonly used to model their behavior. We will also discuss a novel extension used to model power amplifiers in antenna systems which may be under the influence of mutual coupling, something which is becoming increasingly important with the current drive toward compact antenna arrays operating at millimeter-wave frequencies. This is followed by a discussion on oscillator phase noise and the underlying Wiener process commonly used for modeling. We then move on to discuss quantization noise in data converters and how to model the granular noise from the quantization process and the distortion arising from clipping.

In the second part of this chapter, we will introduce and discuss a stochastic framework for modeling of radio hardware impairments in terms of second-order statistics. Analytical expressions for the second order statistics is derived using the Bussgang theorem and the moment theorem for Gaussians. This approach is aimed toward applications in which oversampled time-domain signals are either not available or simply not useful, such as link- or system-level simulations. An example of numerical simulations for an OFDM-based massive MU-MIMO case is presented.

4.1 RF POWER AMPLIFIERS

Power amplifiers are known as the most nonlinear subsystem component of any radio transmitter. As such, modeling and understanding their behavior in order to compensate for the nonlinear distortion it produces in order not to violate the requirements set by the agreed upon standards is crucial.

3GPP defines two important metrics commonly used to quantify the linearity performance, which are error vector magnitude (EVM) and adjacent channel leakage ratio (ACLR). EVM specifically accounts for the performance between a base station (BS) and its scheduled users as it measured the in-band quality of the links. ACLR accounts for the interference leaked into the adjacent frequency channels which is crucial for the co-existence between networks to function. ACLR is a crucial metric for which the regulatory requirements are quite stringent in order to reduce the interference between networks and other applications in the same frequency band.

As power efficiency is highly important for reducing the environmental impact and the total cost of operation and the power amplifier being one of the major consumers of power, different efficiency enhancement technologies have been developed over the years. Amplifier topologies such as the Chireix [6] or the Doherty [8] amplifier, are commonly used, although the latter is more favored in recent scientific literature and in most industrial applications. This is in great measure due to its rather low implementation complexity compared to the Chireix amplifier and the performance in terms of power efficiency.

In more recent years, generalizations of the Doherty concept has been developed, supporting high efficiency over larger dynamic range. This further strengthens the Doherty amplifier in its position as the most popular amplifier topology. However, the improved efficiency performance comes at the cost of nonlinear behavior, [15], which may impact the overall system design depending on the transmit power.

The modeling needs vary in different applications and amplifier topologies, but in general the effects commonly covered in the literature are:

- Nonlinear transfer function and compression.
- Dispersive effects from impedance matching and bias networks.
- Electron traps from material artifacts.
- Long-term thermal memory effects due to heating and cooling.
- Load-modulation via mutual coupling in between antennas.

We will now briefly introduce the Volterra series and discuss some of the more common subsets of the Volterra series used in the scientific literature. Strengths and weaknesses of each model are also discussed and illustrated using measurements. Further on, we will discuss some of the more non-conventional models using orthogonal basis functions. We will end the section on power amplifier modeling with a discussion of global and local basis functions. It should be noted that the models discussed here are all acting in complex baseband and, thus, this carries no information regarding RF harmonics and other passband specific aspects.

4.1.1 THE VOLTERRA SERIES

The Volterra series dates back to 1887 [24] and may be considered as one of the most powerful tools for modeling not only radio power amplifiers, but nonlinear time invariant systems in general. The full Volterra series may be viewed as a superset of polynomial-based nonlinear models containing permutations of the input signal. As many of these are not always relevant in all modeling scenarios, many efforts have been made on pruning the full Volterra series in order to tailor it for specific amplifier topologies.

Much of the very early work on the Volterra series was performed over real-valued variables for passband systems. However, due to the quadrature nature of digital communication systems and the analytical manners by which these are modeled, the natural step of extending the Volterra series to complex variables was taken, [2].

The discrete-time, complex-valued Volterra series for causal systems with symmetric kernels ($\theta_{i,j} = \theta_{j,i}$) of nonlinear order P and memory-depth M can be written as

$$\begin{aligned} y[n] &= \sum_{p=1,3,5,\ldots}^{P} \sum_{m_1=0}^{M} \sum_{m_2=m_1}^{M} \cdots \sum_{m_{(p+1)/2}=m_{(p-1)/2}}^{M} \\ &\times \sum_{m_{(p+3)/3}=0}^{M} \cdots \sum_{m_p=m_{p-1}}^{M} \theta_{p,m_1,\ldots,m_p} \times \prod_{l=1}^{(p+1)/2} x[n-m_l] \prod_{j=(p+3)/2}^{p} \bar{x}[n-m_j]. \end{aligned} \tag{4.1}$$

We may note here that in the Volterra series, as stated in Eq. (4.1), there will always be an even number of conjugate terms accompanied by an odd amount of non-conjugate counterparts. This causes the frequent occurrence of terms in the form of $x[n-m_i]|x[n-m_j]|^p$, which plays an important role in power amplifier modeling, as the behavior is for a large part dependent on the signal envelope.

One beneficial aspect of any model based on the Volterra series is that they are linear in the parameters. This means that the model may be written as a simple matrix-vector multiplication,

$$\mathbf{y} = \mathbf{X}\boldsymbol{\theta}, \tag{4.2}$$

in which

$$\mathbf{X}(n,l) = \prod_{l=1}^{(p+1)/2} x[n - m_l] \prod_{j=(p+3)/2}^{p} \bar{x}[n - m_j] \tag{4.3}$$

is the nth row and lth column of the regression matrix and $\boldsymbol{\theta}$ is a vector containing the coefficients $\theta_{p,m_1,\dots,m_p}$. This linear system of equations yields $\mathbf{y} = \left[y[0], \dots, y[N]\right]^T$, which is the vector of output samples. Due to these models being parameter linear, the identification process is generally quite simple, provided that the regression matrix is numerically sound and not ill-conditioned.

The mean squared error (MSE)-optimal parameter estimate can be found as

$$\hat{\boldsymbol{\theta}} = \mathbf{X}^{+}\mathbf{y} \tag{4.4}$$

where we are making use of the Moore–Penrose pseudo-inverse, denoted $\mathbf{X}^{+}$.

4.1.2 COMMON SUBSETS OF THE VOLTERRA SERIES

The Volterra series as presented in Eq. (4.1) contains many terms which may not be providing useful degrees of freedom for accurate modeling. In order to avoid over- or noise-fitting by introducing unnecessary basis functions, some reductions of the Volterra series have been proposed over the years. We will now cover some of the more important ones.

4.1.2.1 Static Polynomial

The smallest and simplest subset of the Volterra series is that of the static polynomial. This model contains only terms at time n and can therefore not accurately model systems which are dispersive in nature. This excludes numerically accurate modeling of power amplifiers operating at a large fractional bandwidth, as these often exhibit dynamic memory effects, making the nonlinear behavior of the amplifier strongly frequency dependent. For narrow-band applications, however, the static polynomial model may very well be sufficient.

The static polynomial as it occurs in most literature is simply defined as

$$y[n] = \sum_{p=1}^{P} \theta_p x[n]|x[n]|^{2p-1} \tag{4.5}$$

where θ_p are the model parameters, sometime referred to as *kernels*. One very important observation regarding the static model, is that its asymmetries in the distortion spectrum may not be captured with this model as there is no memory involved. The output distortion spectrum will therefore always be symmetric around the center frequency.

Third-Order Static Polynomial

Most models covered in this chapter are aimed at accurate numerical predictive simulations or at providing a basis for DPD-algorithms. However, at certain points, simplified models may provide valuable insight in terms of analytical results. One special case of the static polynomial most suitable for this is that of the third-order static polynomial model. This is simply written as

$$y[n] = \theta_1 x[n] + \theta_2 x[n]|x[n]|^2. \tag{4.6}$$

Examples on how to use this model in conjunction with the moment theorem for Gaussians in order to analyze the impact of power amplifier distortion in massive MIMO arrays are shown in [18]. It may seem simplistic, but for amplifier operating in a reasonably nonlinear region, the third-order term is most commonly the more dominant one, even if higher-order terms are present. Thus, this may serve as a decent first-order approximation of a PA for analytical exercises, leading to valuable insight without having to struggle with tedious analytical manipulation.

4.1.2.2 A Note on Odd–Even and Odd Orders

Before moving on to more advanced behavioral models, we need to examine an issue which often occurs throughout the scientific literature. In regards to power amplifier behavioral modeling, there are mainly two different conventions used when defining the nonlinear orders of the model. These two conventions are commonly referred to as *odd–even-order* or just simply *odd-order* models. We will discuss this here using the static polynomial as an example, but the same reasoning applies for most Volterra-based models. The model which is written as

$$y[n] = \sum_{p=1}^{P} \theta_p x[n]|x[n]|^{2(p-1)} \tag{4.7}$$

is most commonly referred to as odd order, both since $2(p-1)$ is an odd number and since $x[n]|x[n]|^{2(p-1)}$ is a permutation of $x[n]$ an odd number of times. Another, slightly different convention commonly which is slightly less likely to occur in the scientific literature is written as

$$y[n] = \sum_{p=1}^{P} \theta_p x[n]|x[n]|^{p-1}, \tag{4.8}$$

in which the basis functions are referred to as odd–even, since the term $x[n]|x[n]|^{p-1}$ always is an odd permutation over $x[n]$, even if the power of $|x[n]|$ is allowed to be odd instead of solely even.

It has been claimed in the past that an odd–even model provides more degrees of freedom, [7], and should therefore provide a richer description of the amplifier behavior. However, this is a claim that has been rebuked by the authors of [13], who show that the odd–even- and odd-order models have the same modeling properties and the odd–even orders do not provide any richer description of the amplifier. This is a result of expanding the term $|x|$ in terms of $|x|^{2p}$, and showing that the series is uniformly convergent. The difference between orders in terms of modeling accuracy is in fact proportionality to the model truncation error.

Moving on, we will in this chapter stick to the more common odd-order version in the following sections as this is the conventional ordering in most publications.

4.1.2.3 Memory Polynomial

The first natural extension of the static polynomial is to add memory terms in order to model dispersive effects introducing a frequency selective behavior. The memory polynomial (MP) model as described in [12], can be written in its complex baseband form as

$$y[n] = \sum_{p=1}^{P} \sum_{m=1}^{M} \theta_{p,m} x[n-m] |x[n-m]|^{2(p-1)}, \tag{4.9}$$

where P is the nonlinear order and M is the memory depth. From Eq. (4.9), we see that the MP is in principle a nonlinear finite impulse response (FIR) filter. As described in [20], the MP is an extension to the common Hammerstein model in which a nonlinear function is followed by a linear filter.

In the MP model only diagonal memory terms from the Volterra series are included, which in some cases may limit the modeling capabilities, particularly in the case that the amplifier in question displays a strong, nonlinear memory for which cross-terms will be necessary. For simpler applications and for analytical manipulation, this model is often good enough and easy to use.

4.1.2.4 Generalized Memory Polynomial

The generalized memory polynomial (GMP) was first presented in [20]. GMP extends the aforementioned MP by introducing cross-terms in the memory-domain [12], which is written as

$$\begin{aligned} y[n] &= \sum_{p=1}^{P} \sum_{m=1}^{M} \alpha_{p,m} x[n-m] |x[n-m]|^{2(p-1)} \\ &+ \sum_{p=1}^{P} \sum_{m=1}^{M} \sum_{l=1}^{L} \beta_{p,m,l} x[n-m] |x[n-m-l]|^{2(p-1)} \\ &+ \sum_{p=1}^{P} \sum_{m=1}^{M} \sum_{l=1}^{L} \gamma_{p,m,l} x[n-m] |x[n-m+l]|^{2(p-1)} \end{aligned} \tag{4.10}$$

where P is the nonlinear order, M is the memory depth and L is the number of leading and lagging cross-terms. In this model, the memory terms span an L-dimensional strip in the memory-space of the Volterra series. Reference [20] also introduces a more general way of creating the model using index arrays in order to facilitate pruning of the parameter space in order to achieve the desired complexity/performance trade-off.

4.1.3 GLOBAL VS. LOCAL BASIS FUNCTIONS

All of the models examined so far are using so-called global basis functions. This is in the sense of basis functions acting upon the entire range of the envelope. This may in some cases cause numerical issues, in particular when using high-order polynomial basis functions in conjunction with signals, which has a large dynamic range combined with low probabilities of occurring peaks. This will end up causing a poor numerical fit as the number of samples collected around the peak power will be small, which tends to have a negative impact particularly on the high-order terms as their estimator variance rapidly increases with order.

One straightforward solution to this numerical issue is to use so-called local basis functions. By this, we mean dividing the envelope-space into segments which then get associated with their own specific set of basis functions. Let $\mathcal{P}$ be the envelope-range which we divide into Ω-ordered sections as

$$\mathcal{P} = \bigcup_{\omega=1}^{\Omega} \mathcal{P}_u \tag{4.11}$$

where $\mathcal{P}_u \cap \mathcal{P}_{u'} = \varnothing$ for $u \neq u'$. We can now use this to form modified versions of the models previously discussed. For example, examining the static polynomial, we can write

$$y[n] = \sum_{p=1}^{P} \sum_{\omega=1}^{\Omega} \mathbf{I}_{|x[n]| \in \mathcal{P}_\omega} \theta_p x[n] |x[n]|^{2(p-1)}, \tag{4.12}$$

in which $\mathbf{I}$ is the indicator function. It may be difficult to see from the equation above, but this yields a regression matrix which is largely sparse, e.g. the majority of elements will be zero. This may increase the overall numerical complexity of the model, but as the regression matrix will have a sparse structure, this complexity may be reduced using sparse matrix methods for computation.

In Fig. 4.1, we see one example of a static polynomial fitted with local basis functions on data generated via a memoryless polynomial. As the number of sections increases, we increase the modeling accuracy at the cost of added complexity.

Quite simple examples of local basis functions may be in the form of a look-up table (LUT) in which elements of the table only act upon its own specific region of the envelope. Another example which allows for switching between local cross-terms in the memory-space, is the so-called vector-switching model [1].

4.1.4 EXPERIMENTAL MODEL VALIDATION

Concluding the discussion on power amplifier modeling, the numerical accuracy will be examined using a set of measurements performed on a Gallium-Nitride (GaN)-based class-AB amplifier using the web-lab system have been used [14]. The models considered in this chapter will be the static polynomial, the GMP, and a vector-switching extension of the GMP. For easy quantification of the model performance, we need a set of metrics.

4.1.4.1 Quantifying Modeling Performance

In order to make a sound selection of the model to suit the application in mind, we need to examine how these behave using measurement data. For this purpose, we need to consider a couple of important figures of merit. One of the most commonly used figures of merit is the normalized mean squared error (NMSE). Here, we compare the measured and modeled output over the entire measurement bandwidth. The modeling error is then normalized with the measured output variance in order to make the figure

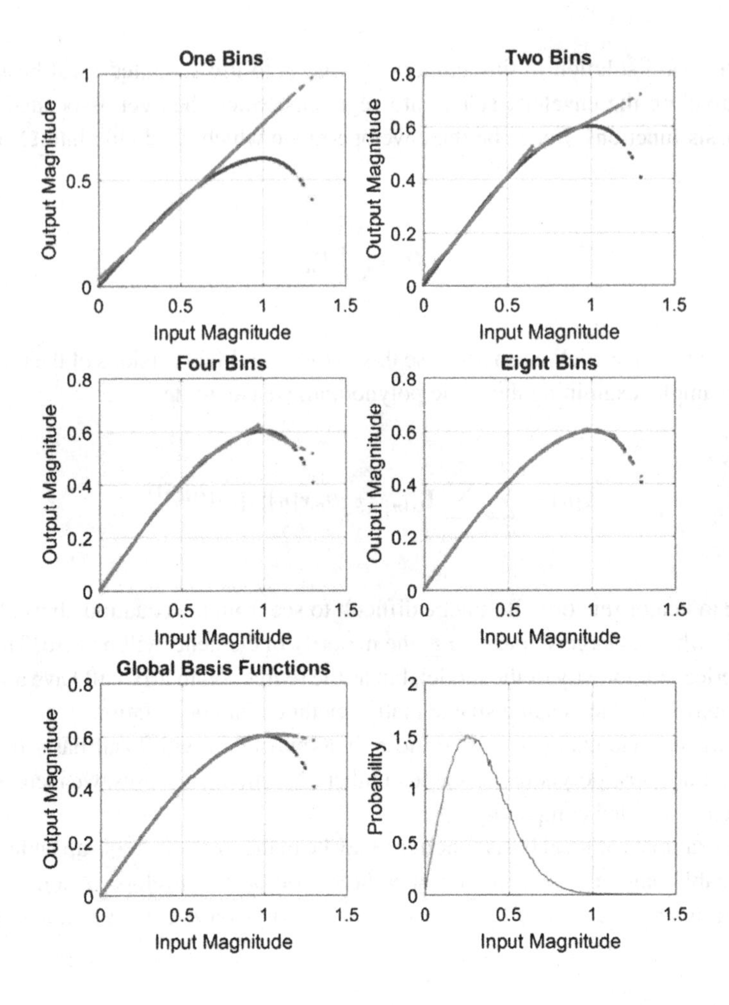

FIGURE 4.1

Illustration of local vs. global basis functions. An illustration of how local basis functions fit to a static model reference. Here, piecewise linear functions are used over different numbers of envelope bins.

power independent. NMSE is commonly defined as

$$\text{NMSE} = \frac{\sum_{n=0}^{N} |y_{\text{meas}}[n] - y_{\text{mod}}[n]|^2}{\sum_{n=0}^{N} |y_{\text{meas}}[n]|^2} \tag{4.13}$$

Table 4.1 Model validation results

PA model	NMSE (dB)	ACEPR (dB)
Static polynomial	−28.5	−29.0
GMP	−32.5	−33.5
VS-GMP	−37.0	−38.0

where y_{meas} is the measured output and y_{mod} is the modeled output. Compared to conventional MSE, the NMSE removes any power dependencies, enabling comparison over different power ranges if necessary.

The second figure of merit useful to assess the modeling performance is the adjacent channel error power ratio (ACEPR), which provides a measure of how accurate the model predicts the generated adjacent channel distortion. The ACEPR is defined as

$$\text{ACEPR} = \frac{\displaystyle\int_{\text{Adj. band}} |y_{\text{meas}}(f) - y_{\text{mod}}(f)|^2 \, df}{\displaystyle\int_{\text{Adj. band}} |y_{\text{meas}}(f)|^2 \, df}. \tag{4.14}$$

We may in ACEPR observe similarities to NMSE, but with the emphasis put on modeling of the distortion outside of the assigned channel.

In order to evaluate the modeling performance, we need two sets of measurement data—one for identification and one for evaluation. This is required to provide a fair comparison such that the model is not fitted to one specific signal realization. In this experiment, we use filtered complex Gaussian noise which commonly has statistical properties very similar to any wide-band OFDM-based system due to the central limit theorem. Since we are not evaluating a postdecoding metric such as bit error rate (BER), Gaussian noise is sufficient.

Fig. 4.2 depicts the measured power spectral density (PSD) and amplitude-to-amplitude modulation (AMAM) of the identification and evaluation data in terms of instantaneous normalized gain.

The model parameters are then found using the Moore–Penrose pseudo-inverse as described in Eqs. (4.3) and (4.4). After performing the identification, each model runs with the evaluation data which is compared to the measurements. The results in terms of NMSE and ACEPR are presented in Table 4.1.

In order to further illustrate the modeling capacities for each case, we may examine the AMAM characteristics and PSD from measurement data. Fig. 4.3 illustrates the measured and simulated AMAM. Fig. 4.4 shows the corresponding simulated PSD for the models.

As we may observe, the static polynomial shows no dispersion effects due to its inherent lack of memory. The GMP however, is capable of modeling these effects as shown. As this particular amplifier is quite nonlinear, Fig. 4.3 illustrates the manner in which with both the static polynomial and the GMP with global basis functions are having certain difficulties fitting to the data near peak power. This is however not the case for the VS-GMP, which further improves the modeling accuracy by applying local basis functions. In each region of the envelope, a low-order GMP ($P = 3$) is applied. Further

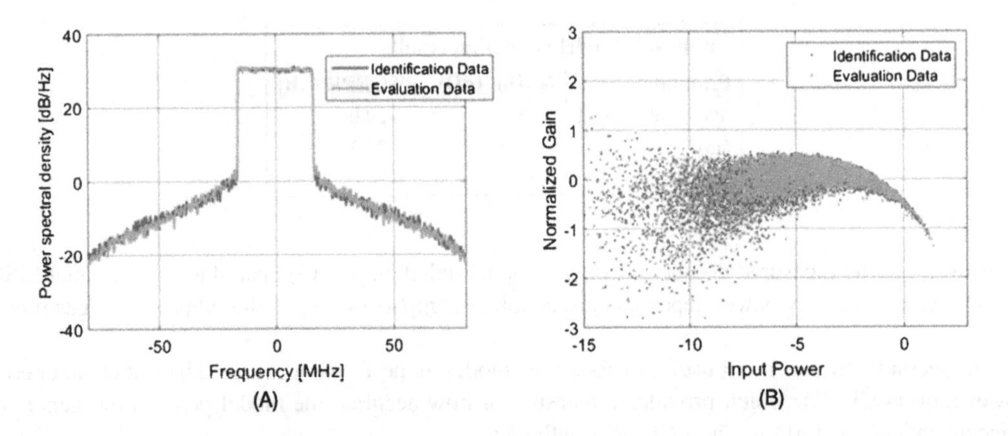

FIGURE 4.2

Measured PSD (A) and AMAM (B) characteristics of the identification and evaluation data used in the comparison between models.

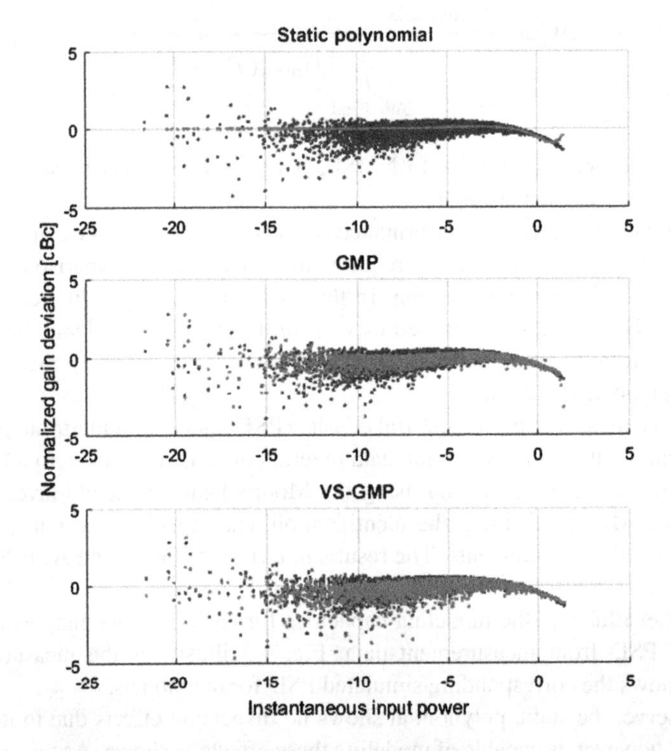

FIGURE 4.3

Comparison between measurement and models. Comparative simulations between the static polynomial model and the GMP, both with global and local basis functions. Blue (dark gray in print version) curves denote measurement data, red (light gray in print version) curves denote predicted output from the models.

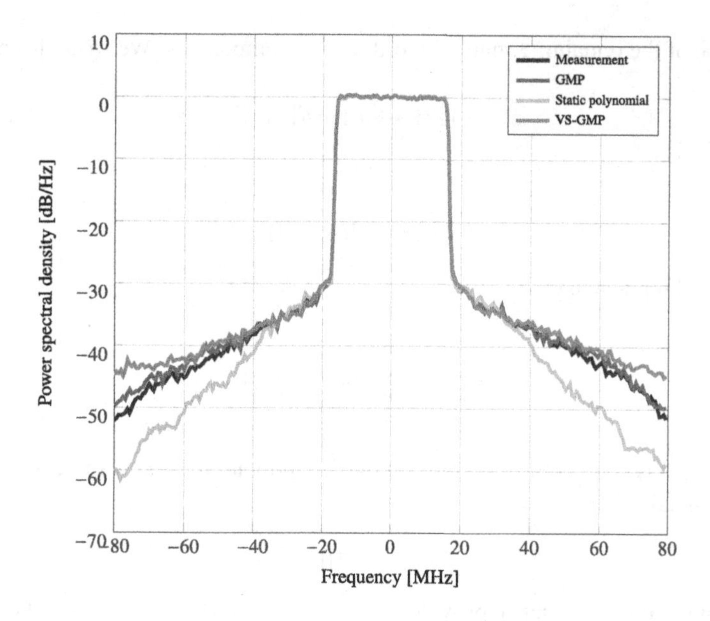

FIGURE 4.4

Comparative simulations between the static polynomial model and the GMP using global and local basis functions.

practical comparisons between models using global and local basis functions may be found in [1]. The numerical results are summarized in Table 4.1.

4.1.5 MUTUALLY ORTHOGONAL BASIS FUNCTIONS

A less common, yet useful model for power amplifier modeling is constructed by using mutually orthogonal basis functions. Consider the expansion of a nonlinear function using the basis functions H_p,

$$y[n] = \sum_{p=1}^{P} \theta_p H_p(x[n]). \tag{4.15}$$

These basis functions can be designed to be mutually orthogonal with respect to the statistics of our input signal. That implies, for all p, k, that the property

$$\mathbb{E}\left[H_p(x[n])H_k^*(x[n])\right] = 0 \tag{4.16}$$

needs to hold. For the case of a complex Gaussian distributed input signal, we can achieve this mutual orthogonality by using the complex Itô-generalization of the Hermitian polynomials, at least for a given input power σ_x^2.

The mutual orthogonality between each basis function may also be beneficial as it provides a description in which the linear term, e.g. the useful signal part, is orthogonal to the distortion as this is a linear combination of the remaining, mutually orthogonal components. We write this as

$$y[n] = s[n] + d[n] \tag{4.17}$$

where

$$s[n] = \theta_1 H_1(x[n]) \tag{4.18}$$

is the linearly scaled input signal and

$$d[n] = \sum_{p=2}^{P} \theta_p H_p(x[n]) \tag{4.19}$$

is the distortion which is expressed using the complex Itô polynomials; now these are mutually orthogonal to the input signal,

$$\mathbb{E}[s[n]d^*[n]] = 0. \tag{4.20}$$

The property of mutual orthogonality provides a clearer path to determining the statistical properties of the distortion in terms of its covariance matrix, in a similar manner as the linear stochastic model discussed in Chapter 4.4.

This approach provides a simple way of analyzing of the spatial properties of the distortion, in the context of MU-MIMO [17]. Here, it is shown that

$$\frac{K^3 + K^2}{2} L^2 \upsilon(f) \tag{4.21}$$

where K is the number of users or transmit-directions, L is the number of significant taps of the channel and $\upsilon(f)$ is a constant depending on the excess bandwidth. A necessary condition for the distortion to behave omni-directionally is

$$\frac{K^3 + K^2}{2} L^2 \upsilon(f) \geq M. \tag{4.22}$$

As Fig. 4.5 illustrates, we see that, as the number of users increases from $K = 1$ to $K = 3$ and 10, the number of dominant directions increases with $\frac{K^3+K^2}{2}$ (as this is modeled using a frequency-flat, narrow-band channel with $L^2\upsilon(f) = 1$). A few further observations may be made from Fig. 4.5 is that the radiated adjacent power is at worst in parity with the single-antenna case. This occurs only in the case of $K = 1$ and only towards the user served. Everywhere else, any potential victim user will receive less distortion compared to the corresponding single-antenna case.

As K then increases, the adjacent channel power is divided over more directions than that of the in-band signal. Thus, it never reaches the same radiated power as in the single-antenna case. This indicates that, in most practical cases, the impact of the radiated distortion will be less in a large array than in the case of a single-antenna system. More discussion and rigorous derivation of the radiation patterns can be found in [17].

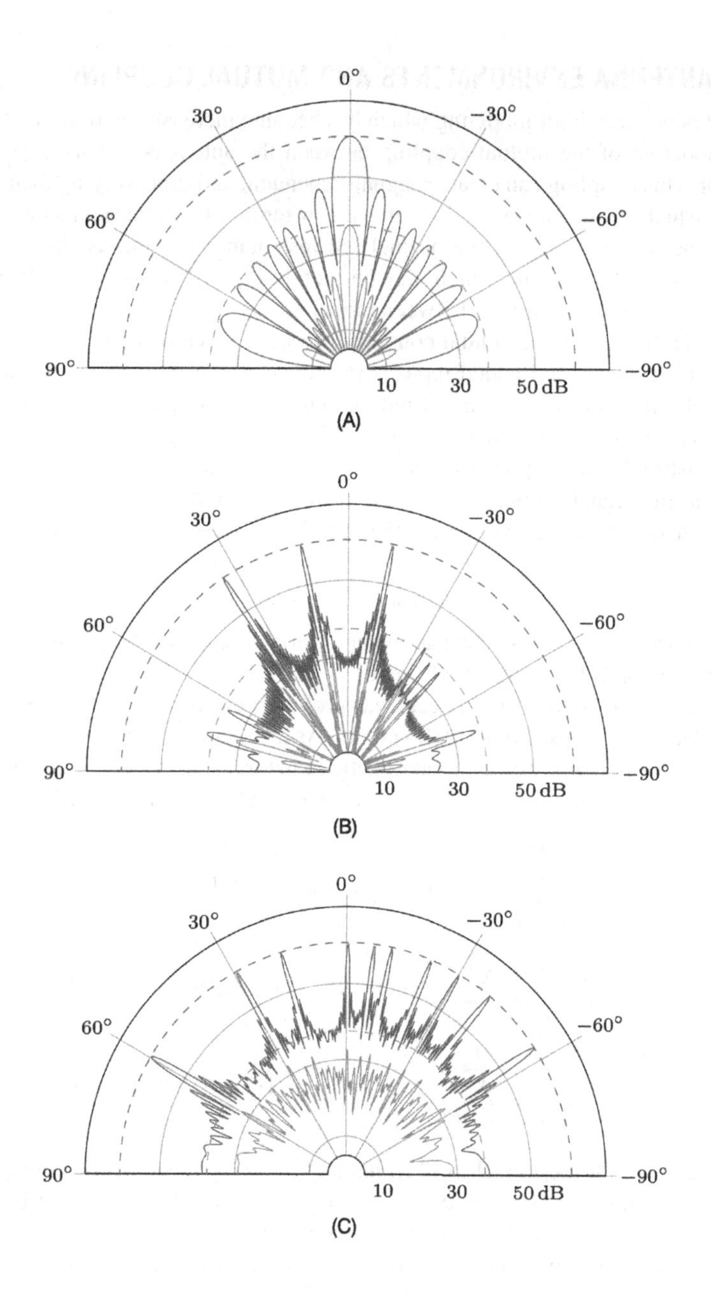

FIGURE 4.5

Radiation patterns from uniform linear arrays. Blue (dark gray in print version) curves indicate the in-band radiated power. Red (light gray in print version) curves indicate the power radiated in the adjacent channel. The dashed lines correspond to the single-antenna case with the same ACLR. (A) Radiated pattern for in- and adjacent-band in the case of $K = 1$ and $M = 16$, (B) radiated pattern for in- and adjacent-band in the case of $K = 3$ and $M = 128$, (C) radiated pattern for in- and adjacent-band in the case of $K = 10$ and $M = 128$.

4.1.6 MULTI-ANTENNA ENVIRONMENTS AND MUTUAL COUPLING

Another aspect of power amplifier modeling which has become increasingly important in multiantenna applications is modeling of the mutual coupling between the antennas and its impact on the power amplifier behavior. This coupling can create a signal-dependent and time-varying load-impedance seen by the amplifier, which in turn changes the behavior in terms of the instantaneous transducer gain. Such effects may be seen by plotting the normalized instantaneous gain, as shown in Fig. 4.7. The large variation of instantaneous gain, which may be seen in Fig. 4.7, is caused by load-modulation of the power amplifier via mutual coupling between branches.

In order to model the effects of mutual coupling, we need to view the power amplifier as a dual-input system in which the second added input is the reflected wave stemming from the transmitted signals from neighboring transmitters, transfered via mutual coupling between antennas as illustrated in Fig. 4.6. Here, $x_1, \ldots, x_M$ are the M transmit signals and $y_1, \ldots, y_M$ the M output signals affected by the nonlinear distortion of the power amplifiers. We also need the variables $y_{r,1}, \ldots, y_{r,M}$, which are mixtures of transmit signals being reflected back into each amplifier via the coupling.

At every time instant n, we may compute the reflected wave to the antenna port m as the scalar product

$$y_{r,m} = \mathbf{y}^T \boldsymbol{\lambda}_m, \tag{4.23}$$

in which $\boldsymbol{\lambda}_m = [\lambda_m 1m, \ldots, \lambda_{Mm}]$ is the parameter vector describing the mutual coupling and $\mathbf{y} = \left[y_1, \ldots, y_M\right]$ is the vector of the transmitted signals.

The first approach to modeling such effects was proposed by Root et al. [22], through a model referred to as the polyharmonic distortion (PHD) model. As a large-signal extension to the S-parameters, and thus nonlinear, this model is, however, quasistatic in nature. This has led to the development of an extension based on the memory polynomial, as derived and validated in [9]. We have

$$\begin{aligned} y_m[n] \quad &= \sum_{p=1}^{P}\sum_{m=1}^{M_1} \alpha_{p,m} x_m[n-m_1]|x_m[n-m_1]|^{2p-1} \\ &+ \sum_{p=1}^{P}\sum_{m_1=1}^{M_1}\sum_{m_2=1}^{M_2} \beta_{p,m,l} y_{r,m}[n-m_2]|x_m[n-m_1]|^{2p-1} \\ &+ \sum_{p=1}^{P}\sum_{m_1=1}^{M_1}\sum_{m_2=1}^{M_2} \gamma_{p,m,l} x_m^2[n-m_1] y_{r,m}^*[n-m_2]|x_m[n-m_2]|. \end{aligned} \tag{4.24}$$

It shares some features with the model in [22], namely the occurrence of the conjugate terms in the reflected wave ($y_{r,m}^*$). The conjugate term occurs due to non-analyticity of the describing function. This goes against conventional Volterra theory in which the number of conjugates in each basis function is always even and accompanied by an uneven and larger number of non-conjugated terms, forming basis functions in the form $x[n]|x[n]|^p$.

The model stated in Eq. (4.24) is parameter linear in the time domain and is suitable for wide-band applications as it supports memory in both incident and reflected waves. As memory is included also in the reflected term, the description of the crosstalk network stated in Eq. (4.23) is no longer sufficient. The modification is, however, rather small as the coupling coefficients $\boldsymbol{\lambda}$ are linear and may be modeled by conventional FIR-filters.

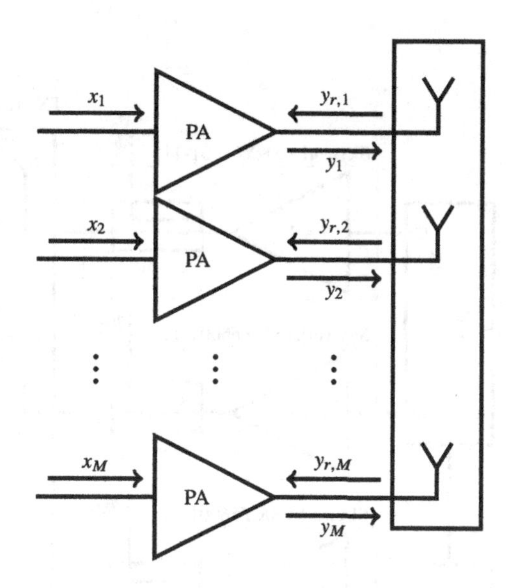

FIGURE 4.6

System model for the dual-input model. The system model for an active antenna system with nonlinear amplifiers. The reflected signal caused by mutual coupling causes a time-varying behavior which is not captured by conventional models.

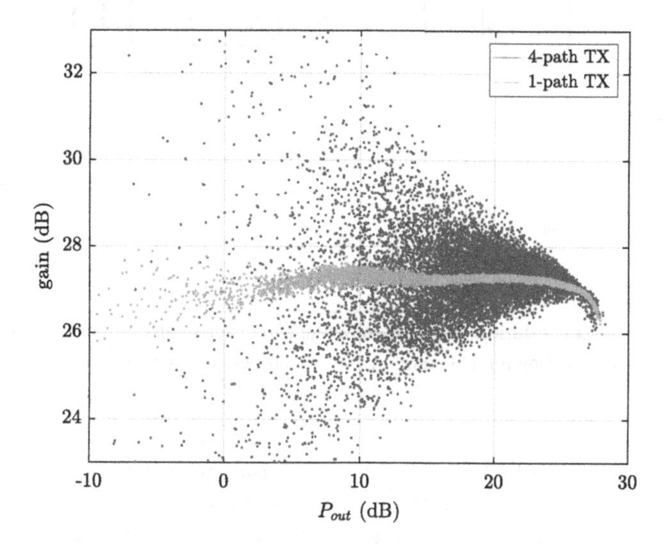

FIGURE 4.7

Measured instantaneous gain. Normalized instantaneous gain of a power amplifier with (blue; dark gray in print version) and without (yellow; light gray in print version) the impact of mutual coupling.

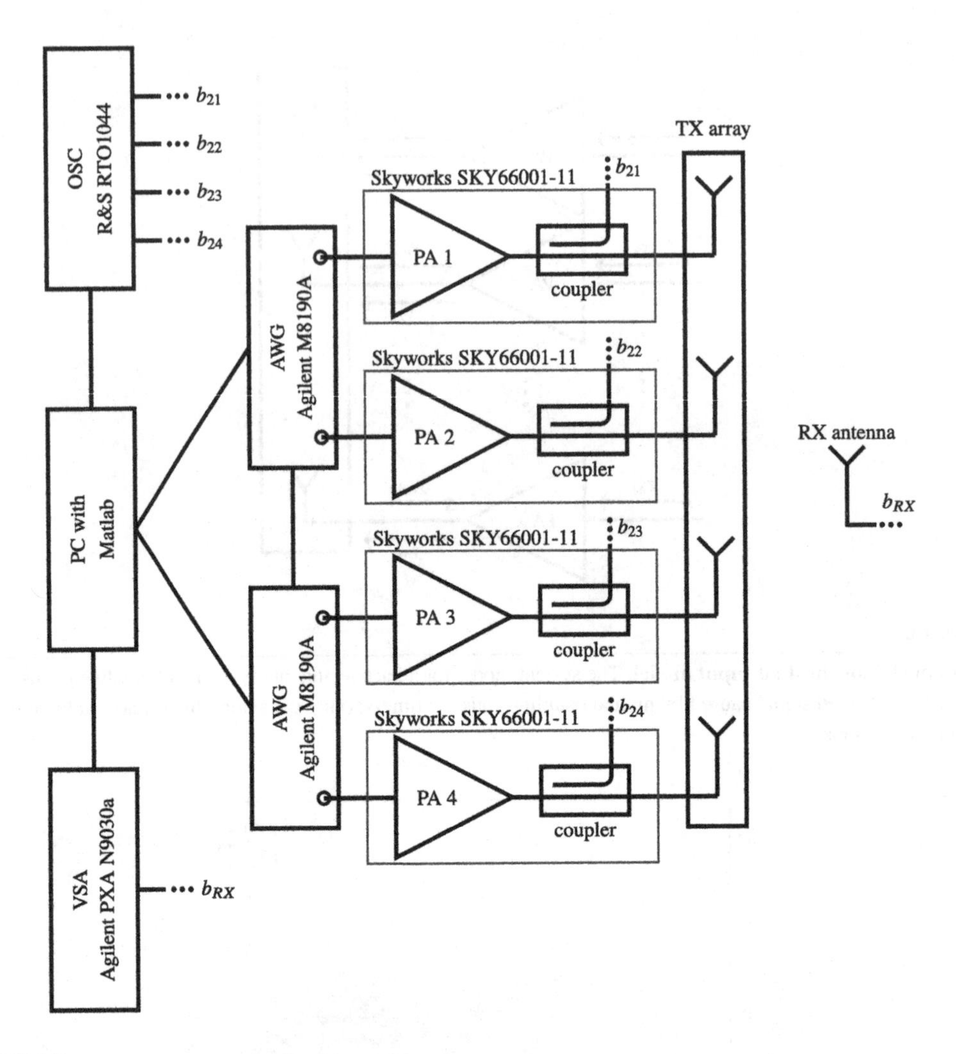

FIGURE 4.8

Measurement setup for model validation. The system model for an active antenna system with nonlinear amplifiers. The reflected signal caused by mutual coupling causes a time-varying behavior which is not captured by conventional models.

The model described in Eq. (4.24) is validated via measurements presented in [9]. These measurements were performed both conducted at the amplifier output reference plane, as well as over the air using a reference receiver. Using wide-band arbitrary waveform generators as shown in Fig. 4.8, four power amplifier modules were connected to two different configurations of a 2×2 patch antenna. The measured characteristics of these arrays are shown in Fig. 4.9. As shown, the two arrays have, via variations in patch separation, different amounts of coupling. The spectrum plots shown in Fig. 4.9 are for the case of strong mutual coupling.

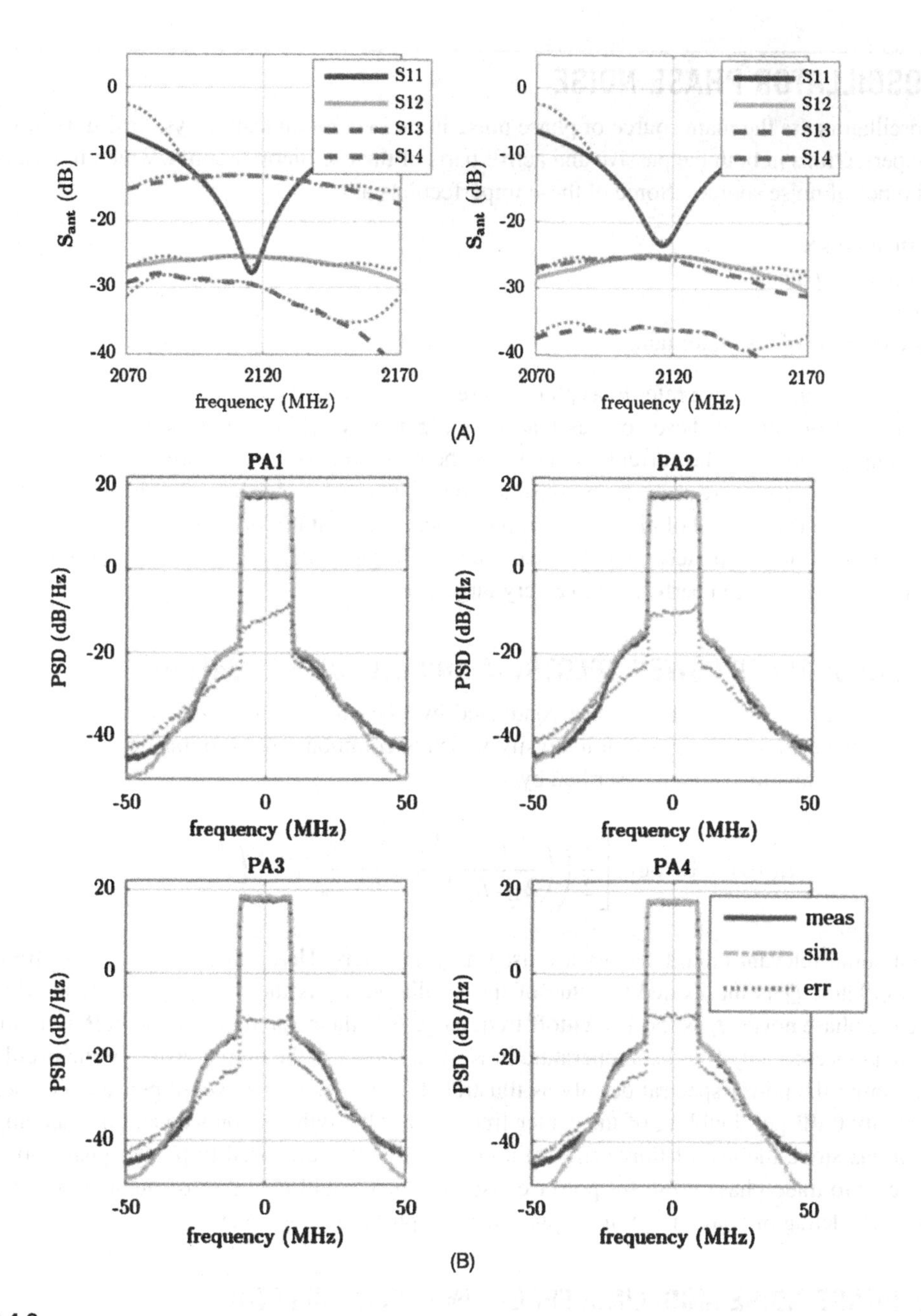

FIGURE 4.9

Antenna array characteristics and measured power spectral density (A) Measured S-parameter characteristics of the two different 2x2 patch-antenna arrays used in the validation process. To the left, the antenna spacing is 0.5λ and to the right 0.7λ, (B) measured and simulated output power spectrum from each of the four power amplifiers, followed by the model error spectrum.

4.2 OSCILLATOR PHASE NOISE

Local oscillators are the main source of phase noise in radio communication systems. It mainly stems from imperfections in both the passive and active parts of the oscillator circuitry, which in turn inflicts several types of noise sources. Some of these imperfections are:

- Thermal noise.
- Flicker or $1/f$ noise.
- Shot noise.
- Finite Q-value of resonator tank.

Together, this forms what we refer to as phase noise.

The impact of most of these sources may be mitigated by different means, for example using a PLL or simply increasing the current pumped into the resonator tank. There are drawbacks to each of these methods such as increased power consumption or increment of other noise in other parts of the spectrum. And with the trend of phase noise rapidly increasing with frequency, it is of utter importance that these effects are accurately modeled and studied. In particular, analysis of phase noise and its impact on OFDM has been performed to a very large extent.

4.2.1 PHASE-NOISE POWER SPECTRUM AND LEESON'S EQUATION

One of the early studies on phase noise conducted by Leeson in 1966, resulted in an empirical expression for phase-noise power spectral density which is commonly used to this day [16]. The power spectrum approximation is often described by

$$L(f_m) = 10\log_{10}\left[\frac{1}{2}\left(\left(\frac{f_0}{2Q_l f_m}\right)^2 + 1\right)\left(\frac{f_c}{f_m} + 1\right)\left(\frac{FkT}{P_s}\right)\right], \tag{4.25}$$

in which some relevant circuit factors are used as parameters. Here, f_0 is the operating frequency of the oscillator, Q is the loaded Q-value of the oscillator, f_m is the offset frequency at which we observe the phase noise, f_c is the $1/f$ cutoff frequency, F is the noise figure of the buffer amplifier, k is Boltzmann's constant, T is the temperature in Kelvin, and P_s is the output power of the oscillator.

Examining the power spectral density as illustrated in Fig. 4.10, the overall phase noise increases with roughly 6 dB per doubling of the center frequency. Although Leeson's equation is an empirical model, it has stood the test of time and is still commonly cited and used to predict phase-noise performance or to trade phase noise for power consumption. We will now discuss an approach used for simplistic modeling of phase noise in the presence of a phase-noise tracker.

4.2.2 PHASE-NOISE MODELING: FREE-RUNNING OSCILLATOR

As it is common to describe a communication system in discrete time due to the nature of its digital implementation, we will stay with phase-noise modeling in the discrete-time domain here. The scientific literature is however not limited to discrete time modeling as a Wiener-process may be formed in continuous time as well.

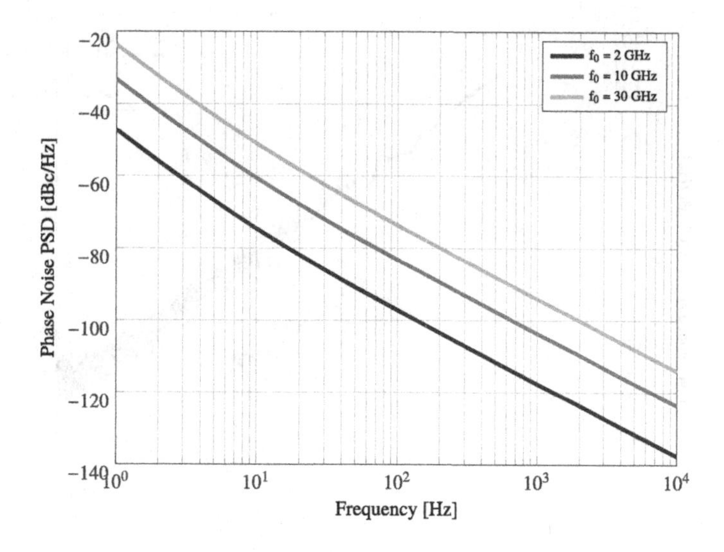

FIGURE 4.10

Phase-noise power spectral density. Phase-noise power spectral density according to Leeson's approximation for different center frequencies, $f_0 = 2$, 10, and 30 GHz.

Phase noise is multiplicative in its nature and applied as

$$y[n] = x[n]e^{i\varphi[n]}, \tag{4.26}$$

in which the phase-noise component φ is commonly modeled using a random walk process,

$$\varphi[n] = \varphi[n-1] + w[n]. \tag{4.27}$$

Just as in the continuous time case, this is an accumulative process in which the variance of φ grows over time. In the presence of a phase-noise tracker, either in the shape of a phase-locked loop or a tracker at the receiver, we may model the residual phase noise as an autoregressive process,

$$\varphi[n] = \lambda\varphi[n-1] + w[n]. \tag{4.28}$$

Here, $w \sim \mathcal{N}(0, \sigma_w^2)$ is the innovation noise and $0 < \lambda < 1$ is a constant determined mainly by the PLL. This phase-noise model is a wide-sense stationary (WSS) process which has the autocorrelation function

$$\mathbb{E}\{\varphi[n]\varphi[n-l]\} = \begin{cases} \dfrac{4\pi^2 f_c^2 t_s \kappa \lambda^{|l|}}{1-\lambda^2} & \text{if } m = m', \\ 0 & \text{else,} \end{cases} \tag{4.29}$$

where f_c is the center frequency, t_s is the sample-time, and κ is a constant depending on the quality of the oscillator and $\varphi \sim \mathcal{N}(0, 4\pi^2 f_c^2 t_s \kappa)$. Examples of the simulated power spectral density for three

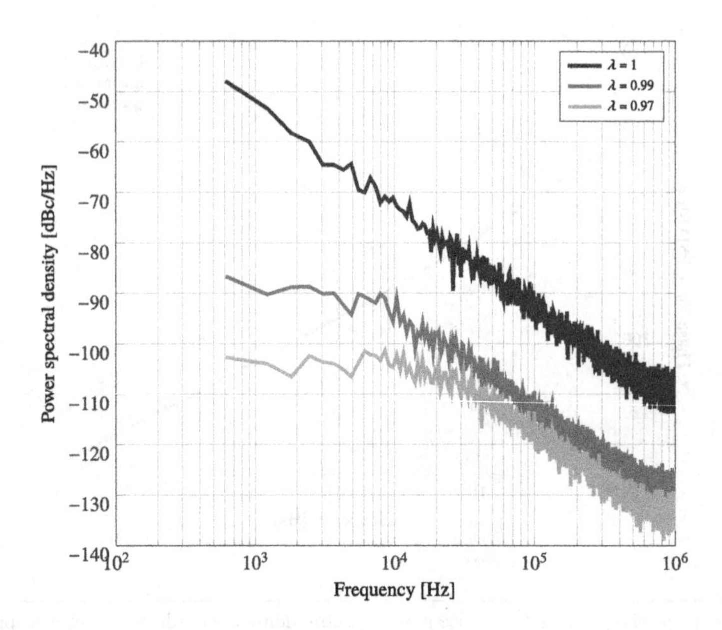

FIGURE 4.11

Phase-noise power spectral density. Simulated phase-noise spectral density using the model in (4.26)–(4.28) with three different values of λ.

different values of λ are shown in Fig. 4.11. Notice the flattening of the spectrum near the carrier frequency as λ decreases.

In practical applications, decreasing λ is analog to increasing the PLL-bandwidth which improves the phase noise around the carrier. However, this tends to increase the white noise part at higher offset frequencies, something which is not captured by this particular model.

4.2.3 PHASE-NOISE MODELING: PHASE-LOCKED LOOP

The phase noise of the PLL-based oscillator consists of three main noise sources, i.e., noises from the reference oscillator θ_{ref}, the phase–frequency detector and the loop filter θ_{LP}, and the voltage controlled oscillator (VCO) θ_{VCO}, as shown in Fig. 4.12. The Laplace transform of the phase noise of the PLL-based oscillator is given as [26]

$$\theta_{\text{out}} = \frac{N_{\text{D}} K_{\text{VCO}} Z(s)\left(K_{\text{D}}\theta_{\text{ref}} + \theta_{\text{LP}}\right) + sN_{\text{D}}\theta_{\text{VCO}}}{sN_{\text{D}} + K_{\text{D}} K_{\text{VCO}} Z(s)}, \tag{4.30}$$

where K_{D} denotes the gain of the phase–frequency detector, K_{VCO} represents the sensitivity of the VCO, $Z(s)$ represents the loop filter, and $1/N_{\text{D}}$ is the frequency divider. The noise sources include both white noise (thermal noise) and colored noise (flicker noise). The PSD of the noise source can be

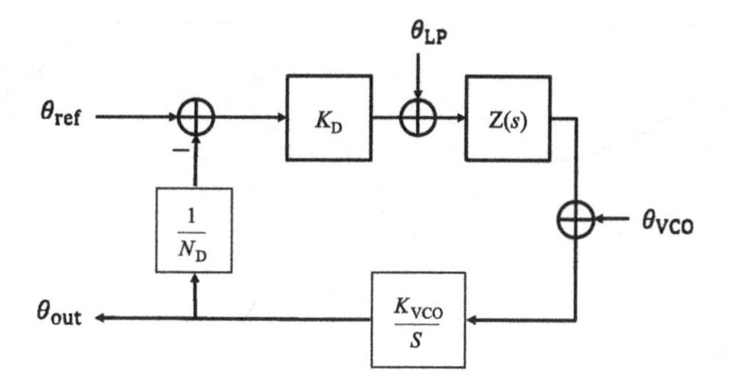

FIGURE 4.12

Phase-noise model of PLL-based oscillator. Blockdiagram of the phase-noise model for a PLL-based oscillator.

modeled as [26]

$$S(f) = S_0 \frac{1+\left(\frac{f}{f_z}\right)^2}{1+\left(\frac{f}{f_p}\right)^2} + S_0 \frac{1+\left(\frac{f_{zo}}{f_z}\right)^2}{1+\left(\frac{f_{zo}}{f_p}\right)^2}\left(1+\left(\frac{f_{zo}}{f_{po}}\right)^k\right)\frac{1}{1+\left(\frac{f}{f_{po}}\right)^k}, \tag{4.31}$$

where S_0 is the power density at zero frequency, f_p and f_z are the (3-dB) pole and corner frequencies of the thermal noise, respectively, $k = 1$ and 3 for flick noises of the PLL and of the oscillator, respectively, and f_{po} and f_{zo} are the pole and corner frequencies of the flicker noise, respectively. The detailed modeling parameters are listed in Table 4-2 of [26]. As an example, Fig. 4.13 shows the estimated PSDs of the carrier phase noise at 6, 28 and 60 GHz, respectively.

4.3 DATA CONVERTERS

Data converters are known as bottlenecks in digital communication systems as the transceiver bandwidths increases, while at the same time demands on dynamic range increase via denser modulation schemes. Amongst the sources of impairment related to data converters, the one most commonly modeled is the quantization noise. As data converters are mixed-signal components, issues with nonlinearities caused by either un-even quantization levels or output buffer amplifiers have also been modeled in the past, quite often using a Volterra series approach [4].

4.3.1 MODELING OF QUANTIZATION NOISE

The modeling of quantization noise is an area well mapped out and the study on quantization is therefore a quite mature field. Some of the most important foundations can be found in [25,23]. Part of this has led to a commonly used simplification of modeling a quantizer by adding uniform noise:

$$\mathcal{Q}(x) = x + w \tag{4.32}$$

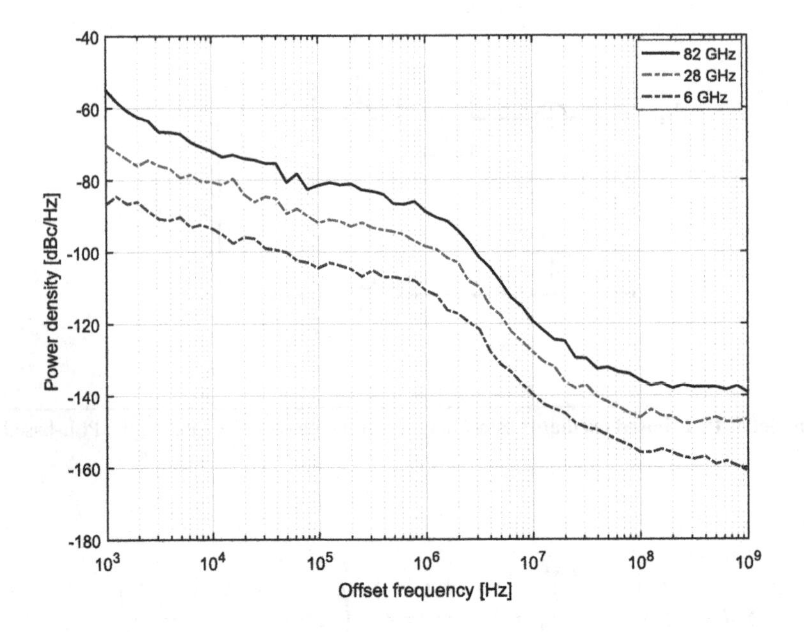

FIGURE 4.13

The PSD of the carrier phase noise. Simulated PSD of the phase noise generated by the model in Fig. 4.12, parameterized for three different center frequencies.

where $\mathcal{Q}$ denotes the quantizer function and w is an additive noise process with parameters determined by the quantizer properties and the input signal statistics. Under certain constraints, the signal to quantization noise ratio (SQNR) is modeled as

$$\text{SQNR} \approx 6.02q \tag{4.33}$$

where q is the number of bits used. This is, however, not always a good approximation depending on the input signal statistics. More exact conditions for when the quantization noise is uniform have been presented in [23]. For the general case using a DAC or ADC with q bits, we have a discrete set of 2^q possible outputs from the quantizer. Most commonly, uniform mid-rise quantizers are assumed; however, other alternatives exist. Given a finite range of the possible input signal, the quantizer can be modeled as

$$\mathcal{Q}(x) = \begin{cases} \frac{\Delta}{2}(1-2^q) & \text{if } x < -\frac{\Delta}{2}2^q, \\ \Delta\lfloor\frac{x}{\Delta}\rfloor + \frac{\Delta}{2} & \text{if } |x| < \frac{\Delta}{2}2^q, \\ \frac{\Delta}{2}(2^q-1) & \text{if } x > \frac{\Delta}{2}2^q, \end{cases} \tag{4.34}$$

in which Δ denotes the LSB size. For a quadrature signal, represented by a complex random variable, we define the input–output relation for the set of baseband DAC's or ADC's by

$$f_{\text{DAC}}(x) = \mathcal{Q}(\text{Re}x + i\text{Im}x) = \mathcal{Q}(\text{Re}x) + i\mathcal{Q}(\text{Im}x). \tag{4.35}$$

As exact modeling of the quantization noise may be difficult and linear, stochastic approaches as proposed in [10] may be sufficient, but we will leave the discussion for Section 4.7.

4.4 STATISTICAL MODELING

In certain applications such as system-level performance assessment, it may be difficult to make use of behavioral models as they often operate on oversampled time-domain signals. In most communication system engineering applications, the signals are modeled as stochastic random variables with given second-order statistics described with a covariance matrix. As such a signal passes through a nonlinear function, for example a building block, the statistical properties changes. With advanced multiantenna systems making use of the spatial properties of the communication channel, the importance of knowing the second-order statistics increases as it determines the spatial properties of both the transmit signal as well as the hardware impairments.

4.4.1 THE BUSSGANG THEOREM AND THE SYSTEM MODEL

Determining the second-order statistics of the distortion is straightforward using the Bussgang theorem. Let $f(\cdot)$ be a nonlinear function and $x \sim \mathcal{N}(0, \sigma_x^2)$. The input–output cross-covariance is equal to the input autocorrelation up to a constant C [5], as

$$R_{f(x),x} = C R_x. \tag{4.36}$$

This property may be used in order to formulate a linear stochastic model which describes the output of a nonlinear system when the input is modeled using a Gaussian random variable x. The model is written as

$$f(x) = \alpha x + d \tag{4.37}$$

where $d \sim \mathcal{N}(0, \sigma_d^2)$ is the distortion noise which is orthogonal to the scaled input signal αx and α is the gain constant, determined by

$$\alpha = \frac{\mathbb{E}\left[x\bar{y}\right]}{\sigma_x^2}. \tag{4.38}$$

From this, we have $\mathbb{E}\left[\alpha x\bar{d}\right] = 0$. For the multiantenna case, considering a system with M antennas, we may model this in matrix-vector format as

$$\mathbf{y} \;=\; \mathbf{G}\mathbf{x} + \mathbf{e} \tag{4.39}$$

where $\mathbf{G}$ is the complex gain matrix and $\mathbf{e}$ is the distortion noise. The gain matrix is defined as

$$\mathbf{G} = \operatorname{diag}\left(\mathbb{E}\left[x_1\bar{y}_1\right], \ldots, \mathbb{E}\left[x_M\bar{y}_M\right]\right). \tag{4.40}$$

The remaining distortion noise, which is orthogonal to $\mathbf{Gx}$, is then distributed as

$$\mathbf{e} \sim \mathbb{C}\mathcal{N}\left(\mathbf{0}, \mathbf{C_e}\right) \tag{4.41}$$

where

$$\mathbf{C_e} = \mathbf{C_y} - \mathbf{G}\mathbf{C_x}\mathbf{G}^H, \tag{4.42}$$

which forces $\mathbf{e}$ and $\mathbf{x}_n$ to be mutually orthogonal. This description is equivalent to the model discussed in Section 4.1.5 as they have exactly the same properties, which stems from using mutually orthogonal basis functions in the shape of Itô–Hermite polynomials.

4.5 STOCHASTIC MODELING OF POWER AMPLIFIERS

In the case of power amplifiers, derivation of the distortion covariance is straightforward by repeatedly using the moment theorem for Gaussians [21]. This may, however, be tedious, as the number of cross-products and higher-order moments tends to explode. For this reason, we will stick to the canonical third-order model here. Using the common third-order polynomial as foundation, written as

$$y[n] = \theta_1 x[n] + \theta_2 x[n]|x[n]|^2, \tag{4.43}$$

we would, as illustrated in [10], be able to derive the model in the following form:

$$\mathbf{y} \;=\; \mathbf{G}^{\mathrm{PA}}\mathbf{x} + \mathbf{e}^{\mathrm{PA}}. \tag{4.44}$$

Here $\mathbf{G}^{\mathrm{PA}} = \mathbf{I}_{MN} \otimes \operatorname{diag}\left(\mathbf{g}^{\mathrm{PA}}\right) \in \mathbb{C}^{MN \times MN}$, or more explicitly

$$\mathbf{G}^{\mathrm{PA}} = \mathbf{I}_{MN} \otimes (\theta_1 \mathbf{I}_{MN} + 2\theta_2 \operatorname{diag}(\mathbf{C_x})). \tag{4.45}$$

The distortion covariance due to orthogonality will become

$$\mathbf{C}_{\mathbf{e}^{\mathrm{PA}}} = \mathbf{C_y} - \mathbf{G}^{\mathrm{PA}}\mathbf{C_x}(\mathbf{G}^{\mathrm{PA}})^H \tag{4.46}$$

in which $\mathbf{G}^{\mathrm{PA}}$ is given by Eq. (4.45), $\mathbf{C_x}$ is the transmit covariance matrix which is given by precoder design and $\mathbf{C_y}$ is the covariance matrix of the distorted output. This has to be manually computed using the moment theorem for Gaussians, which in the end results in

$$\begin{aligned}
\mathbf{C_y} &= |\theta_1|\mathbf{C_x} + 2|\theta_2|^2 \mathbf{C_x} \circ |\mathbf{C_x}|^2 \\
&+ 2\bar{\theta}_1\theta_2 \left(\mathbf{I}_{MN} \otimes \operatorname{diag}(\mathbf{C}_{\mathbf{x}_n})\right)\mathbf{C_x} + 2\bar{\theta}_2\theta_1 \mathbf{C_x}\left(\mathbf{I}_{MN} \otimes \operatorname{diag}(\mathbf{C}_{\mathbf{x}_n})\right) \\
&+ 2|\theta_2|^2 \left(\mathbf{I}_{MN} \otimes \operatorname{diag}(\mathbf{C}_{\mathbf{x}_n})\right) \times \mathbf{C_x}\left(\mathbf{I}_{MN} \otimes \operatorname{diag}(\mathbf{C}_{\mathbf{x}_n})\right)
\end{aligned} \tag{4.47}$$

where $\circ$ is the element-wise product. A full derivation of this can be found in [10].

4.6 OSCILLATOR PHASE NOISE

In the case of phase noise, formulating a stochastic model in the same format as for the power amplifier is not trivial as phase noise is in itself not generated through a nonlinear function but is in fact already described as a stochastic process. Therefore, we may not use the same straightforward method using the Bussgang theorem. However, we may do a signal decomposition using LMMSE Estimation in order to formulate a model in the linear same format.

We recall that the behavioral model for the phase noise is described by

$$y[n] = x[n]e^{i\varphi[n]} \tag{4.48}$$

where $\varphi[n] = \lambda\varphi[n-1] + w[n]$. The stochastic model for the phase noise, in this case, is written as

$$\mathbf{x} = \mathbf{G}^{\text{PN}}\mathbf{w} + \mathbf{e}^{\text{PN}} \tag{4.49}$$

where

$$\mathbf{G}^{\text{PN}} = e^{-\frac{2\pi^2 f_c^2 t_s \kappa}{1-\lambda^2}} \mathbf{I}_{MN}. \tag{4.50}$$

In the case of phase noise with separate oscillators using PLL synchronization, the covariance of the error-term becomes

$$\mathbf{C}_{\mathbf{e}^{\text{PN}}} = \mathbf{C}_{\mathbf{y}} - e^{-\frac{4\pi^2 f_c^2 t_s \kappa}{1-\lambda^2}} \mathbf{C}_{\mathbf{x}} \tag{4.51}$$

in which the output covariance is

$$\mathbf{C}_{\mathbf{y}} = \text{diag}\,(\boldsymbol{\psi})\,\text{diag}\,(\mathbf{C}_{\mathbf{x}}) + e^{-\frac{2\pi^2 f_c^2 t_s \kappa}{1-\lambda^2}}\,\text{nondiag}\,(\mathbf{C}_{\mathbf{x}}). \tag{4.52}$$

Here, $\boldsymbol{\psi} = \text{vec}\left(\left[\psi_0, \ldots, \psi_{N-1}\right]\right)$ which for each ψ_m may be computed as

$$\psi_m = e^{-\frac{4\pi^2 f_c^2 t_s \kappa}{1-\lambda^2}\left(1-\lambda^{\left|(m+\frac{N}{2})\text{mod}N - \frac{N}{2}\right|}\right)}. \tag{4.53}$$

It is of interest to notice that the gain and covariance matrix depend only on PLL, the oscillator parameter λ, and κ.

4.7 STOCHASTIC MODELING OF DATA CONVERTERS

Finally, before moving on to concatenate the model-blocks into a complete transmitter model, we need a description of the quantization noise from the data converters. Just as with power amplifier distortion and phase noise, we model the process in a linear fashion. We have

$$\mathbf{x} \;=\; \mathbf{G}^{\text{DAC}}\mathbf{w} + \mathbf{e}^{\text{DAC}}. \tag{4.54}$$

The gain matrix for a DAC, denoted

$$\mathbf{G}^{\text{DAC}} = \mathbf{I}_{MN} \otimes \text{diag}\left(\mathbf{g}^{\text{DAC}}\right), \tag{4.55}$$

will element-wise be computed as [11]

$$\mathbf{g}^{\mathrm{DAC}} = \frac{\Delta}{\sqrt{\pi}} \operatorname{diag}\left(\mathbf{C}_{\mathbf{x}_n}\right)^{-\frac{1}{2}} \sum_{i=1}^{2^q-1} e^{-\Delta^2 (i-2^{q-1})^2 (\mathbf{C}_{\mathbf{x}_n})^{-1}}. \tag{4.56}$$

Now again, following the orthogonality we get the error covariance matrix:

$$\mathbf{C}_{\mathbf{e}^{\mathrm{DAC}}} = \mathbf{C}_{\mathbf{y}} - \mathbf{G}^{\mathrm{DAC}} \mathbf{C}_{\mathbf{x}} \mathbf{G}^{\mathrm{DAC}}. \tag{4.57}$$

The quantized output covariance matrix, from [11, Eq. 10], is then computed as

$$\begin{aligned} \mathbf{C}_{\mathbf{y}} \approx\ & \frac{\Delta^2}{2}(2^q-1)^2 \mathbf{I}_{BN} - 4\Delta^2 \sum_{i=1}^{2^q-1} \left(i - 2^{q-1}\right) \\ \times\ & \Phi\left(\sqrt{2} \operatorname{diag}\left(\mathbf{C}_{\mathbf{x}}\right)^{-\frac{1}{2}} (i-2^{q-1})\right) \\ +\ & \mathbf{G}^{\mathrm{DAC}} \operatorname{nondiag}\left(\mathbf{C}_{\mathbf{x}}\right) \mathbf{G}^{\mathrm{DAC}}. \end{aligned} \tag{4.58}$$

With the individual models for each of the subsystem component, a more complete transmitter model may be assembled.

4.8 MODEL CONCATENATION AND SIMULATIONS

To assemble the chain forming a complete transmitter chain model, we may use the linear properties of each subsystem model to our advantage. If we consider the linear models for two different (vector) nonlinear systems

$$\begin{aligned} \mathbf{y}_1 &= \mathbf{G}_1 \mathbf{x} + \mathbf{e}_1 \\ \mathbf{y}_2 &= \mathbf{G}_2 \mathbf{x} + \mathbf{e}_2 \end{aligned} \tag{4.59}$$

the concatenated model for the series non-linearity becomes

$$\mathbf{y} = \mathbf{G}_2(\mathbf{G}_1 \mathbf{x} + \mathbf{e}_1) + \mathbf{e}_2. \tag{4.60}$$

Using this step repeatedly, we may find the transmitter-model as derived in [10]

$$\begin{aligned} \mathbf{y} &= \mathbf{G}^{\mathrm{PA}} \left(\mathbf{G}^{\mathrm{PN}} \left(\mathbf{G}^{\mathrm{DAC}} \mathbf{z} + \mathbf{e}^{\mathrm{DAC}} \right) + \mathbf{e}^{\mathrm{PN}} \right) \mathbf{w} + \mathbf{e}^{\mathrm{PA}} \\ &= \underbrace{\mathbf{G}^{\mathrm{PA}} \mathbf{G}^{\mathrm{PN}} \mathbf{G}^{\mathrm{DAC}} \mathbf{z}}_{\text{Scaled and rotated signal}} + \underbrace{\mathbf{G}^{\mathrm{PA}} \mathbf{G}^{\mathrm{PN}} \mathbf{e}^{\mathrm{DAC}} + \mathbf{G}^{\mathrm{PA}} \mathbf{e}^{\mathrm{PN}} + \mathbf{e}^{\mathrm{PA}}}_{\text{Distortion noise}}, \end{aligned} \tag{4.61}$$

from which we may observe an equivalent gain matrix, $\mathbf{G}^{\mathrm{PA}} \mathbf{G}^{\mathrm{PN}} \mathbf{G}^{\mathrm{DAC}}$ and an accumulative noise process with rescaled contributions from each subsystem component. Using the concatenated model, we may now describe the signal-to-interference, distortion and noise ratio (SINDR).

4.8.1 SIGNAL-TO-INTERFERENCE AND NOISE RATIO

As described in more detail in [10], the SINDR for an OFDM system at the kth UE and the lth subcarrier may be approximated by means of the concatenated model discussed in previous section. The SINDR is

$$\gamma_{k,l} \approx \frac{\exp\left(-\frac{2\pi f_c t_s \kappa}{1-\lambda^2}\right)\left|\hat{\mathbf{h}}_{k,l}^T \operatorname{diag}(\mathbf{g}^{\text{PA}})\operatorname{diag}(\mathbf{g}^{\text{DAC}})\hat{\mathbf{p}}_{k,l}^T\right|^2}{I_{k,l}+E_{k,l}+N_0} \tag{4.62}$$

in which we have that

$$I_{k,l} = \exp\left(-\frac{2\pi f_c t_s \kappa}{1-\lambda^2}\right)\sum_{v\neq k}\left|\hat{\mathbf{h}}_{k,l}^T \operatorname{diag}(\mathbf{g}^{\text{PA}})\operatorname{diag}(\mathbf{g}^{\text{DAC}})\hat{\mathbf{p}}_{v,l}^T\right|^2 \tag{4.63}$$

is the interference power and

$$E_{k,l} = \Bigg[(\mathbf{F}_N \otimes \mathbf{I}_M)\left(\mathbf{G}^{\text{PA}}\left(\exp\left(-\frac{2\pi f_c t_s \kappa}{1-\lambda^2}\right)\times\left(\mathbf{C_v}-\mathbf{G}^{\text{DAC}}\mathbf{C_z}\mathbf{G}^{\text{DAC}}\right)\right.\right. \tag{4.64}$$

$$\left.\left.+\left(\mathbf{C_w}-\exp\left(-\frac{2\pi f_c t_s \kappa}{1-\lambda^2}\right)\mathbf{C_v}\right)\left(\mathbf{G}^{\text{DAC}}\right)^H+\mathbf{C_x}-\mathbf{G}^{\text{DAC}}\mathbf{C_w}\left(\mathbf{G}^{\text{DAC}}\right)^H\right)\left(\mathbf{I}_M\otimes\mathbf{F}_N^H\right)\Bigg]_{k+lK,k+lK}$$

is the received distortion power on the lth subcarrier and the kth user. Here, $\mathbf{F}_N$ is the $N \times N$ DFT-matrix for which $\mathbf{F}_N\mathbf{F}_N^H = \mathbf{I}_N$.

Looking at the interference power, comparing to conventional signal-to-interference-and-noise ratio (SINR)-analysis of massive MU-MIMO systems, further degeneration stemming from non-ideal hardware may be observed.

4.8.2 SIMULATIONS

In order to illustrate the usage of the stochastic transceiver model discussed in this chapter, we will present some basic link-simulations. These make use of the stochastic models in order to emulate an OFDM-based massive MU-MIMO array transmitter and we compare the results with simulations using conventional models. We will consider an OFDM-system with frequency selective linear precoding using maximum ratio transmission (MRT) and zero-forcing (ZF). The precoding matrices are defined as

$$\mathbf{P}_k^{\text{MRT}} = \alpha_{\text{MRT}}\frac{1}{M}\mathbf{H}_k^H \tag{4.65}$$

$$\mathbf{P}_k^{\text{ZF}} = \alpha_{\text{ZF}}\frac{1}{M}\mathbf{H}_k\left(\mathbf{H}_k\mathbf{H}_k^H\right)^{-1} \tag{4.66}$$

where

$$\mathbf{H}_k = \sum_{t=0}^{T}\hat{\mathbf{H}}_t e^{-jk\frac{2\pi}{N}t} \tag{4.67}$$

is the $M \times K$ frequency-domain matrix describing the channel at subcarrier k. α_{MRT} and α_{ZF} are chosen such that $\mathbb{E}\left[|\mathbf{x}_n|^2\right] = 1$ for the precoded vector $\mathbf{x}_n$. The full set of parameters are found in Table 4.2. We consider an OFDM-based MU-MIMO system with M antennas and K users and a Rayleigh inde-

Table 4.2 Simulation parameters

Parameter	Setting
Carrier frequency	3 GHz
Carrier type	OFDM
Number of occupied subcarriers	1200
FFT size	4096
Carrier bandwidth	20 MHz
Channel parameter	**Setting**
Channel type	Rayleigh IID in spherical coordinates
Number of channel taps	5
Power delay profile	Exponential
DAC parameter	**Setting**
q	6 bits
LSB (Δ)	0.0081
Phase-noise parameter	**Setting**
λ	0.99
κ	$5 \cdot 10^{-17}$
Power amplifier parameter	**Setting**
θ_1	1
θ_2	$-0.03491 + i0.00565$

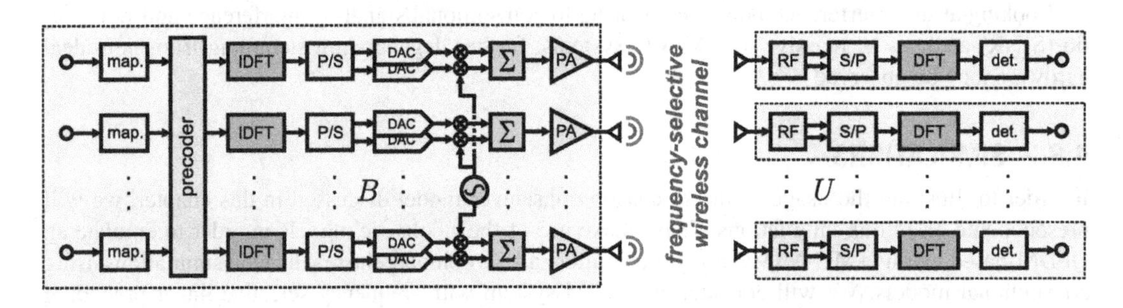

FIGURE 4.14

Complex baseband system model. The complex baseband system model including a base-station equipped with M antennas, each consisting of DAC's, local oscillators and power amplifiers. This is followed by the radio channel and K served users.

pendent identically distributed (IID) channel. Fig. 4.14 shows the system model including impairment models for quantization noise, phase noise and power amplifier distortion.

4.8.3 SIMULATION RESULTS

The simulation results are summarized in Figs. 4.15–4.17. Fig. 4.15 shows the simulated power spectral density of each model block as well as the composite transmitter model, compared to the corresponding

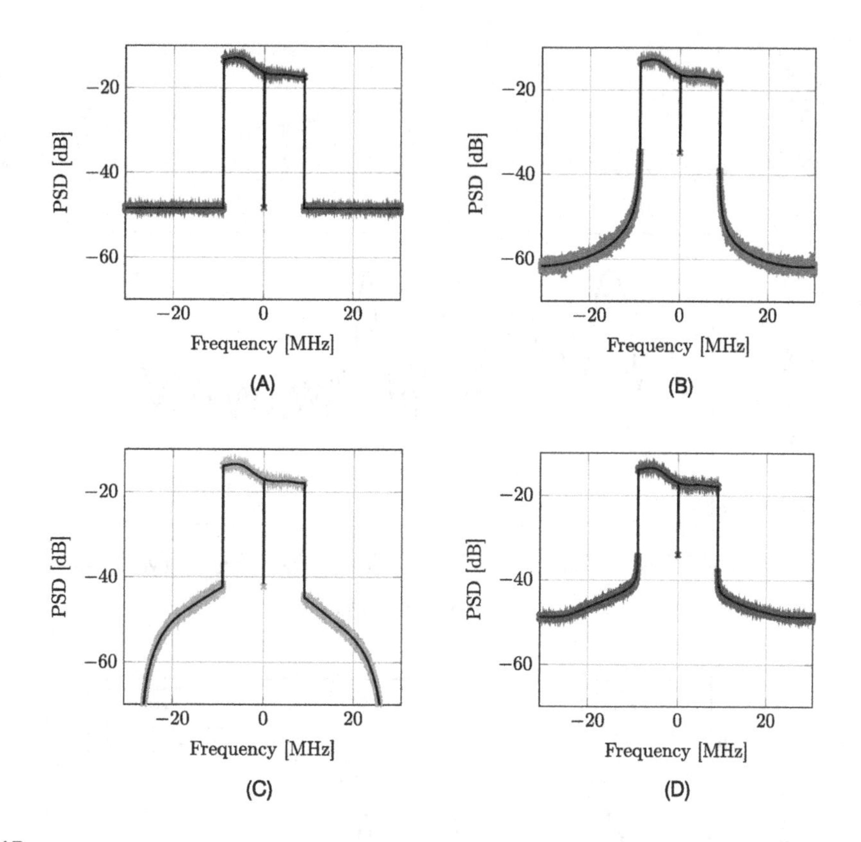

FIGURE 4.15

Output power spectral density from the statistical models. PSD of the transmitted signal for the case when each hardware impairment in the RF-chain are considered separately, and when all of the hardware impairments are considered together. The markers correspond to simulation results whereas the black lines are the analytical outcome. (A) Finite-resolution DACs only (ideal LO and linear PA). (B) Nonideal LO only (infinite-resolution DAC and linear PA). (C) Nonlinear PAs only (infinite-resolution DACs and ideal LO). (D) Concatenated impairment model (finite-resolution DACs, nonideal LO, and nonlinear PA).

behavioral models. As may be seen, the stochastic model accurately predicts the power spectral density in not only each model block, but also in the case of the complete transmit model. The PSD is accurately predicted both in- and out-of band, which is otherwise not the case for most linear and stochastic models which today is widely used in the literature, [3].

Fig. 4.16 shows the radiated power of both the transmit signal (A) and the adjacent channel power (B). Here we may spot similar effects to those noted in Section 4.1.5. In a similar manner, the inband distortion takes on a similar radiation pattern as the transmit signal in this case.

Finally, Fig. 4.17 shows the performance in terms of uncoded BER with and without impairment models. As may be observed, the radio hardware impairments has a negative impact on the uncoded BER performance for both ZF and MRT.

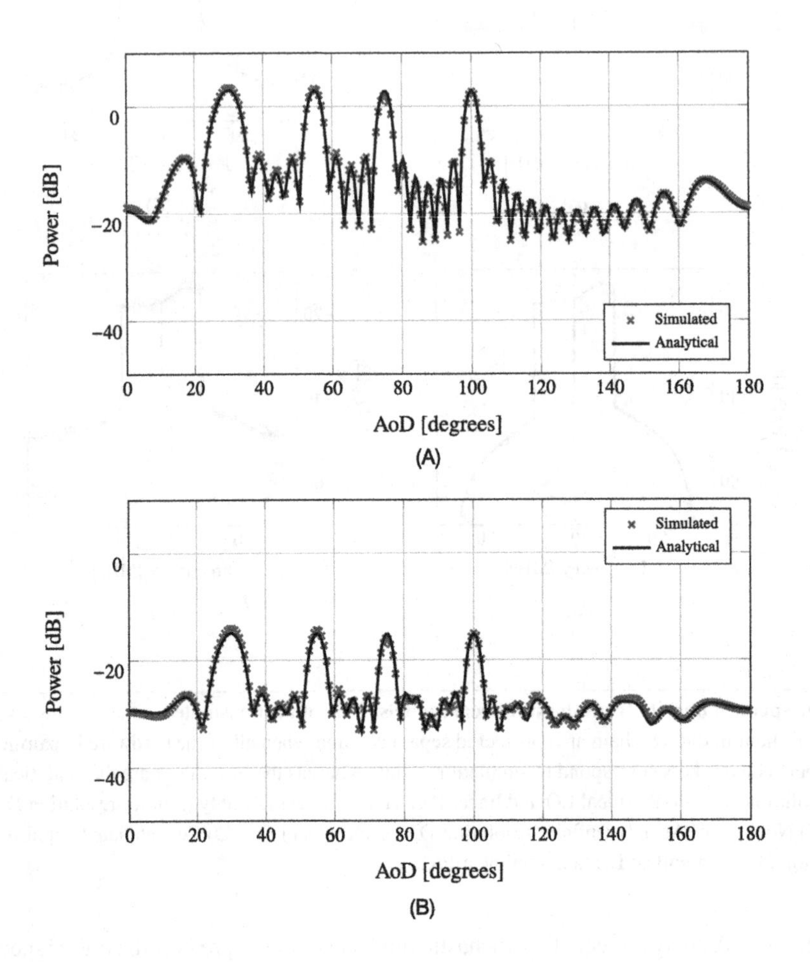

FIGURE 4.16

Simulated radiation patterns using stochastic impairment models. Simulated radiation patterns of the desired signal (A) and the impairment component (B), consisting of power amplifier distortion, phase noise, and quantization noise.

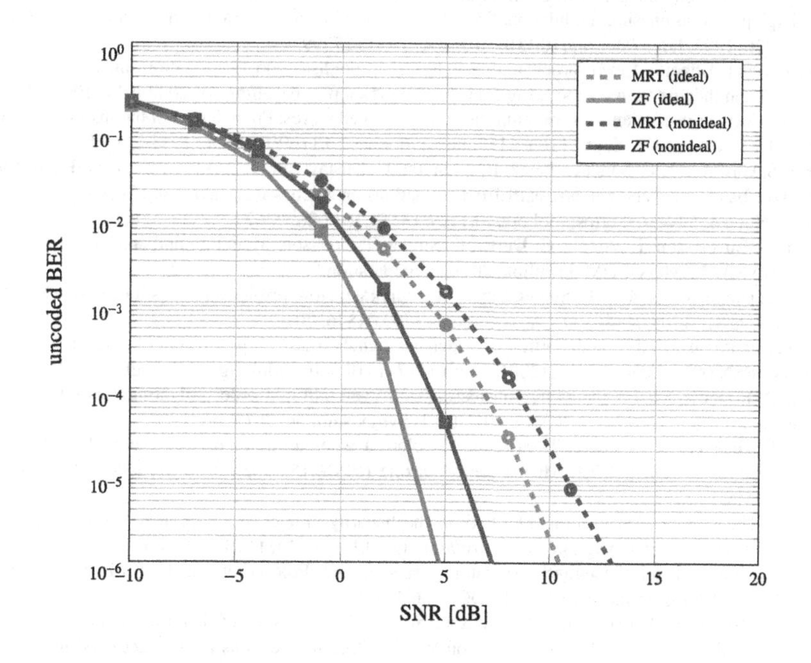

FIGURE 4.17

Link-performance assessment using stochastic impairment models. Comparison of uncoded BER using linear precoding, with and without the impact of hardware impairments.

REFERENCES

[1] S. Afsardoost, T. Eriksson, C. Fager, Digital predistortion using a vector-switched model, IEEE Transactions on Microwave Theory and Techniques (ISSN 0018-9480) 60 (4) (2012, April) 1166–1174, https://doi.org/10.1109/TMTT.2012.2184295.

[2] S. Benedetto, E. Biglieri, R. Daffara, Modeling and performance evaluation of nonlinear satellite links-a Volterra series approach, IEEE Transactions on Aerospace and Electronic Systems (ISSN 0018-9251) AES-15 (4) (1979, July) 494–507, https://doi.org/10.1109/TAES.1979.308734.

[3] E. Björnson, J. Hoydis, M. Kountouris, M. Debbah, Massive MIMO systems with non-ideal hardware: energy efficiency, estimation, and capacity limits, IEEE Transactions on Information Theory 60 (11) (2014) 7112–7139.

[4] N. Bjorsell, P. Suchánek, P. Handel, D. Ronnow, Measuring Volterra kernels of analog-to-digital converters using a stepped three-tone scan, IEEE Transactions on Instrumentation and Measurement 57 (4) (2008) 666–671.

[5] J.J. Bussgang, Crosscorrelation functions of amplitude-distorted gaussian signals, 1952.

[6] H. Chireix, High power outphasing modulation, Proceedings of the Institute of Radio Engineers (ISSN 0731-5996) 23 (11) (1935, Nov.) 1370–1392, https://doi.org/10.1109/JRPROC.1935.227299.

[7] L. Ding, G.T. Zhou, Effects of even-order nonlinear terms on predistortion linearization, in: Digital Signal Processing Workshop, 2002 and the 2nd Signal Processing Education Workshop. Proceedings of 2002 IEEE 10th, 2002, pp. 1–6.

[8] W.H. Doherty, A new high efficiency power amplifier for modulated waves, Proceedings of the Institute of Radio Engineers (ISSN 0731-5996) 24 (9) (1936, Sept.) 1163–1182, https://doi.org/10.1109/JRPROC.1936.228468.

[9] K. Hausmair, S. Gustafsson, C. Sánchez-Pérez, P.N. Landin, U. Gustavsson, T. Eriksson, C. Fager, Prediction of nonlinear distortion in wideband active antenna arrays, IEEE Transactions on Microwave Theory and Techniques (ISSN 1557-9670) 65 (11) (2017, Nov.) 4550–4563, https://doi.org/10.1109/TMTT.2017.2699962.

[10] S. Jacobsson, U. Gustavsson, D. Astely, G. Durisi, C. Studer, T. Eriksson, Modeling and analysis of transmit-RF impairments in massive MU-MIMO OFDM downlink, 2018, in preparation.

[11] S. Jacobsson, G. Durisi, M. Coldrey, C. Studer, Linear precoding with low-resolution DACs for massive MU-MIMO-OFDM downlink, CoRR, arXiv:1709.04846, http://arxiv.org/abs/1709.04846, 2017.

[12] J. Kim, K. Konstantinou, Digital predistortion of wideband signals based on power amplifier model with memory, Electronics Letters (ISSN 0013-5194) 37 (23) (2001, Nov.) 1417–1418, https://doi.org/10.1049/el:20010940.

[13] P.N. Landin, D. Rönnow, RF PA modeling considering odd–even and odd order polynomials, in: Communications and Vehicular Technology in the Benelux (SCVT), 2015 IEEE Symposium on, 2015, pp. 1–6.

[14] P.N. Landin, S. Gustafsson, C. Fager, T. Eriksson, Weblab: a web-based setup for PA digital predistortion and characterization [application notes], IEEE Microwave Magazine (ISSN 1527-3342) 16 (1) (2015, Feb.) 138–140, https://doi.org/10.1109/MMM.2014.2367857.

[15] P.M. Lavrador, T.R. Cunha, P.M. Cabral, J.C. Pedro, The linearity-efficiency compromise, IEEE Microwave Magazine (ISSN 1527-3342) 11 (5) (2010, Aug.) 44–58, https://doi.org/10.1109/MMM.2010.937100.

[16] D.B. Leeson, A simple model of feedback oscillator noise spectrum, Proceedings of the IEEE (ISSN 0018-9219) 54 (2) (1966, Feb.) 329–330, https://doi.org/10.1109/PROC.1966.4682.

[17] C. Mollén, U. Gustavsson, T. Eriksson, E.G. Larsson, Spatial characteristics of distortion radiated from antenna arrays with transceiver nonlinearities, IEEE Transactions on Wireless Communications (2018), accepted for publication, arXiv: 1711.02439, http://arxiv.org/abs/1711.02439, 2017.

[18] C. Mollén, U. Gustavsson, T. Eriksson, E.G. Larsson, Out-of-band radiation measure for MIMO arrays with beamformed transmission, in: 2016 IEEE International Conference on Communications (ICC), 2016, May, pp. 1–6.

[19] G.E. Moore, Cramming more components onto integrated circuits, Electronics 38 (8) (1965) 114.

[20] D.R. Morgan, Z. Ma, J. Kim, M.G. Zierdt, J. Pastalan, A generalized memory polynomial model for digital predistortion of RF power amplifiers, IEEE Transactions on Signal Processing 54 (10) (2006) 3852–3860.

[21] I.S. Reed, On a moment theorem for complex Gaussian processes, IEEE Transactions on Information Theory 8 (3) (1962, Apr.) 194–195, https://doi.org/10.1109/TIT.1962.1057719.

[22] D. Root, et al., Polyharmonic distortion modeling, IEEE Microwave Magazine 7 (3) (2006) 44–57.

[23] A. Sripad, D. Snyder, A necessary and sufficient condition for quantization errors to be uniform and white, IEEE Transactions on Acoustics, Speech, and Signal Processing (ISSN 0096-3518) 25 (5) (1977, Oct.) 442–448, https://doi.org/10.1109/TASSP.1977.1162977.

[24] V. Volterra, Sopra le funzioni che dipendono da altre funzioni, Tip. della R. Accademia dei Lincei, 1887.

[25] B. Widrow, I. Kollár, Quantization Noise: Roundoff Error in Digital Computation, Signal Processing, Control, and Communications, Cambridge University Press, Cambridge, UK, ISBN 9780521886710, 2008.

[26] P. Zetterberg, A. Wolfgang, A. Westlund, et al., Initial multi-node and antenna transmitter and receiver architectures and schemes, mmMAGIC Deliverale D5.1, 2016.

CHAPTER

MULTICARRIER WAVEFORMS

5

The waveform is a core component of any communication technology. Broadly speaking, there are two main categories of waveforms: i) single-carrier waveforms, ii) multicarrier waveforms. 2G and 3G cellular systems (GSM, UMTS, CDMA2000) as well as ZigBee and Bluetooth adopted single-carrier waveforms. The 4G cellular system (LTE) employs a multicarrier waveform. Multicarrier waveforms are also used in asymmetric digital subscriber line (ADSL), wireless local area network (WLAN), WiMAX, digital audio broadcast (DAB), and digital video broadcast-terrestrial (DVB-T) standards. Typically, single-carrier waveforms have a low peak-to-average power ratio (PAPR), which makes them power efficient—suitable for coverage limited scenarios and extending battery life of user equipment. On the other hand, multicarrier waveforms provide high spectral efficiency, flexible resource allocation in the frequency domain, and possibly easy integration with multiantenna technology. These are the key drivers for 5G NR.

NR will support various use cases with different deployments in frequencies from below 1 GHz to 100 GHz. Although single-carrier waveforms can be interesting for massive IoT devices (with extended battery life) and for operation in high carrier frequencies (where transmission loss is high), the multicarrier waveforms have been seen as main candidates of 5G due to the reasons mentioned above. The 3GPP evaluated several multicarrier waveforms as well as single carrier waveforms. It was observed that different waveforms have their pros and cons, which makes different waveforms suitable in different scenarios. However, the 3GPP also acknowledged the fact that the 5G NR radio interface based on multiple waveforms will make the overall system design complex. Considering overall performance, system requirements, and the need for a single waveform, the 3GPP concluded in favor of OFDM for both uplink and downlink transmissions. Moreover, the DFT-Spread OFDM (DFTS-OFDM) waveform, which has single-carrier properties, has been kept as an option for uplink transmission in coverage limited scenarios.

This chapter presents various multicarrier waveforms and provides a comparison that has led to the down selection of OFDM for 5G NR. The chapter is structured as follows. Section 5.1 introduces state-of-the-art OFDM-based and filter bank multicarrier (FBMC)-based waveforms and Section 5.2 introduces the DFTS-OFDM waveform. Section 5.3 describes the waveform design requirements for NR. Section 5.4 presents the key indicators for waveform comparisons, based on which Section 5.5 compares performance of the multicarrier waveforms.

5.1 MULTICARRIER WAVEFORMS

The main principle behind multicarrier waveforms is to split a high rate data stream into multiple low rate streams that are transmitted simultaneously over a number of carriers (termed subcarriers). By doing so, the symbol duration for each low rate subcarrier increases and therefore the relative amount of time dispersion caused by a multipath (frequency selective) channel decreases. This makes

5G Physical Layer. https://doi.org/10.1016/B978-0-12-814578-4.00010-2

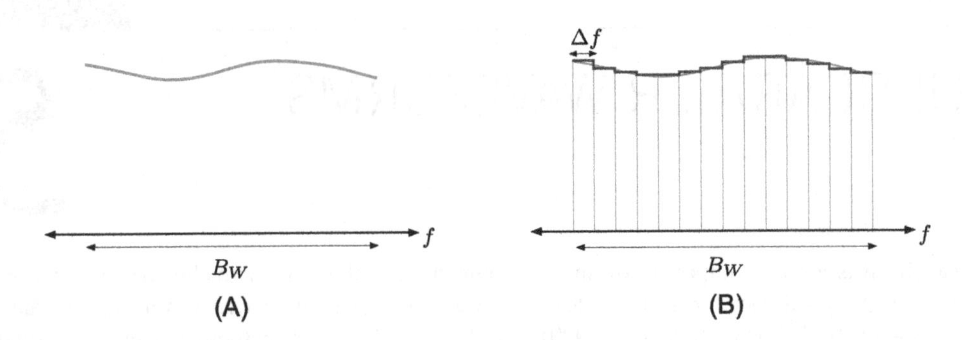

FIGURE 5.1

(A) Frequency-selective channel. (B) Frequency flat subchannels.

a multicarrier waveform robust to the intersymbol interference caused by the multipath channel. Let us look at this in the frequency domain. A frequency-selective fading channel is split into multiple (approximately) frequency flat channels and each subcarrier now experiences frequency flat fading, as illustrated in Fig. 5.1. This avoids complex time domain equalization, making the multicarrier transmissions suitable for high data rate wireless communication.

Multicarrier waveforms provide the freedom of multiplexing multiple users in the frequency domain (allocating different subcarriers to different users) and spatial domain (mapping different subcarriers to different antennas). Furthermore, multicarrier waveforms are robust to narrowband interference compared to single-carrier waveforms. In a multicarrier system, narrowband interference only corrupts a small number of subcarriers, whereas in a single-carrier transmission an entire link can fail due to narrowband interference.

5.1.1 THE PRINCIPLE OF ORTHOGONALITY

Mathematically, a multicarrier waveform in baseband can be expressed as

$$\tilde{x}(t) = \sum_{s\in\mathbb{Z}} \sum_{i\in I} X_{s,i}\, \tilde{p}_i(t - sT), \tag{5.1}$$

where s denotes the waveform symbol index, $\mathbb{Z}$ is a set of integers, i denotes the subcarrier index, $I \in \{i = 1, 2, \dots, N\}$ contains indices of the subcarriers, N is the total number of subcarriers, $X_{i,s} \in \mathbb{C}$ is a complex modulated (QAM or phase shift keying (PSK)) symbol mapped to the subcarrier i and the symbol index s, and $\tilde{p}_i(t)$ is a pulse shape (or prototype filter) used for the subcarrier i. Typically, $\tilde{p}_i(t)$ is a bandpass filter centered at the subcarrier frequency f_i. For example, for OFDM, $\tilde{p}_i(t) = e^{j2\pi f_i t}$ for $0 \le t \le T$ and $\tilde{p}_i(t) = 0$ elsewhere. In a multicarrier system, the number of active subcarriers is usually chosen smaller than the total number of subcarriers to relax filtering operations, i.e., $X_{i,s} = 0$ for some i at the band edges. The modulation and demodulation operations for a multicarrier system are illustrated in Fig. 5.2.

The demodulation operation is based on matched filtering, which is optimal for linear and additive noise channels in the sense of maximizing signal-to-noise ratio (SNR) of the demodulated (reconstructed) signal. The matched filter demodulation comprises the convolution of the received signal $\tilde{r}(t)$

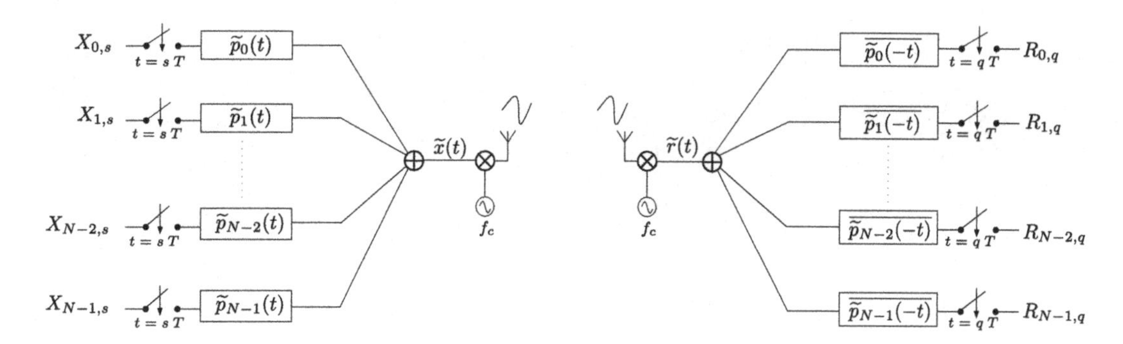

FIGURE 5.2

Modulation and demodulation process in a multicarrier system.

with the complex conjugated time-reversed version of the transmitter prototype filter (or pulse shape), followed by sampling at a rate equal to $\frac{1}{T}$. For simplicity, let us assume that $\widetilde{r}(t) = \widetilde{x}(t)$ and denote $R_{l,q}$ as demodulated (reconstructed) QAM (or PSK) symbol from a waveform symbol q and a subcarrier l. The demodulation operation is then given by

$$\begin{aligned} R_{l,q} &= \widetilde{r}(t) * \overline{\widetilde{p}_l(-t)}\Big|_{t=qT} \\ &\overset{(a)}{=} \Big\langle \widetilde{r}(t), \widetilde{p}_l(t-qT) \Big\rangle \\ &\overset{(b)}{=} \sum_{s\in\mathbb{Z}} \sum_{i\in I} X_{s,i} \Big\langle \widetilde{p}_i(t-sT), \widetilde{p}_l(t-qT) \Big\rangle, \end{aligned} \tag{5.2}$$

where $\overline{\widetilde{p}_i(t)}$ denotes complex conjugation of $\widetilde{p}_i(t)$; $*$ is the convolution operator; $\langle v_1(t), v_2(t) \rangle := \int_{\infty}^{\infty} v_1(t) v_2{}^*(t) dt$ denotes the inner product between $v_1(t)$ and $v_2(t)$; (a) follows from the definition of the convolution operation (i.e., expressing convolution in terms of correlation); and (b) follows from the assumption that $\widetilde{r}(t) = \widetilde{x}(t)$. From (5.2), we can see that the perfect reconstruction of the transmitted QAM symbols ($R_{l,q} = X_{l,q}$) requires the following orthogonality condition to be fulfilled:

$$\Big\langle \widetilde{p}_i(t-sT), \widetilde{p}_l(t-qT) \Big\rangle = \delta_{i,l}\, \delta_{s,q}, \tag{5.3}$$

where $\delta_{a,b}$ is the Kronecker delta function, defined as $\delta_{a,b} = 1$ when $a = b$ and $\delta_{a,b} = 0$ when $a \neq b$.

A multicarrier waveform that satisfies the orthogonality condition in (5.3) is called an orthogonal waveform; for example, the orthogonal frequency division multiplexing (OFDM) waveform. Let us look at the prototype filters used for OFDM and see how they fulfill the orthogonality condition. For an ith subcarrier in OFDM, the prototype filter is given by

$$\widetilde{p}_i(t) := \widetilde{p}(t)\, e^{j2\pi f_i t}, \tag{5.4}$$

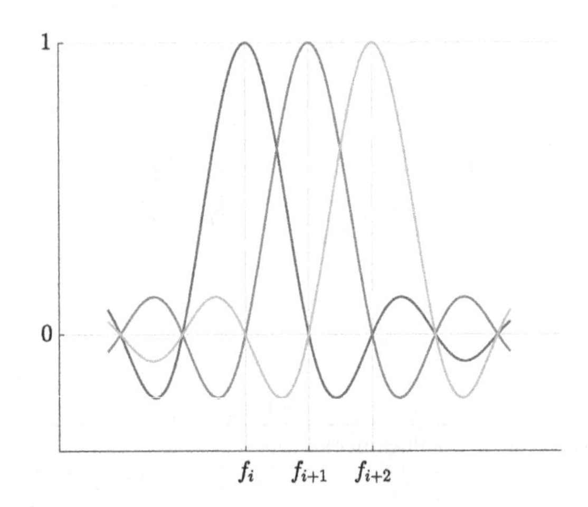

FIGURE 5.3

Frequency domain illustration of OFDM subcarriers.

where $f_i = \frac{i}{T}$ and $\widetilde{p}(t)$ is a rectangular filter with length T defined as

$$\widetilde{p}(t) := \begin{cases} \frac{1}{\sqrt{T}} & \text{if } 0 \leq t \leq T \\ 0 & \text{elsewhere} \end{cases}, \tag{5.5}$$

where the scaling $\frac{1}{\sqrt{T}}$ is simply chosen to get unit energy, i.e., $\int_{-\infty}^{+\infty} |\widetilde{p}(t)|^2 \, dt = 1$. Without loss of generality, the inner product (cf. (5.3)) between the prototype filters is given by

$$\begin{aligned} \langle \widetilde{p}_i(t), \widetilde{p}_l(t) \rangle &= \int_0^T \widetilde{p}_i(t) \, \overline{\widetilde{p}_l(t)} \, dt \\ &= \int_0^T \widetilde{p}(t) \, e^{j2\pi f_i t} \, \widetilde{p}(t) \, e^{-j2\pi f_l t} \, dt \\ &= \int_0^T e^{j2\pi (f_i - f_l)t} \, dt \\ &= \int_0^T e^{j2\pi \frac{i-l}{T} t} \, dt, \end{aligned}$$

which is nonzero (and equal to 1) if and only if $i = l$. We observe that it is the selection of $f_i = \frac{i}{T}$ that makes the OFDM waveform orthogonal. A frequency domain illustration of the OFDM subcarriers is shown in Fig. 5.3. The spacing between consecutive subcarriers is $f_{i+1} - f_i = \frac{i+1}{T} - \frac{i}{T} = \frac{1}{T} := \Delta f$, i.e., the subcarrier spacing Δf is the inverse of the symbol duration T.

When the orthogonality condition is not fulfilled, the waveform is nonorthogonal, for example, the FBMC waveforms. In the FBMC waveforms, the prototype filters of the subcarriers are not orthogonal within the same symbol as well as between the adjacent symbols. Typically, there is interference between subcarriers of the same symbols and between subcarriers of the adjacent different symbols (due to overlapping waveform symbols in the time domain). The inner product (based on (5.3)) takes the following form:

$$\left\langle \widetilde{p}_i(t - sT), \widetilde{p}_l(t - qT) \right\rangle = \begin{cases} 1 & \text{if } i = l \text{ and } s = q \\ \varepsilon_{i,l} & \text{if } i \neq l \text{ and } s = q, \\ \beta_{i,l,s,q} & \text{else,} \end{cases} \tag{5.6}$$

where $\varepsilon_{i,l}$ corresponds to an interference between subcarriers of the same symbol and $\beta_{i,l,s,q}$ is associated with an interference between subcarriers of two different symbols. These two types of interference are known as self-interferences—$\varepsilon_{i,l}$ is referred to as self-intercarrier interference and $\beta_{i,l,s,q}$ is referred to as self-intersymbol interference. The structure of the FBMC waveforms will be discussed in Section 5.1.3.

5.1.2 OFDM-BASED WAVEFORMS

There are several OFDM-based waveforms. In the following, we will present four major variants of the OFDM waveform, namely, CP-OFDM, windowed OFDM (W-OFDM), filtered OFDM (F-OFDM), and universally filtered OFDM (UF-OFDM).

A general discrete-time OFDM-based waveform is given by

$$x[n] = \left(\sum_{s \in \mathbb{Z}} \sum_{i \in I} X_{i,s} \, p_i[n - sN_{sc}] \right) * w_{tx}[n], \tag{5.7}$$

where I contains indices of the active subcarriers,[1] N_{sc} is the total number of subcarriers, $X_{i,s} \in \mathbb{C}$ is the modulated symbol corresponding to the subcarrier i and the time index s, $p_i[n] := p[n]\, e^{j2\pi \frac{i}{N_{sc}} n}$ is the prototype filter (or pulse shape) used for the subcarrier i, and $w_{tx}[n]$ is a filter to improve spectral characteristics of the transmitted signal.[2] The modulation and demodulation operations of the OFDM-based waveforms are shown in Fig. 5.4 and Fig. 5.5, respectively.

5.1.2.1 Cyclic Prefix OFDM

CP-OFDM employs a rectangular pulse shape, i.e., $p[n]$ in (5.7) is given by

$$p[n] := \begin{cases} \frac{1}{\sqrt{N_{sc}}} & \text{if } 0 \leq n \leq N_{sc} - 1, \\ 0 & \text{elsewhere,} \end{cases} \tag{5.8}$$

where N_{sc} is the total number of subcarriers. There is no post filtering operation, i.e., $w_{tx}[n] = \delta[n]$ in (5.7). To combat the intersymbol interference (ISI) in a multipath channel, a cyclic prefix (CP)

[1] In a multicarrier system, the number of active subcarriers is usually chosen smaller than the total number of subcarriers to relax filtering operations.

[2] The sharper the spectrum roll-off of the signal, the easier it is to fulfill the out-of-band (OOB) emission requirements specified by the 3GPP for cellular systems. The OOB emission requirements are specified in terms of a spectrum emission mask and an adjacent channel leakage ratio for both base stations and devices.

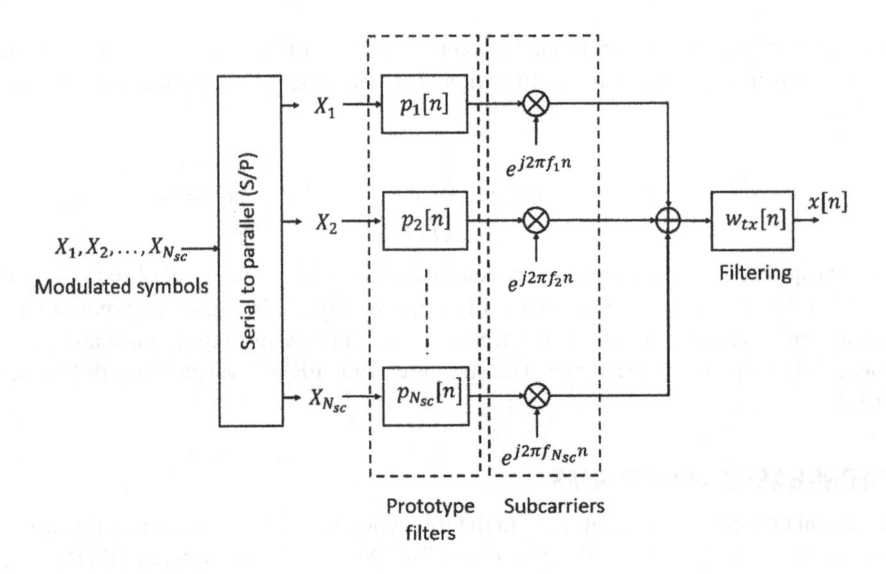

FIGURE 5.4

A general modulation structure of OFDM waveforms.

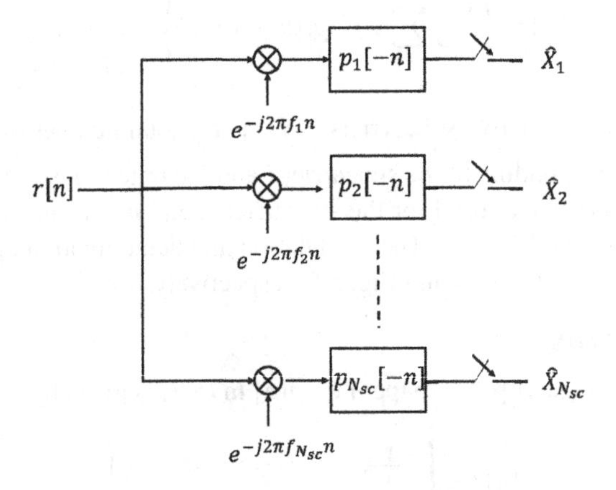

FIGURE 5.5

A general demodulation structure of OFDM waveforms.

is inserted in the OFDM symbol. The CP refers to the cyclic extension of an OFDM symbol, that is, appending the last N_{cp} samples of the OFDM symbol to the front of the symbol as illustrated in Fig. 5.6. If the CP duration is chosen greater than the delay spread of the channel, then the received OFDM signal does not suffer from any ISI. In LTE, the normal CP duration is 7% of the core OFDM symbol (i.e., $N_{sc} = 2048$ samples and $N_{cp} = 144$ samples).

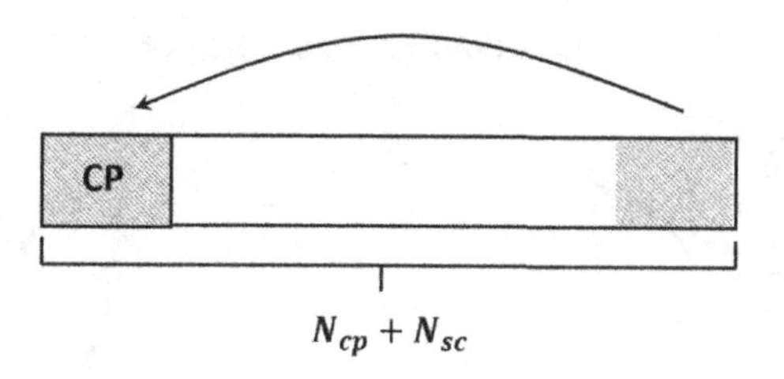

FIGURE 5.6

Cyclic prefix is inserted to avoid intersymbol interference due to the multipath channel.

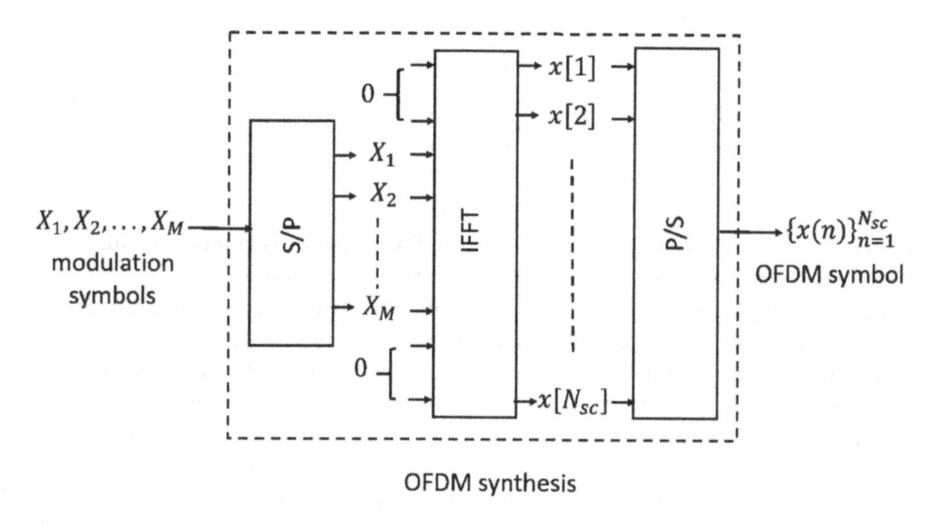

FIGURE 5.7

OFDM waveform is efficiently synthesized via fast Fourier transform.

In practice, CP-OFDM modulator is implemented via the computationally efficient fast Fourier transform (FFT). As shown in Fig. 5.7, M modulated QAM symbols ($\{X_i\}_{i=1}^{M}$) are mapped to orthogonal subcarriers via a serial-to-parallel (S/P) transformation followed by an IFFT of size N and parallel-to-serial (P/S) conversion. In this example, the total number of subcarriers is N and the number of active subcarriers is equal to M. In practice, the number of active subcarriers is kept smaller than the total number of subcarriers to relax filtering operations. This is achieved by zero padding QAM symbols prior to the IFFT operation, as shown in Fig. 5.7.

5.1.2.2 Windowed OFDM

The spectrum of CP-OFDM decays rather slowly. The main reason for its slow decay is the signal discontinuities at the symbol boundaries due to the rectangular pulse shape used in OFDM (cf. (5.8)). To improve the spectral shape of OFDM, a nonrectangular pulse shape $p[n]$ with smooth edges can be employed in (5.7), which we refer to as windowed-OFDM (W-OFDM) waveform. As an example,

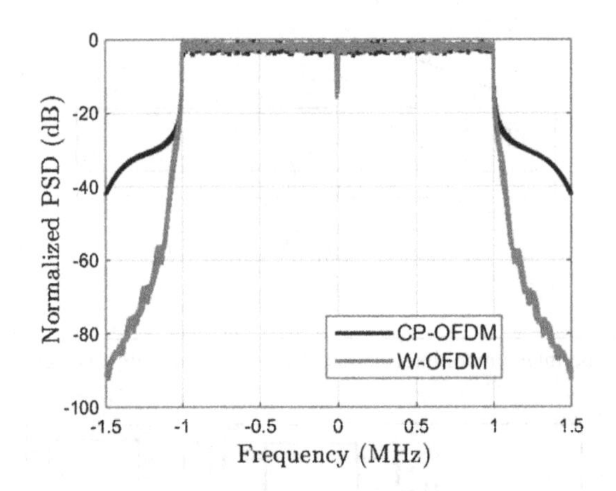

FIGURE 5.8

PSDs of CP-OFDM and W-OFDM waveforms.

the spectrum of W-OFDM (considering a window with linear slopes) is compared with CP-OFDM in Fig. 5.8. We observe that W-OFDM has a much sharper spectrum roll-off.[3]

The windowing can be done either at the transmitter side (to suppress interference to adjacent bands) or at the receiver side (to suppress interference pick-up from adjacent bands) [10,14]. In the following, we explain the transmitter side and the receiver side windowing. The transmitter side windowing is illustrated in Fig. 5.9. In the transmitter side windowing, the boundaries of each OFDM symbol are multiplied with a smooth slope in the time domain, increasing smoothly from 0 to 1 (increasing slope) or decreasing smoothly from 1 to 0 (decreasing slope). The increasing slope is applied at the beginning of the CP, while the decreasing slope is applied after the end of the core OFDM symbol within an extra added cyclic suffix. Fig. 5.9 also shows that the increasing slope of the next OFDM symbol overlaps with the decreasing slope of the previous OFDM symbol. Since the receiver only keeps the samples of the core OFDM symbol, the transmitter side windowing is transparent to the receiver. Next, we discuss the receiver side windowing.

A standard OFDM receiver cuts out the desired OFDM symbol period by applying a rectangular window in the time domain to the received signal and then subsequently applies the FFT for demodulating the subcarriers. Application of the rectangular window in the time domain corresponds to a convolution in the frequency domain with a sinc function. The sinc-like function leads to a high interference pick-up from any adjacent nonorthogonal signals such as OFDM signals with other numerologies. To reduce the interference pick-up, the rectangular window must be replaced by a smooth window function. The receiver side windowing operation is illustrated in Fig. 5.10. A smooth increasing window slope is applied at the boundary between the CP and the core OFDM symbol (half within each); a decreasing smooth window slope is applied at the boundary between the core OFDM symbol and the added cyclic suffix. If the applied window slopes fulfill the Nyquist criterion (i.e., they are

[3]The sharper the spectrum roll-off of the signal, the easier it is to fulfill the out-of-band emission requirements specified by the 3GPP for cellular systems.

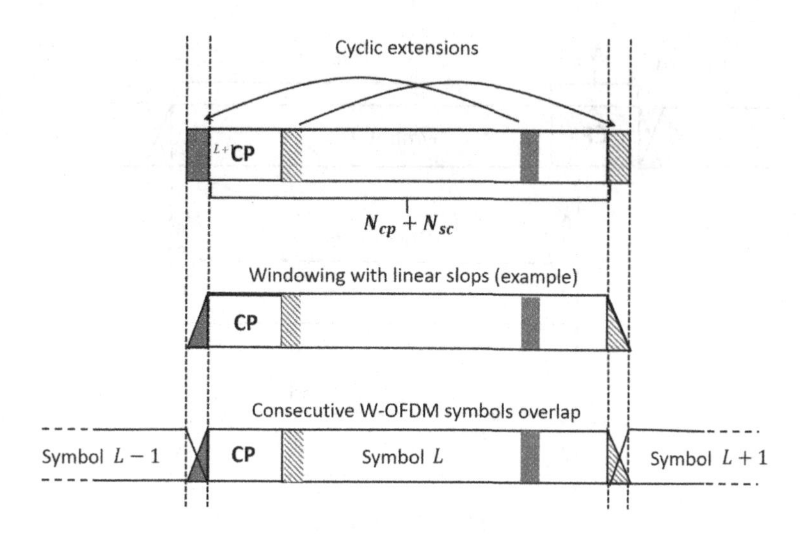

FIGURE 5.9

The transmitter side windowing.

center symmetric), the signal part cut away by the decreasing windowing slope is the same as the remaining signal part after application of the increasing window slope within the CP, since the CP is a copy of the last part of the OFDM symbol. If the windowed CP part is added to the last part of the core OFDM symbol, the core OFDM symbol is restored at its second boundary. The core OFDM symbol can also be restored at the first symbol boundary by applying the same trick. Now the complete OFDM is restored, and its subcarriers are orthogonal again. The FFT is applied to the restored core OFDM symbol as indicated in Fig. 5.10.

5.1.2.3 Filtered OFDM

Like windowing (or pulse shaping), filtering is another technique to improve spectral characteristics or reduce out-of-band (OOB) leakage of OFDM waveform [16]. Filtering (like windowing) is typically done at both transmitter and receiver sides to reduce interference to adjacent bands and reduce interference pick-up from adjacent bands, respectively. Let us refer to the general OFDM waveform modulation structure in (5.7). In filtered OFDM (F-OFDM), the prototype filter (pulse shape) $p[n]$ is rectangular including the CP, as we have in CP-OFDM. In addition, a filter $w_{tx}[n]$ is employed to suppress OOB leakage (cf. Fig. 5.4). The filter $w_{tx}[n]$ is derived as the product of an ideal bandpass filter impulse response and a time domain mask (or a window), that is,

$$w_{tx}[n] = h[n]w[n], \tag{5.9}$$

where $h[n]$ is an ideal bandpass filter with the selected signal bandwidth and $w[n]$ is a window function. Various window functions are common for the construction of FIR filters with different characteristics, for example, Hanning, Hamming, Blackman, Bartlett, and Kaiser window functions.

The filter $w_{tx}[n]$ is dependent on the signal bandwidth, which implies that the filter need to be dynamically designed or selected based on the signal bandwidth. This is different from W-OFDM,

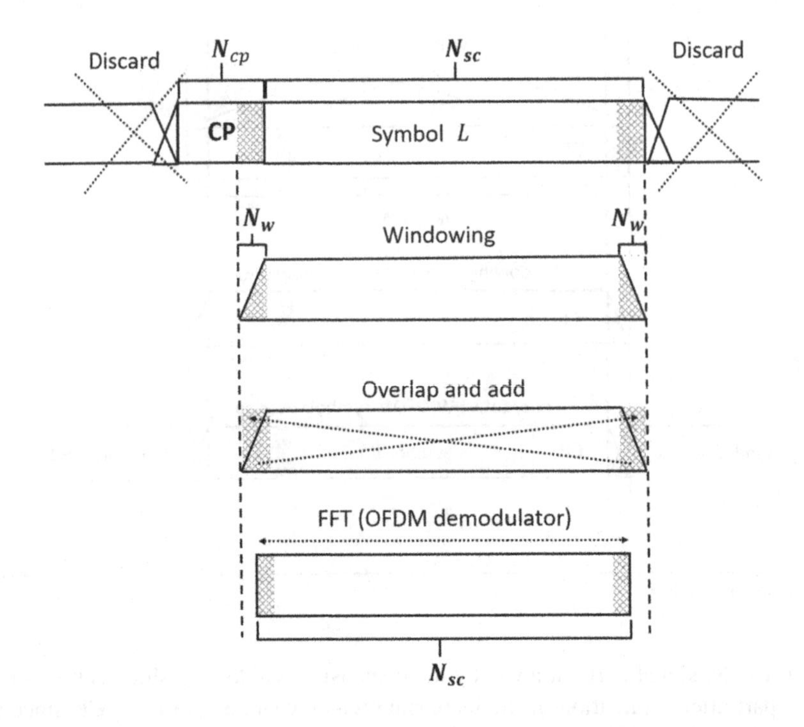

FIGURE 5.10

The receiver side windowing.

whose construction does not change with the signal bandwidth (i.e., the pulse shape $p[n]$ does not change with signal bandwidth). Another aspect of F-OFDM is that when long filters are used for an improved OOB suppression, there can be significant group delay due to the filtering. For example, if filters with length equal to half OFDM symbol are used at the transmitter and at the receiver sides, then the filter processing delay is equal to one OFDM symbol. The large processing delay increases overhead for TDD link direction switching, which is not suitable for latency critical applications.

5.1.2.4 Universally Filtered OFDM

UF-OFDM waveform is synthesized by performing filtering on groups of subcarriers (subbands) to improve the spectral shape (reduce OOB leakage) of each subcarrier group (subband) [12,9]. The subcarrier group-wise (subband) filtering is motivated by the fact that the scheduling operation in cellular systems (e.g., LTE and NR) is done based on subbands (resource blocks[4]). The subband filtering improves spectral characteristics. As an example, the spectrum of W-OFDM is compared with CP-OFDM in Fig. 5.11.

The transmitter structure of UF-OFDM is illustrated in Fig. 5.12. We assume that there are M active subcarriers and the subband filtering is performed on groups of B subcarriers (i.e., each subband

[4]In 4G LTE and 5G NR, a resource block comprises of 12 subcarriers.

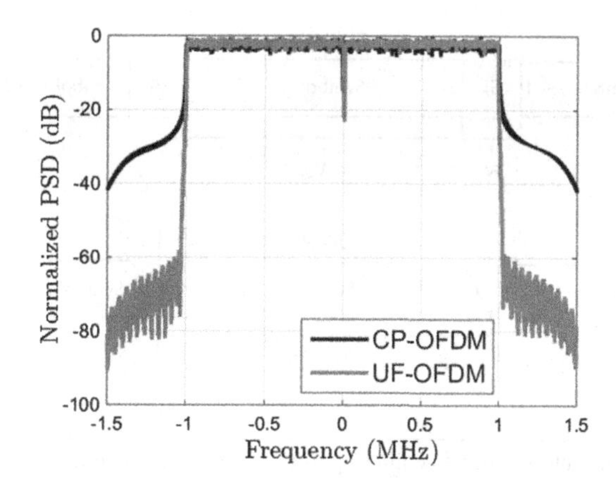

FIGURE 5.11

PSDs of CP-OFDM and UF-OFDM waveforms.

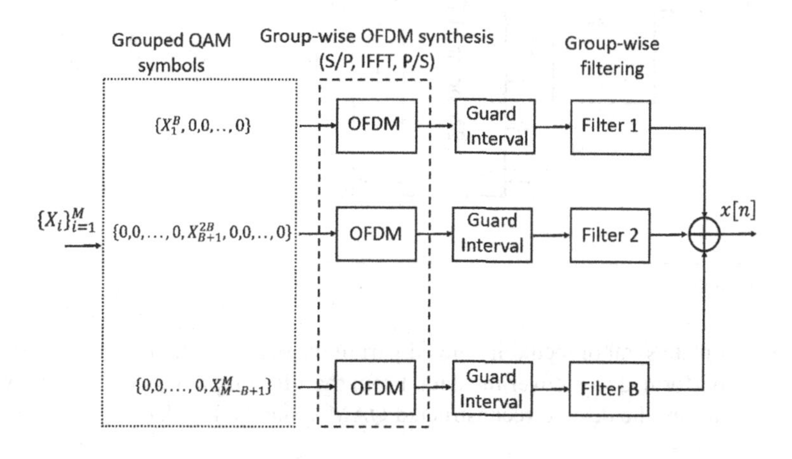

FIGURE 5.12

UF-OFDM transmitter structure.

has B subcarriers). As shown in Fig. 5.12, a set of QAM modulated symbols $\{X_1, X_2, \ldots, X_M\}$ are first divided into groups and zero padded. Each zero padded group follows the same process as we have for the OFDM synthesis, i.e., serial-to-parallel conversion, N_{sc}-point IFFT, and parallel-to-serial conversion. Afterwards, each symbol is padded with zeros in the time domain (i.e., a guard interval (GI) is inserted for each symbol) and filtered. The GI is inserted to prevent intersymbol interference due to the transmit filter delay. For an ISI free UF-OFDM synthesis, the GI length should be at-least equal to the filter length, as illustrated in Fig. 5.13.

The receiver processing of the UF-OFDM waveform is shown in Fig. 5.14. UF-OFDM waveform does not contain the CP, therefore, the cyclic convolution property is not preserved. Unlike the CP-OFDM demodulator that discards the CP part of the symbol before the FFT operation, the UF-OFDM

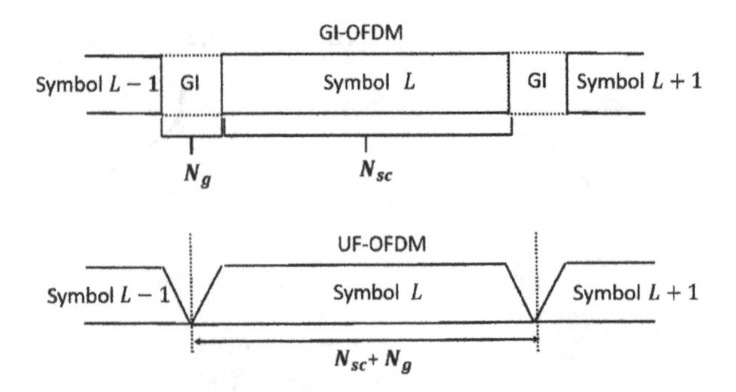

FIGURE 5.13

A guard interval is required for interference-free filtering in UF-OFDM modulator.

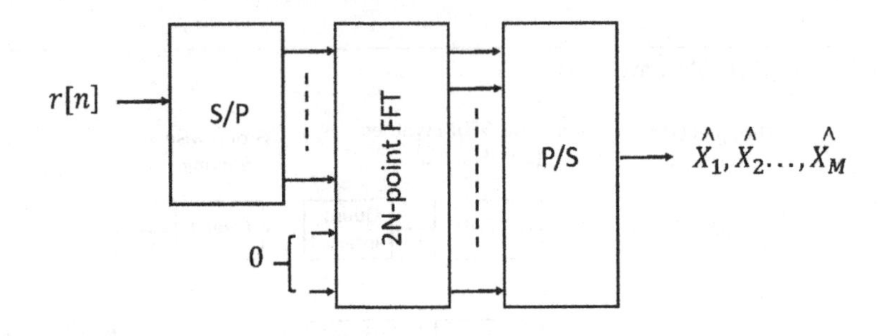

FIGURE 5.14

UF-OFDM receiver structure.

receiver utilizes complete symbol including the GI part in the demodulation. A $2N_{sc}$ point fast Fourier transform (FFT) is performed to recover the data. From the FFT output block (of size $2N_{sc}$), only the even indexed symbols are the desired reconstructed QAM symbols. For details, we refer the reader to [12,11].

5.1.3 FILTER BANK-BASED WAVEFORMS

There is an inverse relation between time and frequency. That is, extending a signal in the time domain makes it shrink in the frequency domain. Based on this principle, the filter bank multicarrier (FBMC) waveforms employ pulse shapes (prototype filters) with a much longer time duration than the OFDM-based waveforms to achieve a more confined spectrum. Spectral confinement is important for efficient utilization of the spectrum, which is a precious resource. However, as the waveform symbols get longer in the time domain, the adjacent waveform symbols overlap. (Non-overlapping symbols would result in loss of data transmission rate.) The FBMC waveforms are not orthogonal. That is, even in the absence of any channel impairment, it is not possible to perfectly reconstruct the QAM symbols transmitted using an FBMC waveform.

A conceptual illustration of the FBMC transmitter is given in Fig. 5.15. We note that conceptually the construction of FBMC is similar to OFDM-based waveforms (e.g., W-OFDM), however, with the

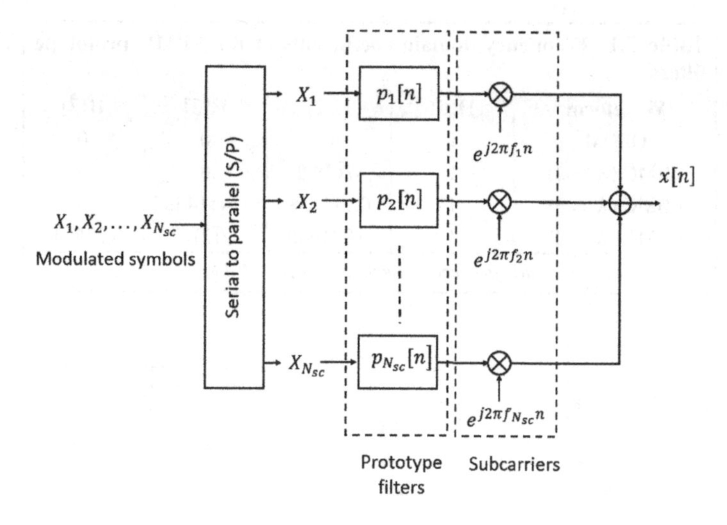

FIGURE 5.15

A conceptual illustration of FBMC modulator.

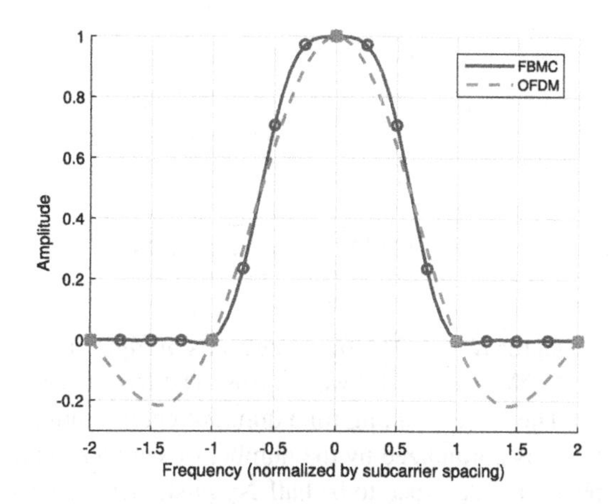

FIGURE 5.16

Frequency responses of the prototype filters of OFDM and FBMC ($K = 4$). The respective frequency domain coefficients are given in Table 5.1.

important difference that the length of the prototype filter is approximately K times greater than the prototype filter (window/pulse) used in the OFDM-based waveforms. For FBMC, the duration of the prototype filter $p_i[n]$ is KN_{sc} samples. The prototype filters for FBMC can be synthesized via frequency domain oversampling of the (rectangular) prototype filter used for OFDM. That is, a prototype filter of length KN_{sc} can be constructed by oversampling the frequency response with a factor K. In Fig. 5.16, we show an example of an oversampled frequency response of the rectangular prototype

Table 5.1 Frequency domain coefficients of the FBMC prototype filters

Waveform	H[0]	H[1]	H[2]	H[3]
OFDM	1	0	0	0
FBMC ($K = 2$)	1	$\sqrt{(2)}/2$	0	0
FBMC ($K = 3$)	1	0.911438	0.411438	0
FBMC ($K = 4$)	1	0.971960	$\sqrt{(2)}/2$	0.235147

The filters have symmetric frequency response, i.e., H[k]=H[-k]

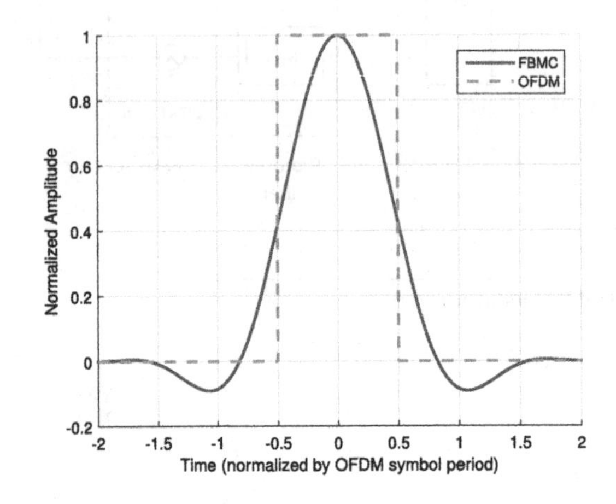

FIGURE 5.17

Impulse responses of the prototype filters used in FBMC ($K = 4$) and OFDM.

filter (used in OFDM) with a factor $K = 4$. The frequency axis in Fig. 5.16 is normalized by $\frac{1}{N_{sc}}$ which is the subcarrier spacing in OFDM and FBMC waveforms. The frequency domain coefficients of the filters are given in Table 5.1. The corresponding time domain representations of these filters are given in Fig. 5.17. The time axis here is normalized by the number of samples in the OFDM symbol without the CP (N_{sc}). The prototype filter is designed to be half-Nyquist, that is, convolution of the transmitter side half-Nyquist and the receiver side half-Nyquist filter is a Nyquist filter that crosses zero (on the time axis) at all the integer multiples of the symbol period [2].

The design, implementation, and characteristics of the prototype filters are discussed in great detail in [2,1]. There are two commonly used methods for implementing FBMC: frequency spreading (FS) and polyphase networks (PPN). In the frequency spreading (FS) method, offset QAM (OQAM) symbols are filtered in the frequency domain, followed by a KN_{sc} point IFFT and an overlap and sum operation. At the receiver side, a sliding window selects KN_{sc} points every $N_{sc}/2$ samples. A KN_{sc} point FFT is performed followed by a matched filtering. In the PPN based FBMC transmitter, an N_{sc} point IFFT is performed on OQAM symbols. The IFFT output (time domain signal) is fed to a polyphase network. At the receiver side, matched filtering is performed followed by a N_{sc} point FFT [2].

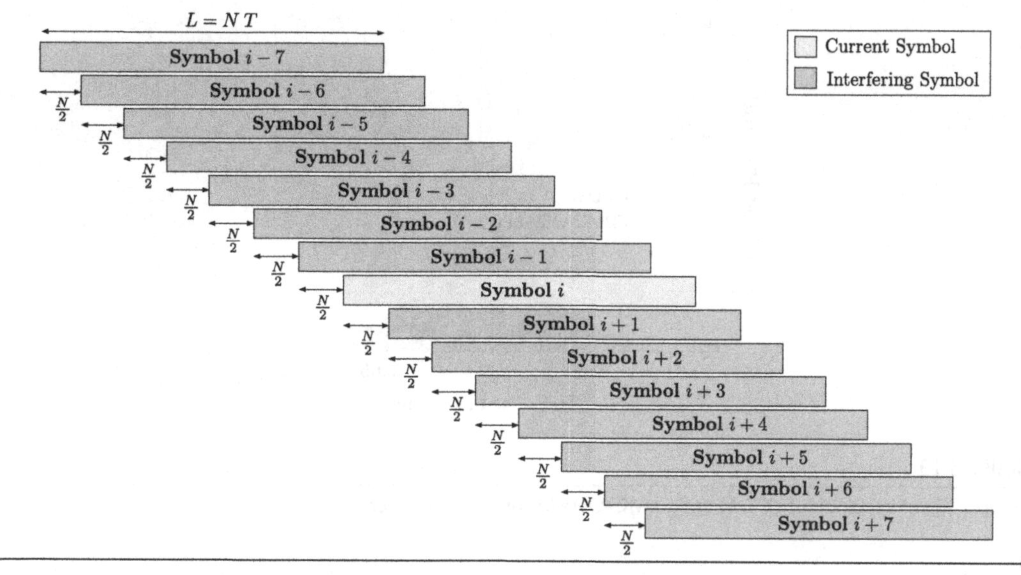

FIGURE 5.18

The overlapping symbol structure of FBMC-OQAM with overlapping factor $K = 4$.

As mentioned earlier, the FBMC waveforms are not orthogonal. There are two reasons for the non-orthogonality:

1. **Intersymbol Interference** An FBMC symbol spans KN_{sc} samples, therefore, the neighboring FBMC symbols overlap as shown in Fig. 5.18. There is a time offset N_{sc} between the adjacent symbols. Due to the overlapping symbols, there is intersymbol interference (ISI) [2]. For a given subcarrier of an FMBC symbol, the ISI primarily comes from the direct adjacent subcarriers of the other overlapping FBMC symbols, i.e., a kth subcarrier in an FBMC symbol primarily suffers interference from subcarriers k and $k-1$ of the overlapping FBMC symbols. The ISI from the subcarriers that are further than the direct neighboring subcarriers is negligible (uniformly upper bounded by -60 dB [2]). There is no ISI between same subcarriers of the overlapping FBMC symbols due to the Nyquist property of the prototype filters [2].
2. **Intercarrier Interference** The subcarriers within an FBMC symbol are not orthogonal either [2]. We refer to this as intercarrier interference (ICI). The ICI primarily comes from the adjacent subcarriers (as in the case of the ISI). For any adjacent subcarriers, there is interference between the real parts or the imaginary parts, but there is no interference between the real and the imaginary parts. To eliminate the ICI between the real parts (or the imaginary parts) of the adjacent subcarriers, they have to be offset by $N_{sc}/2$ samples. This property motivates employing the OQAM modulation, where the subcarriers of the FBMC symbol are modulated with an alternating pattern of real and imaginary valued symbols [2].

Broadly speaking, the FBMC waveforms can be divided into two types—one that employs the offset QAM and one that employs the regular QAM. In the following, we briefly discuss both types.

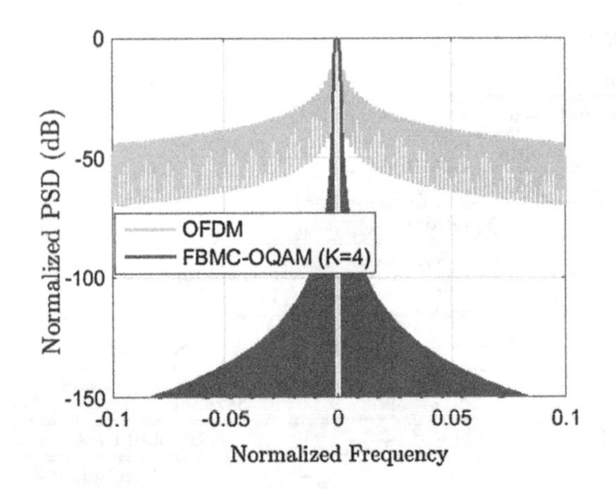

FIGURE 5.19

Comparison of spectrum of OFDM and FBMC-OQAM for one subcarrier.

5.1.3.1 FBMC-OQAM

A discrete-time FBMC-OQAM signal is given by

$$x[n] = \sum_{s\in\mathbb{Z}} \sum_{i\in I} X_{i,s}\, \theta_{i,s}\, p_i\left[n - s\frac{N_{sc}}{2}\right], \tag{5.10}$$

where the set I contains indices of the active subcarriers, N_{sc} is the total number of subcarriers, $X_{i,s} \in \mathbb{R}$ is the modulation symbol of the subcarrier i and the (FBMC) symbol time index s, $\theta_{i,s} := j^{i+s}$ is the phase rotation associated with the offset QAM mapping, $p_i[n] := p[n]\, e^{j2\pi \frac{i}{N_{sc}} n}$ is the filter for the subcarrier i with $p[n]$ as the base filter with length $L = K\, N_{sc}$ (where K is the overlapping/oversampling factor). The overlapping factor K is a parameter of the FBMC-QAM waveform which refers to the property that a given FBMC-OQAM symbol overlaps with $2K - 1$ symbols preceding it and $2K - 1$ symbols following it. Furthermore, $D = \{-2K + 1, ..., 2K - 1\}$ denotes the set of the adjacent FBMC symbols that overlap in time. An illustration of the overlapping symbol structure is given in Fig. 5.18 with a symbol overlapping (or frequency domain oversampling) factor $K = 4$.

Next, we compare spectral shapes of OFDM and FBMC-OQAM waveforms considering the widely adopted FBMC prototype filters [2,1]. The frequency domain coefficients $H[k]$ of these prototype filters are given in Table 5.1 for three different oversampling (overlapping) factors. The frequency response is symmetric around the center frequency (which is necessary for ensuring the Nyquist property), i.e., $H[-k] = H[k]$. (The prototype filter in Fig. 5.16 and Fig. 5.17 is also constructed using the coefficients given in Table 5.1.) In Figs. 5.19–5.22, we compare power spectral densities of OFDM and FBMC-OQAM waveforms with different overlapping (oversampling) factors. The comparisons include the spectrum of a single subcarrier as well as of a group of 300 subcarriers. These results show that FBMC-OQAM has significantly sharper spectral roll-off than OFDM, especially for $K \geq 3$. The overlapping factor $K = 4$ is most commonly used in the literature.

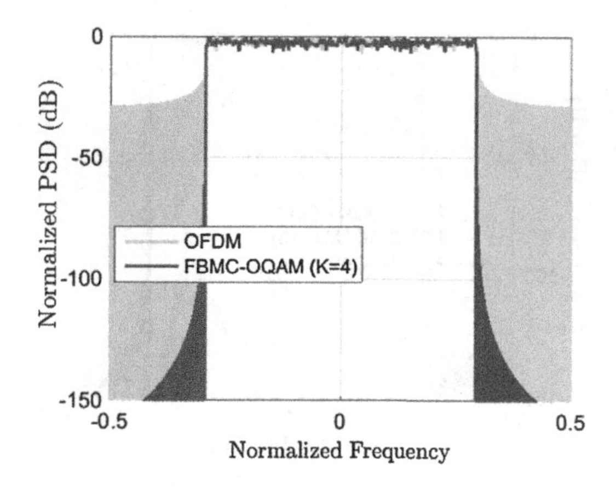

FIGURE 5.20

Comparison of spectrum of OFDM and FBMC-OQAM for 300 subcarriers.

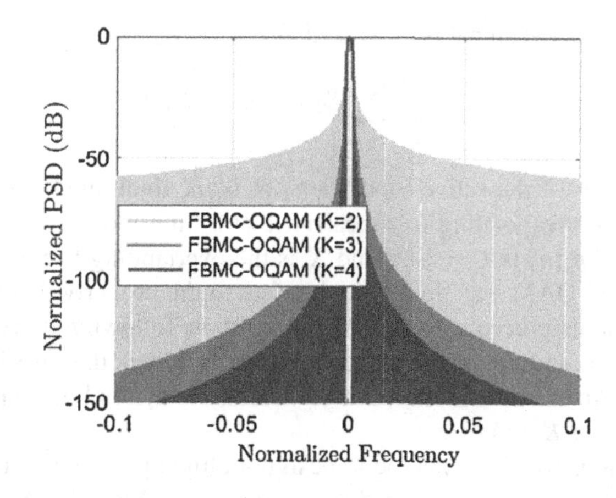

FIGURE 5.21

Comparison of spectrum of one subcarrier of FBMC-OQAM for different overlapping factors.

5.1.3.2 FBMC-QAM

An FBMC waveform that employs a regular QAM (instead of the offset QAM) is referred to as FBMC-QAM. As discussed earlier, there can be significant interference between the real parts (or between the imaginary parts) of the adjacent subcarriers in an FBMC symbol. The offset QAM is one way to reduce the effect of interference from the adjacent subcarriers. Another approach is to design the prototype filters such that the self-interference is minimized or kept to an acceptable level so that the regular QAM can be employed. Typically, the self-interference can be reduced at the cost of degradation in spectral confinement. We will see this shortly in the sequel.

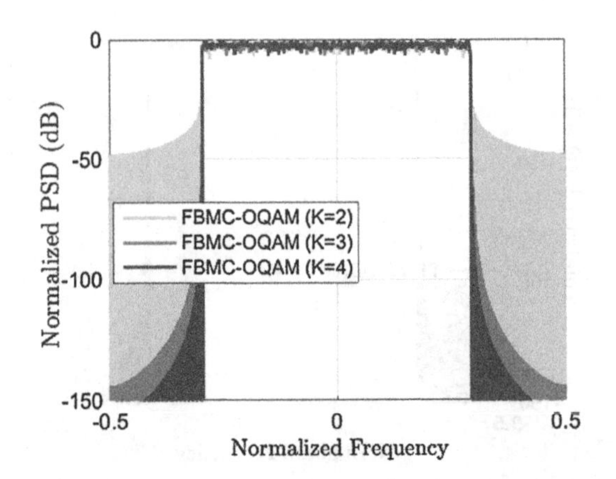

FIGURE 5.22

Comparison of spectrum of 300 subcarriers of FBMC-OQAM for different overlapping factors.

A discrete-time FBMC-QAM signal is expressed as

$$x[n] = \sum_{s \in \mathbb{Z}} \sum_{i \in I} X_{i,s}\, p_i[n - sN_{sc}], \tag{5.11}$$

where I is the set of indices of the active subcarriers, N is the total number of subcarriers, $X_{i,s} \in \mathbb{C}$ is the modulated symbol corresponding to subcarrier i and time index s, and $p_i[n]$ is the filter for subcarrier i. The length of $p_i[n]$ is $L = K\,N$ and K is the overlapping factor. The overlapping factor K is a parameter of FBMC-QAM waveform which refers to the property that a given FBMC symbol overlaps with $K-1$ symbols preceding it and $K-1$ symbols following it. Thus, $D = \{-K+1, ..., K-1\}$ is the set containing indices of the adjacent FBMC symbols that overlap in time considering an overlapping factor K. An example is shown in Fig. 5.23 for a symbol overlapping (or the frequency domain oversampling) factor $K = 4$.

In the recent publications [8,13], a method to design multiple prototype filters (different filters for different subcarriers) has been developed for the synthesis of an FBMC-QAM waveform. To design the prototype filters, a constrained optimization problem is formulated with the objective function being self-interference and the constraints being the spectrum confinement described by falloff rate and the number of nonzero filter taps in the frequency domain. The prototype filters obtained through this method trade-off signal-to-interference ratio caused by the self-interference with spectrum confinement. In [8,13], the authors provide a design example with two prototype filters—one filter is used for the even-numbered subcarrier indices and the other filter is used for the odd-numbered subcarrier indices. That is,

$$p_i[n] := \begin{cases} p_{even}[n]\, \mathrm{e}^{j2\pi \frac{i}{N_{sc}} n} & \text{if } i \text{ is even,} \\ p_{odd}[n]\, \mathrm{e}^{j2\pi \frac{i}{N_{sc}} n} & \text{if } i \text{ is odd.} \end{cases} \tag{5.12}$$

The length of $p_{even}[n]$ and $p_{odd}[n]$ is $L = K\,N_{sc}$. Considering these two prototype filters (see [8,13] for the filter coefficients $p_{even}[n]$ and $p_{odd}[n]$) for FBMC-QAM, we compare its power spectrum

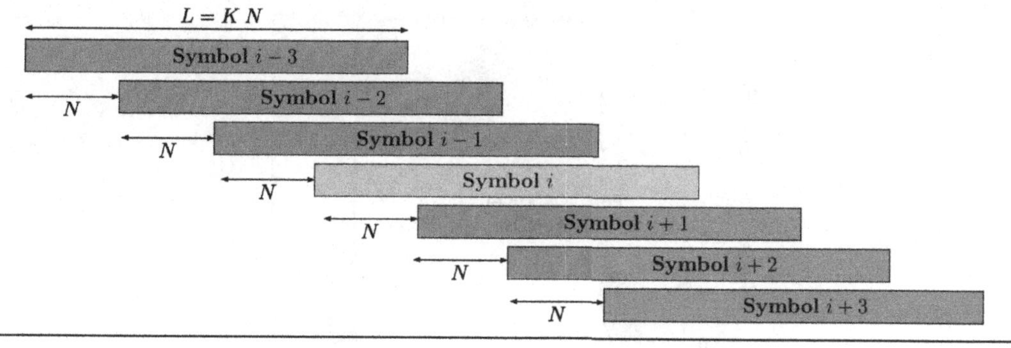

FIGURE 5.23

Overlapping symbol structure of FBMC-QAM with overlapping factor $K = 4$.

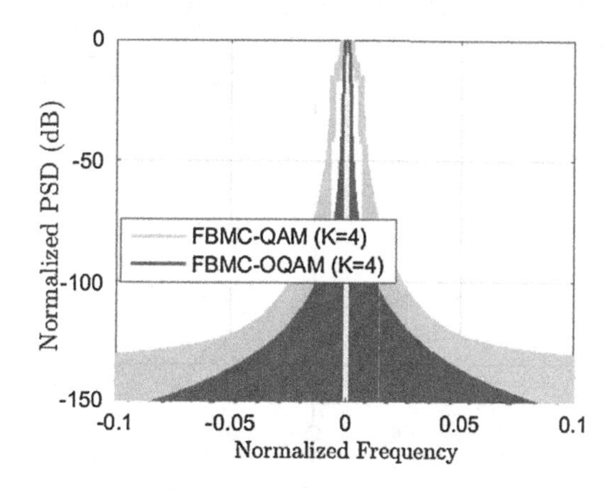

FIGURE 5.24

Comparison of spectrum of FBMC-QAM and FBMC-OQAM for one subcarrier.

with an FBMC-OQAM waveform for a single subcarrier as well for a group of 300 subcarriers in Fig. 5.24 and Fig. 5.25. We observe that the FBMC-OQAM has better spectral characteristics than the FBMC-QAM. A major benefit of FBMC-QAM is that it is better compatible with MIMO transmission due to the utilization of the complex valued QAM symbols.

An important attribute of the FBMC waveforms is that they typically do not employ any CP or GI, unlike the OFDM-based waveforms. The prototype filters in the FBMC waveforms have long decaying tails which can make them robust to the ISI caused by frequency-selective channels. By avoiding the CP overhead, the FBMC waveforms can potentially utilize transmission resources more efficiently than the OFDM-based waveforms, at least in theory. In order to compare the resource usage efficiency, we have sketched the time-frequency lattices for OFDM, FBMC-OQAM, and FBMC-QAM in Fig. 5.26, Fig. 5.27, and Fig. 5.28, respectively. In these figures, Δf denotes the subcarrier spacing

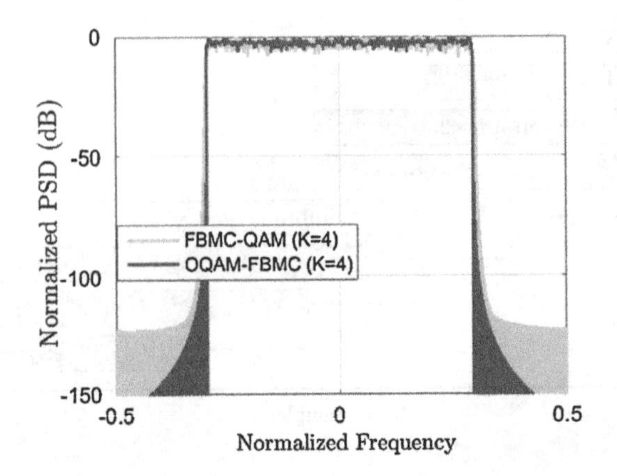

FIGURE 5.25

Comparison of spectrum of FBMC-QAM and FBMC-OQAM for 300 subcarriers.

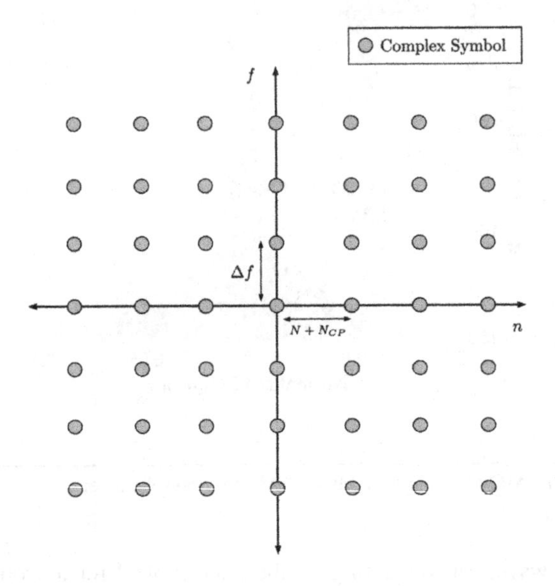

FIGURE 5.26

Time-frequency lattice representation for CP-OFDM.

(which is equal to $\frac{1}{N}$ for all waveforms), N denotes the total number of subcarriers (assumed equal for all waveforms), and N_{CP} is the number of CP samples in CP-OFDM. We observe that FBMC-QAM has one complex symbol per unit area on its lattice and FBMC-OQAM has two real symbols per unit area on its lattice. This means that the two FBMC waveforms have an equal resource usage efficiency. CP-OFDM has $\frac{N}{N+N_{CP}}$ (less than one) complex symbol per unit area, implying a lower resource usage efficiency than the FBMC waveforms. Although FBMC has better time-frequency re-

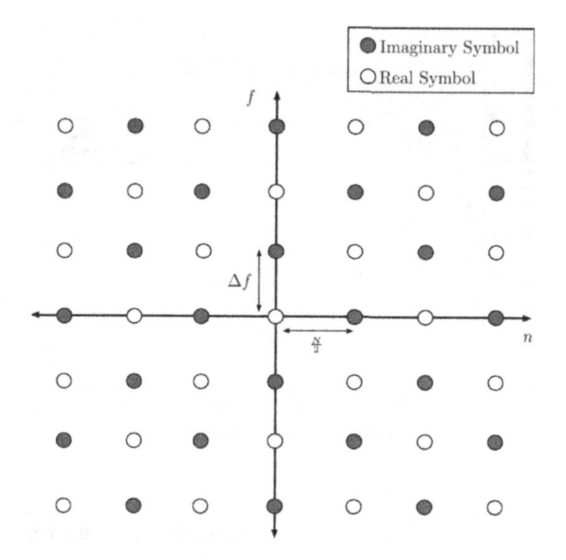

FIGURE 5.27

Time-frequency lattice representation for FBMC-OQAM.

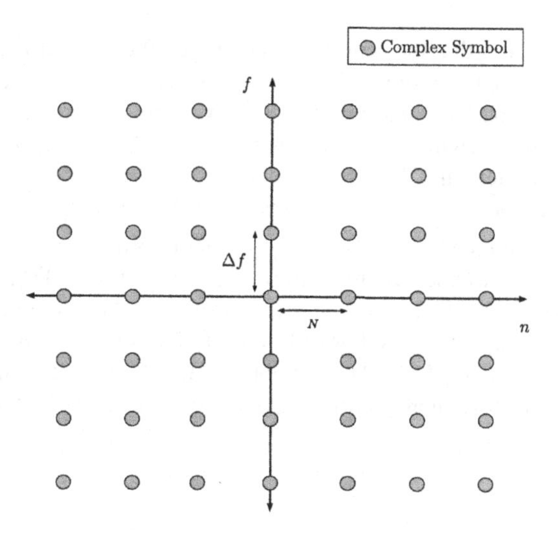

FIGURE 5.28

Time-frequency lattice representation for FBMC-QAM.

source utilization in theory, one cannot simply conclude that CP-OFDM has worse spectral efficiency in a realistic scenario. Here, it is important to keep in mind that the FBMC waveforms suffer from self-interferences (being nonorthogonal waveforms). Moreover, when the channel delay spread is large, the

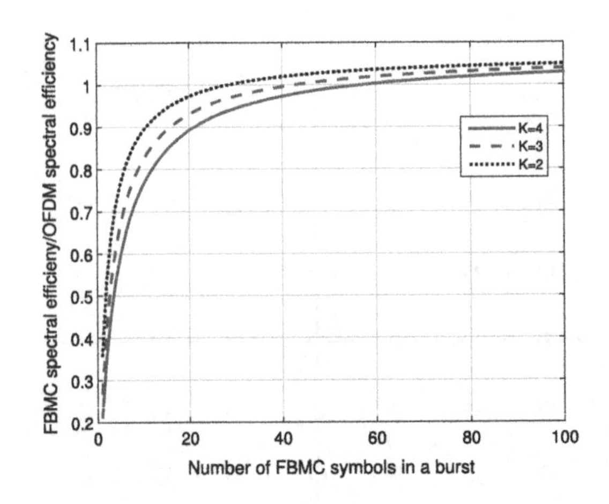

FIGURE 5.29

The ratio of FBMC and OFDM resource usage efficiencies as a function of the number of the waveform symbols per transmission burst.

FBMC waveforms can also have noticeable ISI due to the multipath effect and may require the CP overhead.

Another important aspect is that the time-resource usage efficiency of the FBMC waveforms depends on the transmission burst duration, whereas the time-resource usage efficiency of the OFDM waveforms does not depend on the burst duration. For example, if there are M FBMC symbols in a transmission burst (time slot), then the burst duration has to be $(M+K-1)N_{sc}$ samples due to the fact that each FBMC symbol has length KN_{sc} and the consecutive FBMC symbols overlap with an offset N_{sc}. This implies that the time-resource usage efficiency of FBMC is $\frac{M}{M+K-1}$. For the OFDM waveform the time-resource utilization efficiency is $\frac{N}{N+N_{CP}}$, regardless of the transmission burst duration. In Fig. 5.29, we plot the ratio of the resource utilization efficiencies of the FBMC and OFDM waveforms. We observe that OFDM is more efficient for shorter transmission burst durations, which is especially important for delay sensitive applications. For LTE, a subframe consists of merely 14 OFDM symbols and therefore OFDM is suitable for LTE like assumptions. In NR, a transmission burst can be as short as one OFDM symbol to support latency critical applications (see Section 2.3 for the mini-slot based transmission).

5.2 SINGLE CARRIER DFTS-OFDM

Multicarrier waveforms, including OFDM, have large variations in instantaneous amplitude (and power) due to the superposition of a large number of modulated subcarriers. When there are large variations in the amplitude of the transmit signal, the power amplifiers of the transmitter either operate in a nonlinear region and cause signal distortion (due to nonlinear clipping) or the transmitter dissipates more power in order to operate in a linear region and prevent distortions. In contrast, single-carrier waveforms have small signal variations and therefore higher power efficiency than multicarrier

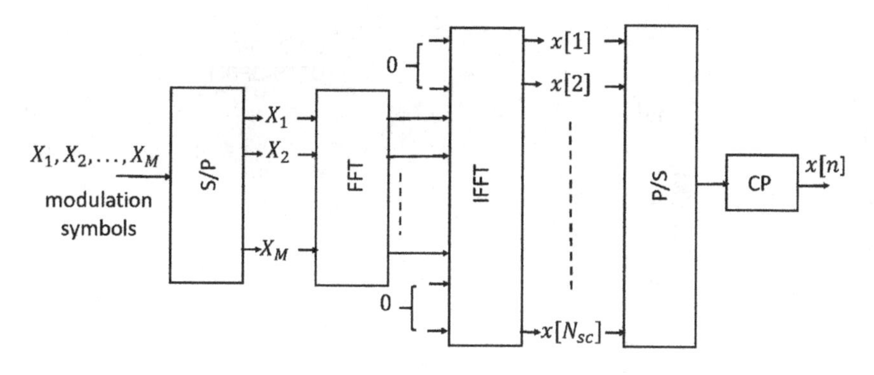

FIGURE 5.30

DFTS-OFDM transmitter structure.

waveforms. For example, there exist constant envelope single-carrier waveforms (e.g., minimum shift keying (MSK), Gaussian minimum shift keying (GMSK)) that allow power amplifiers to run at saturation point without any need for precompensation or postcompensation to account for the clipping. The single carrier PSK modulated waveform also has a constant amplitude. However, pure single-carrier waveforms have their disadvantages—a low spectral efficiency, performance degradation in frequency-selective channels, and lack of flexible resource allocation in the frequency domain. The single-carrier DFT spread OFDM (DFTS-OFDM) waveform aims at combining benefits of the single-carrier and the multicarrier waveforms, with the following key properties:

- Small amplitude variations (single-carrier aspect).
- Flexible resource allocation in the frequency domain (multicarrier aspect).

DFTS-OFDM can be seen as an OFDM waveform with DFT-based precoding (which makes it single-carrier) [7]. The structure of the DFTS-OFDM modulator is shown in Fig. 5.30. An M-point FFT (DFT) is performed on M QAM symbols followed by an N_{sc}-point IFFT with $M < N_{sc}$. A cyclic prefix is inserted as in CP-OFDM to enable low complexity frequency domain equalization at the receiver side. Fig. 5.31 compares the peak-to-average power ratios of CP-OFDM and DFTS-OFDM waveforms via the complementary cumulative distribution function (CCDF) of PAPR, assuming 16 QAM modulation and 1200 subcarriers. CCDF shows the likelihood of PAPR (in terms of percentage) exceeding a given value. We observe a reduced PAPR with DFTS-OFDM. If one employs PSK modulation instead of QAM, PAPR of DFTS-OFDM can be further reduced.

The bandwidth of a DFTS-OFDM signal is given by $\frac{M}{N} f_s$, where f_s is the sampling frequency. The instantaneous signal bandwidth can vary by changing the block size M, to allow for flexible bandwidth assignment. Moreover, users can be multiplexed in the frequency domain by allocating subsets of subcarriers to different users. For example, according to Fig. 5.30, if we want to multiplex two users with equal bandwidths, the first $\frac{M}{2}$ QAM symbols $(X_1, X_2, \ldots, X_{M/2})$ can be associated with user 1 and the last $\frac{M}{2}$ QAM symbols $(X_{M/2}, X_{M/2+1}, \ldots, X_M)$ can be associated with user 2 (assuming M is even). In this example, each user is assigned contiguous subcarriers. Alternatively, each user can also be assigned noncontiguous subcarriers. For example, user 1 and user 2 can be interleaved by assigning odd-numbered subcarriers (i.e., modulation symbols $X_1, X_3, \ldots, X_{M-1}$) to user 1 and even-numbered

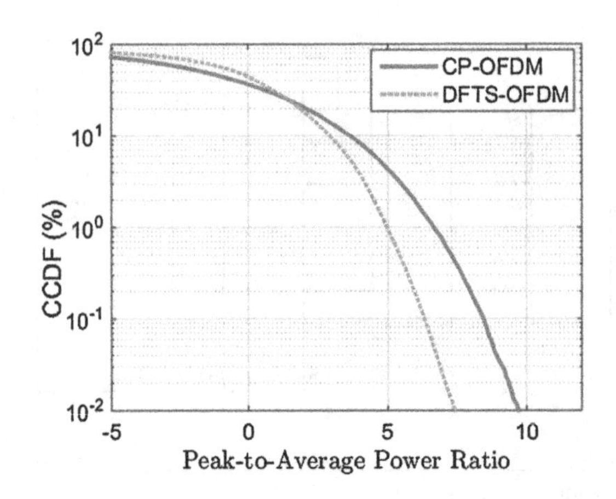

FIGURE 5.31

DFTS-OFDM achieves much reduced PAPR than CP-OFDM.

subcarriers (i.e., modulation symbols $X_2, X_4, \ldots, X_M$) to user 2. When multiple users are multiplexed in DFTS-OFDM waveform, it is often referred to as single-carrier frequency division multiple access (SC-FDMA). If users are assigned contiguous subcarriers, it is called localized frequency division multiple access (LFDMA). When each user is assigned noncontiguous subcarriers that are uniformly distributed across its bandwidth, it is commonly known as interleaved frequency division multiple access (IFDMA).

DFTS-OFDM is used in 4G LTE for the uplink transmissions due to its improved power efficiency. It is also an optional waveform for the 5G NR uplink. However, there are some shortcomings of employing DFTS-OFDM, which are discussed in Chapter 6.

5.3 WAVEFORM DESIGN REQUIREMENTS FOR 5G NR

5G radio access will support various applications (eMBB, URLLC, massive machine type communications (mMTC)), a wide range of frequencies (from sub-1 GHz to 100 GHz), diverse deployments, and various link types (uplink, downlink, device-to-device link, backhaul link). In the following, we provide an overview of important waveform design requirements for different scenarios.

eMBB has high requirements on throughput, user density, and low latency. Both small and large cell deployments are expected for eMBB services. Typically, larger cells in low carrier frequencies in the licensed spectrum and small cells at higher carrier frequencies (e.g., millimeter-wave) in both the licensed and the unlicensed spectrum. For the downlink transmission, a high spectral efficiency and maximum utilization of MIMO are key requirements for the NR waveform. For the uplink transmissions in the large cell deployments, the cell edge users can be power limited and may not be able to achieve high spectral efficiencies. In this case, the power efficiency of the waveform is important. In the small cell deployments, the distance between the users and the base station can significantly reduce but not necessarily the maximum transmission power of the users. In this case, the power efficiency and the MIMO compatibility are important design requirements.

At very high carrier frequencies (e.g., millimeter-wave), analog or hybrid beam-forming is expected, at least in the early stage of the NR deployments (see Chapter 7 for details). In such systems, the number of digital chains is small compared to the number of antennas and part of the beam-forming is realized via the RF beam-forming. With the RF beam-forming, it is not possible to multiplex users in the frequency domain (frequency division multiplexing (FDM)) unless they all are served by the same beam. In this case, the users should be multiplexed in the time domain (time division multiplexing (TDM)) and therefore a short transmit time interval (TTI) is important. The NR waveform has to be confined in time to allow for the short transmissions.

For the URLLC services, the ultra-reliability and very low latency are important. Beam-forming is an important tool for improving the reliability. To achieve a very low latency, the waveform should be confined in the time domain with a low processing latency. For a fast uplink access, an asynchronous transmission may be beneficial, which means that the waveform should be able to efficiently support the asynchronous access. The retransmission mechanisms are important for enhancing the link reliability. In order to reduce latency due to the HARQ, the waveform should efficiently support the short TTIs.

The mMTC use cases target at providing connectivity to a massive number of devices with strict requirements on battery life and processing complexity. mMTC applications are expected in large cell deployments at lower carrier frequencies (e.g., below 6 GHz). mMTC data traffic can be in the form of short data bursts. Moreover, the radio propagation and penetration losses can be high, for example, to provide connectivity to devices in the basement of a building. Hence, the important waveform design requirements for the mMTC are the high power efficiency (in the uplink), the low complexity transceiver processing, the suitability to the short burst durations, and the ability to multiplex large number of users. Asynchronous access can also be important in the uplink.

The D2D communication typically has short range; therefore, the power efficiency is not very important. For the D2D communication, requirements are expected to be high on throughput and therefore the spectral efficiency is important. Since the D2D links are symmetric in nature, it is desirable to have the same waveforms in both directions.

For the backhaul links (base station-to-base station communication), high throughputs and low latencies are important. The wireless backhaul or the self-backhaul is likely with small cell deployments. Hence, the most important waveform design requirements are high spectral efficiency, compatibility with MIMO, and low latency (similar to eMBB). In wireless backhauling, the links are symmetric in nature and, hence, the same waveform is desirable in both directions.

5.4 KEY PERFORMANCE INDICATOR FOR NR WAVEFORM DESIGN

Key characteristics of 5G NR include large channel bandwidths, extreme data rates, ultra-reliability and low latency requirements, harsh propagation conditions, severe RF impairments, massive number of antennas, small sized base stations, and mainly TDD deployments. Considering these aspects, the mmMAGIC project (a European research project[5]) as well as the 3GPP identified key performance

[5]mmMAGIC was an EU project (funded by H2020 program) that developed radio interface concepts and solutions for above 6 GHz mobile radio communications.

indicators (KPIs) for the NR waveform design and evaluated the candidate waveforms for these KPIs. In the following, we present these waveform design KPIs and their importance in different frequency ranges. The key performance indicators for the NR waveform design include:

- **Spectral efficiency** The spectral efficiency is vital to support extreme requirements on data rate, user connection, and traffic densities. In general, the spectral efficiency is more crucial at lower carrier frequencies than at higher frequencies, since the spectrum is not as precious at higher frequencies due to the availability of potentially much larger channel bandwidths.
- **MIMO compatibility** Multiantenna transmission is a driving technology for NR. With the increase in carrier frequency, the number of antenna elements would increase in the base stations as well as in the devices. The use of various MIMO schemes is essential in providing high spectral efficiencies (by enabling SU-MIMO/MU-MIMO) and greater coverage (via beam-forming). Beam-forming is instrumental in overcoming the high transmission losses at very high frequencies (coverage limited scenarios).
- **Low Peak-to-Average-Power-Ratio (PAPR)** A low PAPR is essential for power efficient transmissions from the devices (for example, uplink, sidelink). A low PAPR becomes even more important at very high frequencies. Since small sized low-cost base stations are envisioned at high frequencies, low PAPR is also important for the downlink transmissions.
- **Robustness to channel time selectivity** The robustness to channel time variations is vital in high vehicular speed scenarios. The high speed scenarios are relevant in large cell deployments. The large cell deployments are not expected at very high frequencies due to harsh propagation conditions (coverage limitation). At very high frequencies, the deployments are expected to come in the form of small cells where mobility is not a major concern. However, certain vehicular services may be enabled at very high frequencies.
- **Robustness to channel frequency selectivity** Typically, wideband wireless channels are strongly frequency selective and robustness to frequency selectivity is fundamental to support high throughput communication over wideband channels.
- **Robustness to phase-noise** Phase-noise depends on quality of the local oscillators at the devices and at the base stations. Typically, phase-noise is high due to the UE (transmitter/receiver), since low phase-noise oscillators can be too expensive and power consuming for the devices. Phase-noise robustness is also important for future low-cost base stations. Basically, any link that involves a device and/or low-cost base station puts a high requirement on the phase-noise robustness of the waveform; especially if the communication takes place at high frequencies since phase-noise typically increases with the carrier frequency (see Chapter 4).
- **Robustness to Power Amplifier Non-linearity** Thc impact of the nonlinear PA increases as the signal bandwidth increases (see Chapter 4).
- **Transceiver baseband complexity**, involved in encoding and decoding of the information embedded in a waveform. The baseband complexity is always very important for the devices, especially from the receiver perspective. For NR, the complexity is even a major consideration for base stations, since a base station can be a small sized access node (especially at high frequencies) with limited processing capability. At very high frequencies and large bandwidths, the receiver may also have to cope with severe RF impairments. A low baseband complexity is also important for faster processing and enabling low latency applications. Moreover, the baseband complexity becomes increasingly important as the signal bandwidth increases.

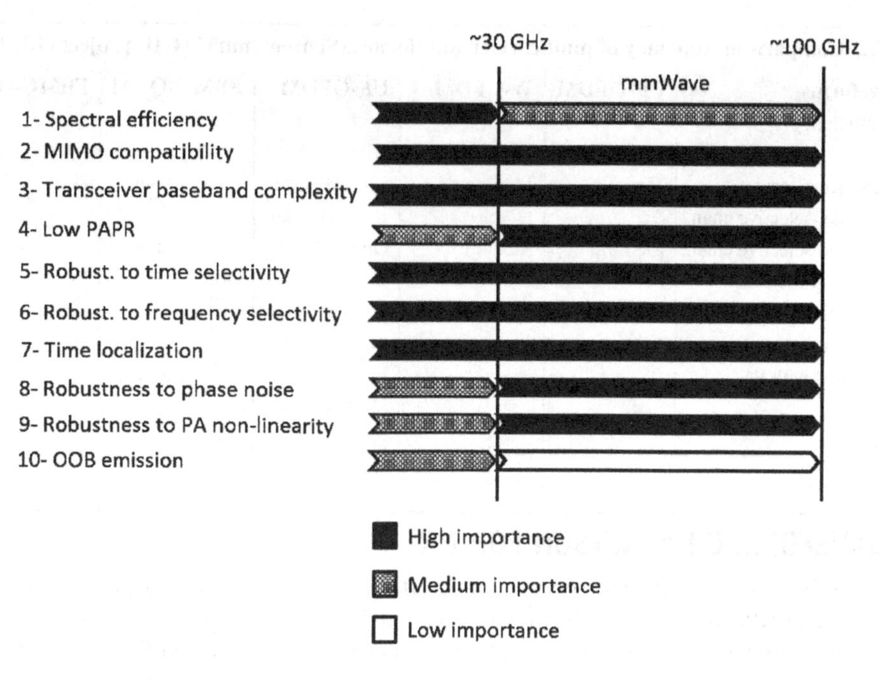

FIGURE 5.32

Importance of waveform performance indicators as a function of carrier frequency.

- **Time localization** Time localization is important to efficiently enable (dynamic) TDD and support low applications. Frequent link direction switching for TDD requires short burst transmissions. A low latency, which is one of the key requirements for both the eMBB and the URLLC services, also requires short transmission time slots. A waveform that is confined in the time domain is suitable for enabling short transmissions.
- **Out-of-band emissions/Frequency localization**: Frequency localization is useful for efficient utilization of spectrum and potential multiplexing of different services (e.g., URLLC, mMTC, eMBB) on a single carrier using different waveform numerologies (see Section 6.3.3). Frequency localization is not a major performance indicator at high frequencies where large channel bandwidths are available. Frequency localization is also relevant for asynchronous access, which can be useful in uplink and sidelink.
- **Flexibility and scalability** Flexibility and scalability of the waveform is important to enable diverse use cases and deployment scenarios.

Fig. 5.32 sketches importance of the waveform KPIs for the NR operation in different frequency ranges. We note that, for millimeter-wave communication, special attention needs to be paid to hardware impairments and power efficiency, whereas frequency localization is not of great importance.

Table 5.2 Comparison summary of multicarrier waveforms. (Source: mmMAGIC project [4,15])

MC-waveforms	CP-OFDM	W-OFDM	UF-OFDM	FBMC-QAM	FBMC-OQAM
Spectral efficiency	High	High	High	High	High
PAPR	High	High	High	High	High
Phase-noise robustness	Medium	Medium	Low/Medium	Low/Medium	Low/Medium
Robust. to freq. selective chan.	High	High	High/Medium	High	High
Robust. to time-selective chan.	Medium	Medium	Open	Low/Medium	Low/Medium
MIMO compatibility	High	High	Open	Open	Low
Time localization	High	High	High	Low	Low
OOB emissions	High	Medium	Medium	Low	Low
OOB emissions with PA	High	High/Medium	High/Medium	High/Medium	High/Medium
Complexity	Low	Low	Medium	High	High
Flexibility	High	High	High	High	High

5.5 WAVEFORM COMPARISON FOR NR

The mmMAGIC project evaluated several multicarrier and single-carrier waveforms based on the waveform design KPIs discussed in the previous section. The project concluded with an overall comparison of the candidate waveforms prior to the NR standardization in the 3GPP. These results served as an important input for the NR standardization. The waveform comparison results are summarized in Table 5.2 for selected multicarrier waveforms. The following color scheme has been used in this table: Green (light gray in print version) refers to a desirable characteristic; red (medium gray in print version) refers to an undesirable characteristic; blue (dark gray in print version) refers to somewhere between desirable and undesirable characteristic. Moreover, "Open", means that further investigations are required to draw any conclusion. The investigations were based on analytical methods as well as simulations. The simulations were performed under common evaluation assumptions in the mmMAGIC waveform simulators. (One of the two simulators is presented in Chapter 9.) The detailed evaluation results are available in the mmMAGIC project report [4] and are summarized in [15]. In the following, we provide the waveform comparison results for selected KPIs (power efficiency, phase-noise robustness, frequency localization, baseband complexity) and refer the reader to [4] for further details.

5.5.1 FREQUENCY LOCALIZATION

CP-OFDM suffers from poor frequency localization/high out-of-band emissions. Several multicarrier waveforms (both OFDM based and FBMC based) aim at improving frequency localization, as shown in Section 5.1.2 and Section 5.1.3. Among these multicarrier waveforms, FBMC-OQAM ($K = 4$) offers the best frequency localization, followed by FBMC-QAM ($K = 4$), and then UF-OFDM and W-OFDM. However, these waveforms behave differently when they are subject to a realistic power amplifier. For example, under the combination of a high transmission power and moderate to large bandwidths, the spectrum roll-off of all waveforms (W-OFDM, UF-OFDM, FBMC) is similar to CP-OFDM, except very close to band edges where FBMC has a steeper roll-off than W-OFDM and UF-OFDM [4].

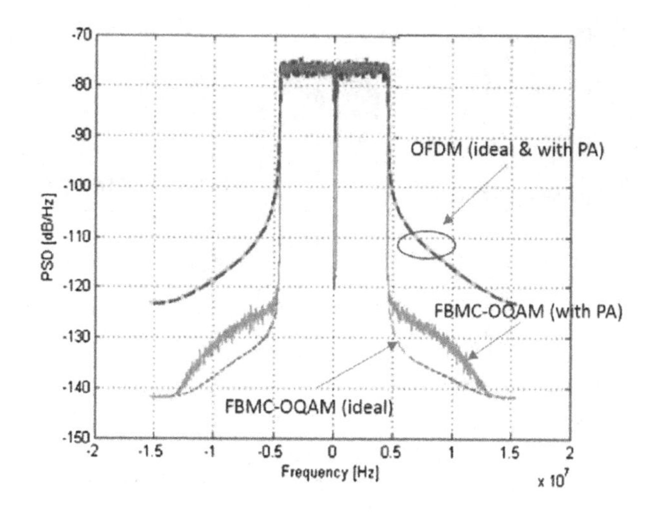

FIGURE 5.33

A PSD comparison of CP-OFDM and FBMC-OQAM with and without PA. Low transmission power and large bandwidth scenario (input power = −21 dBm, output power = 6 dBm, bandwidth = 9 MHz).

In order to show the behavior with respect to nonlinear power amplification, we compare the power spectrum of CP-OFDM and FBMC-OQAM waveforms subject to a power amplifier modeled as a memoryless polynomial in the time domain (as discussed in Chapter 4), according to

$$PA_{output}[n] = PA_{input}[n]p(|PA_{input}[n]|), \tag{5.13}$$

where $p[x] = c_0 + c_1x + c_2x^2 + \cdots + c_kx^k$. For simulations, we have used a tenth order polynomial with coefficients given in [4] and have assumed 2048 subcarriers, 30.72 MHz sampling frequency, and QPSK modulation. For FBMC-OQAM, we have assumed K(overlapping factor) = 4 and the prototype filter given in Table 5.1. In Figs. 5.33–5.35, we compare PSDs of OFDM and FBMC (with and without PA) at different input/output power levels and different signal bandwidths. All PA input/output power values in dBm assume a 50 Ω impedance. We note that the actual transmitter power will be a few (3 to 4) dBs lower due to losses in duplexer/switches. All results are shown without any power back-off. In the following, we summarize the key observations.

- For low transmission power levels or small signal bandwidths, substantially lower OOB emission can be achieved with FBMC compared to OFDM (cf. Fig. 5.33 and Fig. 5.35).
- For high transmission power levels and moderate to high signal bandwidths, the OOB emission is similar for FBMC and OFDM (cf. Fig. 5.35).

In general, for a combination of high transmission power and a large bandwidth, it is unlikely that the sharp spectrum roll-off promised by FBMC and other frequency localized multicarrier waveforms can be realized with state-of-the-art PA technology. In Chapter 8, we provide further evaluations for a power amplifier model with memory.

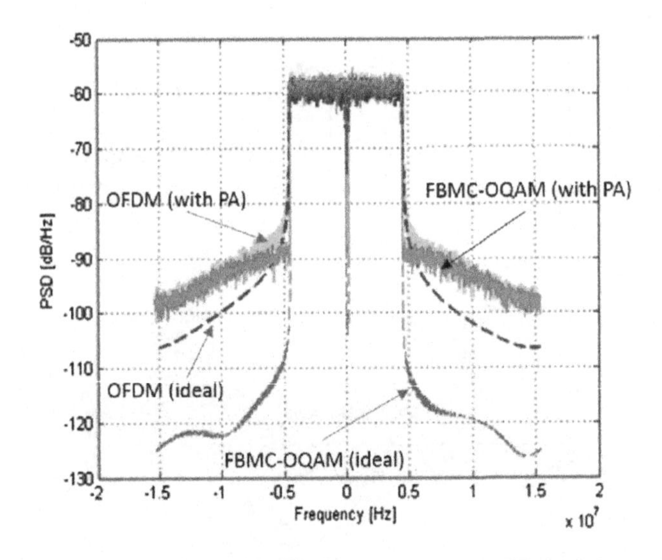

FIGURE 5.34

A PSD comparison of CP-OFDM and FBMC-OQAM with and without PA. High transmission power and large bandwidth scenario (input power = −4 dBm, output power = 24 dBm, bandwidth = 9 MHz).

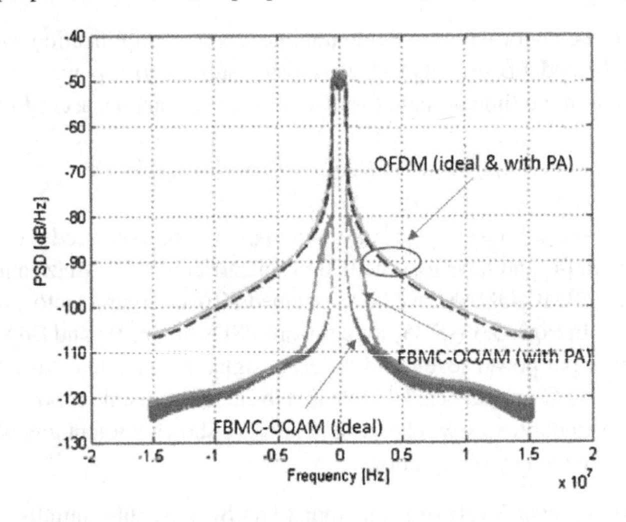

FIGURE 5.35

A PSD comparison of CP-OFDM and FBMC-OQAM with and without PA. High transmission power and small bandwidth scenario (input power = −4 dBm, output power = 24 dBm, bandwidth = 1.08 MHz).

5.5.2 POWER EFFICIENCY

As discussed in Section 5.2, a common drawback of all multicarrier waveforms is their high PAPR (and low power efficiency). In Fig. 5.36, we compare the PAPR of several waveforms, including CP-OFDM,

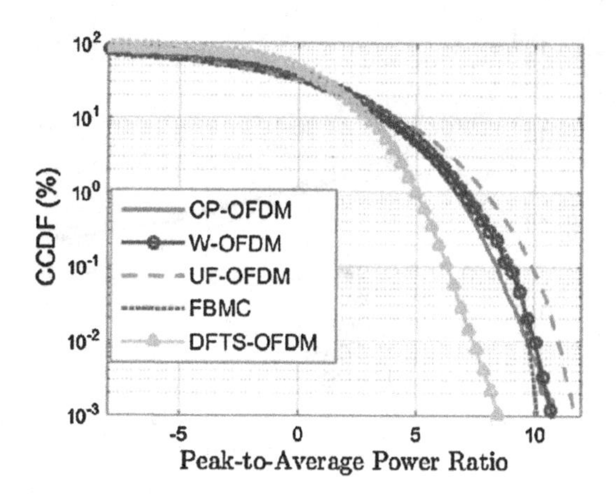

FIGURE 5.36

A comparison of PAPR of multicarrier waveforms and single carrier DFTS-OFDM waveform.

W-OFDM, UF-OFDM, FBMC-OQAM, and DFTS-OFDM (assuming 16 QAM and 1200 subcarriers). We observe that all multicarrier waveforms have similar PAPR except UF-OFDM, which has a higher PAPR. The DFT-based precoding in OFDM (DFTS-OFDM) reduces the PAPR and achieves a higher power efficiency than OFDM. There are various well-known methods to improve the power efficiency of OFDM, which are discussed in Section 6.4.

5.5.3 TIME-VARYING FADING CHANNEL

In Fig. 5.37, we compare performance of several multicarrier waveforms (CP-OFDM, W-OFDM, UF-OFDM, FBMC-OQAM, and FBMC-QAM) in terms of the symbol error rate over a time-varying fading channel with 60 km/h UE speed at 6 GHz carrier frequency (assuming QuaDRiGa channel model). For all waveforms, we assume 16 QAM, 512 subcarriers, and 120 MHz signal bandwidth. As can be seen, CP-OFDM has superior performance in the time-varying fading channel. Further details of this evaluation are given Chapter 9.

5.5.4 BASEBAND COMPLEXITY

We now look at implementation complexity of multicarrier waveforms in terms of the number of real multiplications required for synthesis and demodulation, excluding the computations required for channel estimation. There are different channel estimation techniques used in practice with varying degrees of complexity. We compare complexity of the following multicarrier waveforms: OFDM, W-OFDM, F-OFDM, UF-OFDM, and FBMC. Since FFT/IFFT operations are used in implementations of all of these waveforms, for simplicity we will denote the complexity of an N-point FFT/IFFT operation with $C_{FFT}(N)$. The results presented in the following are based on [3], where the reader can find a detailed analysis including complexity evaluation for the FPGA-based implementations.

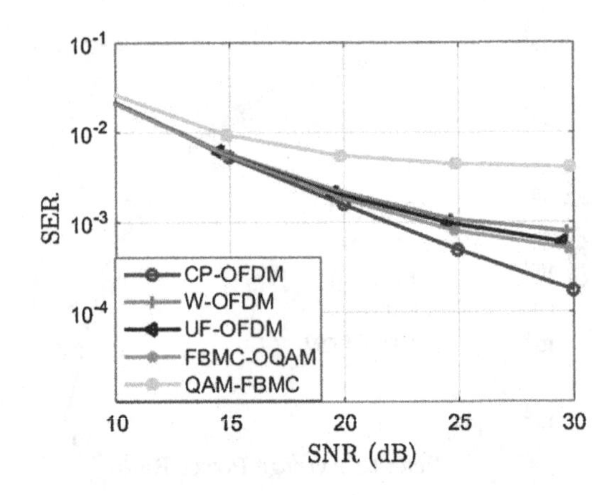

FIGURE 5.37

A comparison of multicarrier waveforms subject to a time-varying fading channel (60 km/h UE speed at 6 GHz carrier frequency).

5.5.4.1 CP-OFDM

Considering the total number of subcarriers N_{sc}, the number of active subcarriers N_{act}, and per subcarrier channel equalization, the number of real multiplications for a CP-OFDM transmitter and receiver are given by

$$C^{Tx}_{OFDM} = C_{FFT}(N_{sc}), \tag{5.14}$$

$$C^{Rx}_{OFDM} = C_{FFT}(N_{sc}) + 4N_{act}, \tag{5.15}$$

where $4N_{act}$ in (5.15) corresponds to one complex multiplication per subcarrier for channel equalization.

5.5.4.2 W-OFDM

The windowing operation in an OFDM transmitter/receiver is merely multiplication of a window function with an OFDM symbol. Consider an OFDM symbol of length $N_{sc} + N_{cp}$ samples with N_{cp} denoting the number of CP samples, the number of real multiplications for the W-OFDM transmitter and receiver are given by

$$C^{Tx}_{W-OFDM} = C^{Tx}_{OFDM} + 4(N_{sc} + N_{cp}) = C_{FFT}(N_{sc}) + 4(N_{sc} + N_{cp}), \tag{5.16}$$

$$C^{Rx}_{W-OFDM} = C^{Rx}_{OFDM} + 4(N_{sc} + N_{cp}) = C_{FFT}(N_{sc}) + 4N_{act} + 4(N_{sc} + N_{cp}). \tag{5.17}$$

The term $4(N_{sc} + N_{cp})$ corresponds to the number of real multiplications required for the transmitter/receiver windowing operation.

5.5.4.3 UF-OFDM

In the UF-OFDM transmitter, the subcarriers are divided into groups (subbands or resource blocks) and then group-wise filtering is performed. Consider a UF-OFDM symbol with N_{sc} total subcarriers, B groups of N_B subcarriers with group-wise filtering, and a filtering operation in the frequency domain. The number of real multiplications per UF-OFDM symbol is given by

$$C_{UF-OFDM}^{Tx} = B\left(C_{FFT}\left(2N_B\right) + C_{FFT}\left(N_B\right) + 8N_B\right) + C_{FFT}\left(2N_{sc}\right). \tag{5.18}$$

The UF-OFDM receiver processing includes windowing in the time domain, transformation of the signal to the frequency domain, filtering in the frequency domain, and equalization in the frequency domain. Since the length of a UF-OFDM symbol is $N_{sc} + N_g$ samples (where N_g is the guard interval; cf. Fig. 6.17), the symbol is zero padded and an FFT of size $2N_{sc}$ is performed while we have a transformation from the time to the frequency domain. Considering these operations, the total number of real multiplications is given by

$$C_{UF-OFDM}^{Rx} = C_{FFT}(2N_{sc}) + 8N_{sc} + 4N_{act}. \tag{5.19}$$

5.5.4.4 FBMC-OQAM

FBMC waveform can be synthesized via two different approaches: a polyphase network (PPN)-based method and a frequency domain filtering (frequency spread)-based method. The two methods have different implementation complexity. For the PPN implementation, the transmitter and receiver side number of real multiplications (assuming an overlapping factor K for FBMC) are given by

$$C_{FBMC-PPN}^{Tx} = 2C_{FFT}(N_{sc}) + 4N_{sc}K + 4N_{act}, \tag{5.20}$$

$$C_{FBMC-PPN}^{Rx} = 2C_{FFT}(N_{sc}) + 4N_{sc}K + 4L_{eq}N_{act}, \tag{5.21}$$

where L_{eq} tap channel equalization is performed per subcarrier.

The complexity of the frequency spread FBMC method is given by

$$C_{FBMC-FS}^{Tx} = 2C_{FFT}(KN_{sc}) + 8N_{act}(K-1), \tag{5.22}$$

$$C_{FBMC-FS}^{Rx} = 2C_{FFT}(KN_{sc}) + 16N_{act}(K-1). \tag{5.23}$$

The transmitter and the receiver side computational complexity of these multicarrier waveforms is compared as a function of the number of subcarriers in Fig. 5.38 and Fig. 5.39, respectively. Here, we have assumed $N_{sc} = N_{act}$ for all waveforms, 7% CP length for OFDM and W-OFDM (i.e., $N_{cp} = .07N_{sc}$), $N_B = 16$ for UF-OFDM, $K = 4$ for FBMC-PPN and FBMC-FS, and $L_{eq} = 3$ for FBMC-PPN. We observe that CP-OFDM has the lowest complexity and the windowing contributes to a minor complexity increase. The complexity of the other multicarrier waveforms is significantly higher than CP-OFDM and the complexity gap enlarges as the number of subcarriers increase. For UF-OFDM, the complexity depends on N_B. As N_B increases, the complexity decreases. We also note that the PPN based FBMC implementation has significantly lower implementation complexity than the frequency spread based implementation.

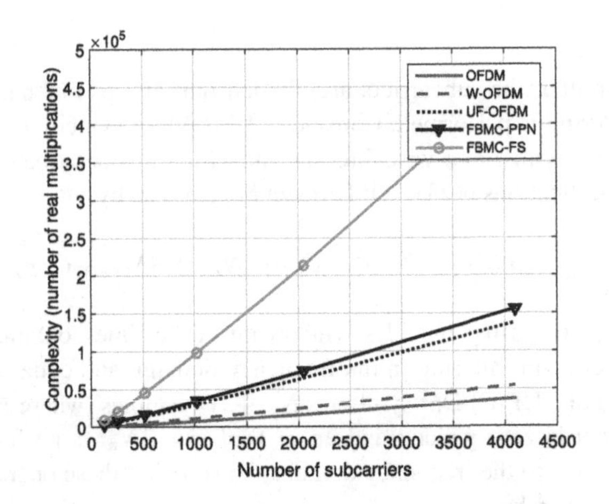

FIGURE 5.38

Transmitter side complexity comparison.

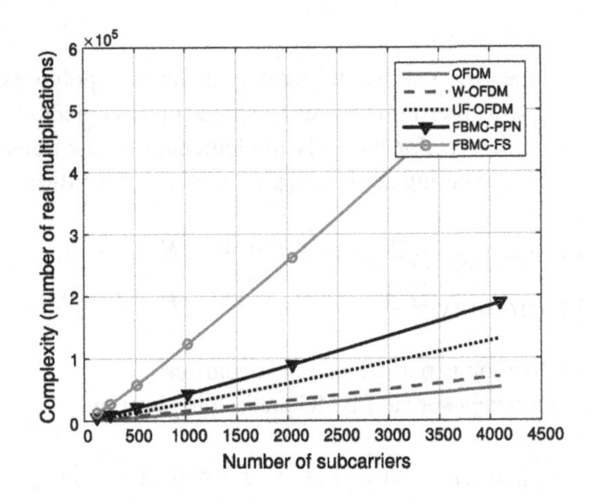

FIGURE 5.39

Receiver side complexity comparison.

5.5.5 PHASE-NOISE ROBUSTNESS COMPARISON

In the following we compare phase-noise robustness of CP-OFDM, FBMC-OQAM, and FBMC-QAM waveforms based on [6]. A baseband signal subject to phase-noise (PN) can be expressed as $r[n] = x[n]\,e^{j\phi[n]}$, where $\phi[n]$ is a random process. Typically, in a phase-locked loop (PLL)-based oscillators $\phi[n]$ is a wide-sense stationary (WSS) random process, and $|\phi[n]| \ll 1$ [5], which leads to the following approximation:

$$r[n] = x[n]\,e^{j\phi[n]} \approx x[n](1 + j\phi[n]). \tag{5.24}$$

5.5.5.1 Phase-Noise Effect in OFDM

Assuming that the received OFDM signal $r[n]$ is subject to the PN, consider the demodulation of a subcarrier l in the 0th OFDM symbol (i.e., $s = 0$) without any loss of generality:

$$
\begin{aligned}
R_{l,0} &= \sum_{n\in\mathbb{Z}} r[n]\,\overline{p_l[n]} \\
&\overset{(a)}{=} \sum_{n\in\mathbb{Z}} \left(\sum_{d\in\mathbb{Z}} \sum_{i\in I} X_{i,d}\, p_i[n-dN] \right) (1 + j\phi[n])\,\overline{p_l[n]} \\
&\overset{(b)}{=} \sum_{i\in I} X_{i,0} \sum_{n\in\mathbb{Z}} p_i[n]\,\overline{p_l[n]}(1 + j\phi[n]) \\
&\overset{(c)}{=} \sum_{i\in I} X_{i,0}\,(\delta_{i,l} + \varphi_{i,l}) \overset{(d)}{=} X_{l,0}\left(1 + \varphi_{l,l}\right) + N_l^{ICI},
\end{aligned}
\tag{5.25}
$$

where (a) follows from (5.7) and (5.24); (b) follows from the property that consecutive OFDM symbols do not interfere in the time domain ($p_i[n - dN]\,\overline{p_l[n]} = 0\ \forall n \in \mathbb{Z}$ when $d \neq 0$), avoiding intersymbol interference (ISI), (c) follows from $\varphi_{i,l} := \sum_{n\in\mathbb{Z}} j\,\phi[n]\,p_i[n]\,\overline{p_l[n]}$ and $\sum_{n\in\mathbb{Z}} p_i[n]\,\overline{p_l[n]} = \delta_{i,l}$, where $\delta_{i,l}$ is the Kronecker delta function; (d) follows by defining $N_l^{ICI} := \sum_{i\in I_{ICI}} X_{i,0}\,\varphi_{i,l}$, and $I_{ICI} = I \setminus \{l\}$ is the set of subcarriers that interfere with subcarrier l due to the PN. According to (5.25), the transmitted symbol X_l is multiplied by the term $(1 + \varphi_{l,l})$. This effect is known as a common phase error (CPE) because it causes an identical phase rotation in all subcarriers (i.e., $\varphi_{l,l} = \varphi_{k,k}$ for any l, k). The additive term N_l^{ICI} refers to an intercarrier interference (ICI), which is different for different subcarriers (that is, $N_l^{ICI} \neq N_k^{ICI}$ for any $l \neq k$).

According to [6], the achievable signal-to-interference ratio (SIR) of the demodulated subcarrier l in the OFDM symbol subject to the PN (in absence of CPE) is given by

$$
\mathrm{SIR}_l = \frac{1}{\left(\int_{-0.5}^{+0.5} S_\phi(\nu)\, W_l^{ICI}(\nu)\, d\nu\right)},
\tag{5.26}
$$

where

$$
W_l^{ICI}(\nu) := \sum_{i\in I_{ICI}} \frac{1}{N^2} \left| \frac{\sin\left(\pi\left(\nu - \frac{i-l}{N}\right)N\right)}{\sin\left(\pi\left(\nu - \frac{i-l}{N}\right)\right)} \right|^2 .
\tag{5.27}
$$

Next, we evaluate the achievable SIR using (5.26) for the PLL based phase-noise model described in Section 4.2.3. The PSDs of the PN model are illustrated in Fig. 5.40 at three oscillator different frequencies: 6 GHz, 28 GHz, and 82 GHz. The achievable SIR as a function of the subcarrier spacing is shown in Fig. 5.41. We observe that the SIR is an increasing function of the subcarrier spacing. This implies that the OFDM system would be more robustness against the PN if a larger subcarrier spacing is used.

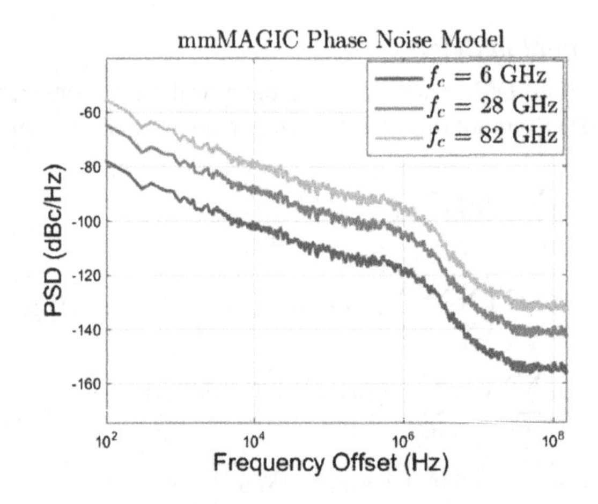

FIGURE 5.40

Power spectrum of the PLL based PN model at three oscillator frequencies: 6 GHz, 28 GHz, and 82 GHz. The model is developed in mmMAGIC (a European research project).

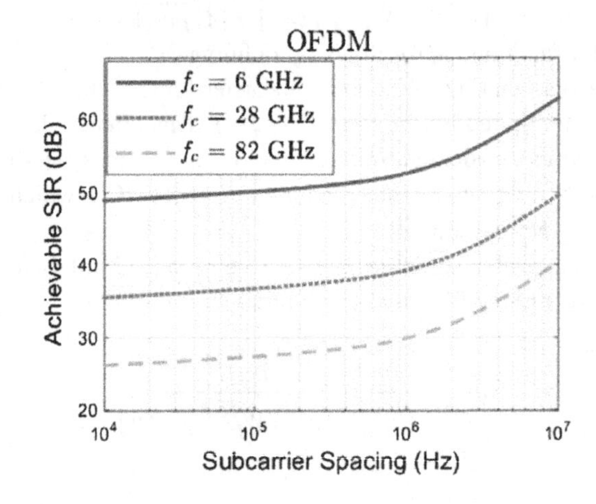

FIGURE 5.41

Achievable signal-to-interference ratio (SIR) for OFDM as a function of its subcarrier spacing at three oscillator frequencies (6 GHz, 28 GHz, and 82 GHz).

5.5.5.2 Phase-Noise Effect in FBMC-QAM

The demodulation of a subcarrier l in the 0th FBMC-QAM symbol (i.e., $s = 0$) subject to the PN is given by

$$R_{l,0} = \sum_{n \in \mathbb{Z}} r[n] \, \overline{p_l[n]}$$

$$\begin{aligned}
&\overset{(a)}{=} \sum_{n\in\mathbb{Z}} \left(\sum_{d\in D} \sum_{i\in I} X_{i,d}\, p_i[n-dN] \right) (1+j\phi[n])\, \overline{p_l[n]} \\
&= \sum_{d\in D} \sum_{i\in I} X_{i,d} \sum_{n\in\mathbb{Z}} p_i[n-dN]\, \overline{p_l[n]}\, (1+j\phi[n]) \\
&\overset{(b)}{=} \sum_{d\in D} \sum_{i\in I} X_{i,d} \left(\lambda_{i,l,d} + \beta_{i,l,d} \right),
\end{aligned} \tag{5.28}$$

where (a) follows from (5.24) and (5.11) and (b) follows by defining

$$\lambda_{i,l,d} := \sum_{n\in\mathbb{Z}} p_i[n-dN]\, \overline{p_l[n]}\,, \tag{5.29}$$

$$\beta_{i,l,d} := \sum_{n\in\mathbb{Z}} j\phi[n]\, p_i[n-dN]\, \overline{p_l[n]}\,. \tag{5.30}$$

The term $\lambda_{i,l,d}$ is due to the self-interference components caused by the overlapping FBMC symbols and the nonorthogonality between the subcarriers, whereas $\beta_{i,l,d}$ is due to the PN; it depends on both the choice of the prototype filters and the PN. We can rewrite (5.28) as

$$R_{l,0} = X_{l,0}\left(1+\beta_{l,l,0}\right) + I_l^{ICI} + I_l^{ISI} + N_l^{ICI} + N_l^{ISI}, \tag{5.31}$$

where we have used the following notations:

$$I_l^{ICI} := \sum_{i\in I_{ICI}} X_{i,0}\, \lambda_{i,l,0}\,, \tag{5.32}$$

$$I_l^{ISI} := \sum_{d\in D_{ISI}} \sum_{i\in I} X_{i,d}\, \lambda_{i,l,d}\,, \tag{5.33}$$

$$N_l^{ICI} := \sum_{i\in I_{ICI}} X_{i,0}\, \beta_{i,l,0}\,, \tag{5.34}$$

$$N_l^{ISI} := \sum_{d\in D_{ISI}} \sum_{i\in I} X_{i,d}\, \beta_{i,l,d}\,, \tag{5.35}$$

where $D_{ISI} = D \setminus \{0\}$ is a set containing the indices of the FBMC-QAM symbols that overlap with the 0th FBMC-QAM symbol. The multiplicative term $(1+\beta_{l,l,0})$ corresponds to the CPE. The CPE has the same value for all even subcarriers and the same value for all odd subcarriers (i.e., $\beta_{k,k,0} = \beta_{l,l,0}$ if $k \bmod 2 = l \bmod 2$ for any l, k). The terms I_l^{ICI} and I_l^{ISI} model the ICI and the ISI due to the self-interferences and the terms N_l^{ICI} and N_l^{ISI} represent the ICI and the ISI due to the PN.

According to [6], the achievable SIR of the demodulated subcarrier l in the FBMC-QAM symbol subject to the PN (in absence of CPE) is given by

$$\mathrm{SIR}_l = \frac{P_X}{P_{I_l^{ICI}} + P_{I_l^{ISI}} + P_{N_l^{ICI}} + P_{N_l^{ISI}}}, \tag{5.36}$$

$$P_{N_l^{ICI}} := P_X \left(\int_{-0.5}^{+0.5} S_\phi(\nu)\, W_l^{ICI}(\nu)\, d\nu \right), \tag{5.37}$$

$$P_{N_l^{ISI}} := P_X \left(\int_{-0.5}^{+0.5} S_\phi(\nu)\, W_l^{ISI}(\nu)\, d\nu \right), \tag{5.38}$$

$$P_{I_l^{ICI}} := P_X\, \delta(\nu)\, W_l^{S-ICI}(\nu), \tag{5.39}$$

$$P_{I_l^{ISI}} := P_X\, \delta(\nu)\, W_l^{S-ISI}(\nu), \tag{5.40}$$

$$W_l^{ICI}(\nu) := \sum_{i \in I_{ICI}} W_{i,l,0}(\nu), \tag{5.41}$$

$$W_l^{ISI}(\nu) := \sum_{d \in D_{ISI}} \sum_{i \in I} W_{i,l,d}(\nu), \tag{5.42}$$

$$W_l^{S-ICI}(\nu) := \sum_{i \in I_{ICI}} W_{i,l,0}^S(\nu), \tag{5.43}$$

$$W_l^{S-ISI}(\nu) := \sum_{d \in D_{ISI}} \sum_{i \in I} W_{i,l,d}^S(\nu), \tag{5.44}$$

$$W_{i,l,d}(\nu) := W_{i,l,d}^S(\nu) = \left| P_i(\nu)\, e^{-j2\pi \nu d N} \circledast \overline{P_l(\nu)} \right|^2, \tag{5.45}$$

where $P_i(\nu) := \mathcal{F}\{p_i[n]\}$, $p_i[n]$ is defined in (5.12), and $\mathcal{F}\{\cdot\}$ is the discrete-time Fourier transform operation.

5.5.5.3 Phase-Noise Effect in FBMC-OQAM

Assuming that the received FBMC-OQAM signal $r[n]$ is subject to the PN, the demodulation of a subcarrier l in the 0th FBMC-OQAM symbol is given by:

$$\begin{aligned}
R_{l,0} &\overset{(a)}{=} \sum_{n \in \mathbb{Z}} \Re r[n]\, \psi_l\, \overline{p_l[n]} \\
&\overset{(b)}{=} \sum_{n \in \mathbb{Z}} \Re \left\{ \left(\sum_{d \in D} \sum_{i \in I} X_{i,d} \theta_{i,d}\, p_i \left[n - d\frac{N}{2} \right] \right) (1 + j\phi[n])\, \psi_l\, \overline{p_l[n]} \right\} \\
&= \sum_{d \in D} \sum_{i \in I} X_{i,d} \sum_{n \in \mathbb{Z}} \Re \left\{ \theta_{i,d}\, p_i \left[n - d\frac{N}{2} \right] \psi_l\, \overline{p_l[n]}\, (1 + j\phi[n]) \right\} \\
&\overset{(c)}{=} \sum_{d \in D} \sum_{i \in I} X_{i,d} \left(\gamma_{i,l,d} + \rho_{i,l,d} \right)
\end{aligned}$$

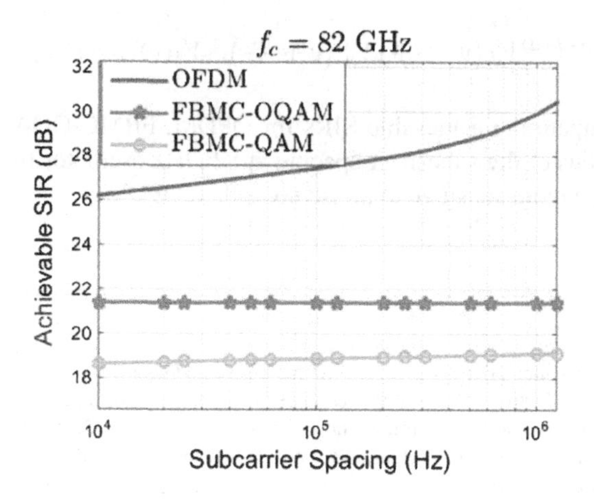

FIGURE 5.42

Achievable signal-to-interference ratio (SIR) for OFDM, FBMC-OQAM ($K = 4$) and FBMC-QAM ($K = 4$) as a function of subcarrier spacing at 82 GHz oscillator frequency.

$$\stackrel{(d)}{=} X_{l,0} + I_l^{ICI} + I_l^{ISI} + N_l^{ICI} + N_l^{ISI}, \tag{5.46}$$

where (a) follows from the offset-demapping which uses the real value operator $\Re\{\cdot\}$ and the phase term $\psi_l = j^{(-l)}$; (b) follows from (5.24) and (5.10); (c) follows by defining

$$\gamma_{i,l,d} := \sum_{n\in\mathbb{Z}} \Re\left\{\theta_{i,d}\, p_i\left[n - d\frac{N}{2}\right] \psi_l\, \overline{p_l[n]}\right\}, \tag{5.47}$$

$$\rho_{i,l,d} := \sum_{n\in\mathbb{Z}} \Re\left\{j\phi[n]\,\theta_{i,d}\, p_i\left[n - d\frac{N}{2}\right] \psi_l\, \overline{p_l[n]}\right\}, \tag{5.48}$$

which are associated with the self-interferences and the PN, respectively; (d) follows from defining I_l^{ICI}, I_l^{ISI} (the same as in (5.32) and (5.33) but replacing $\lambda_{i,l,d}$ by $\gamma_{i,l,d}$), N_l^{ICI} and N_l^{ISI} (the same as in (5.34) and (5.35), but replacing $\beta_{i,l,d}$ by $\rho_{i,l,d}$). It is noteworthy that FBMC-OQAM is not affected by any CPE effect subject to the given PN formulation (i.e., $\forall l \in I\ \rho_{l,l,0} = 0$).

According to [6], the achievable SIR of the demodulated subcarrier l in the FBMC-OQAM symbol subject to the PN is given by (5.36) with the following differences:

$$W_{i,l,d}(\nu) := \frac{1}{4}\left[\left|V_d\left(\nu - \frac{i-l}{N}\right)\right|^2 + \left|V_d\left(\nu + \frac{i-l}{N}\right)\right|^2 - \right.$$
$$\left. - 2\,\Re Y_{i,l,d}(\nu)\right] \tag{5.49}$$

$$W^S_{i,l,d}(\nu) := \frac{1}{4}\left[\left|V_d\left(\nu - \frac{i-l}{N}\right)\right|^2 + \left|V_d\left(\nu + \frac{i-l}{N}\right)\right|^2 + \right.$$
$$\left. + 2\,\Re Y_{i,l,d}(\nu)\right], \tag{5.50}$$

where $Y_{i,l,d}(\nu) := e^{j\pi(i-l+d)} V_d\left(\nu - \frac{i-l}{N}\right) V_d^*\left(\nu + \frac{i-l}{N}\right)$, $V_d(\nu) := P(\nu) \circledast \left(P(\nu)e^{-2\pi\nu\frac{N}{2}d}\right)$, and $P(\nu) := \mathcal{F}\{p[n]\}$.

In Fig. 5.42, we compare the achievable SIRs for OFDM, FBMC-OQAM ($K = 4$), and FBMC-QAM ($K = 4$) as functions of the subcarrier spacing at 82 GHz oscillator frequency. We observe that the FBMC waveforms are more sensitive to phase noise than OFDM.

REFERENCES

[1] M. Bellanger, FS-FBMC: an alternative scheme for filter bank based multicarrier transmission, in: 5th International Symposium on Communications Control and Signal Processing (ISCCSP), 2012, May, pp. 1–4.
[2] M. Bellanger, et al., FBMC physical layer: a primer, http://www.ict-phydyas.org, 2010, 06.
[3] R. Gerzaguet, N. Bartzoudis, L.G. Baltar, V. Berg, J.B. Doré, D. Kténas, O. Font-Bach, X. Mestre, M. Payaró, M. Färber, K. Roth, The 5G candidate waveform race: a comparison of complexity and performance, EURASIP Journal on Wireless Communications and Networking 2017 (1) (2017, Jan.).
[4] J. Luo, A.A. Zaidi, J. Vihriälä, D. Giustiniano, et al., Preliminary radio interface concepts for mm-wave mobile communications. Deliverable D4.1, Millimetre-Wave Based Mobile Radio Access Network for Fifth Generation Integrated Communications (mmMAGIC), https://5g-mmmagic.eu/results, 2016.
[5] A. Mehrotra, Noise analysis of phase-locked loops, in: IEEE/ACM International Conference on Computer Aided Design. ICCAD – 2000. IEEE/ACM Digest of Technical Papers (Cat. No.00CH37140), 2000, Nov., pp. 277–282.
[6] V. Moles-Cases, A.A. Zaidi, X. Chen, T.J. Oechtering, R. Baldemair, A comparison of OFDM, QAM-FBMC, and OQAM-FBMC waveforms subject to phase noise, in: 2017 IEEE International Conference on Communications (ICC), 2017, May, pp. 1–6.
[7] H.G. Myung, J. Lim, D.J. Goodman, Single carrier fdma for uplink wireless transmission, IEEE Vehicular Technology Magazine (ISSN 1556-6072) 1 (3) (2006, Sept.) 30–38, https://doi.org/10.1109/MVT.2006.307304.
[8] H. Nam, M. Choi, C. Kim, D. Hong, S. Choi, A new filter-bank multicarrier system for QAM signal transmission and reception, in: Proc. IEEE ICC, 2014, June, pp. 5227–5232.
[9] F. Schaich, T. Wild, Y. Chen, Waveform contenders for 5G—suitability for short packet and low latency transmissions, in: IEEE Vehicular Technology Conference (VTC Spring), 2014, May, pp. 1–5.
[10] F. Sjoberg, R. Nilsson, M. Isaksson, P. Odling, P.O. Borjesson, Asynchronous zipper, in: IEEE International Conference on Communications (ICC), 1999, pp. 231–235.
[11] T. Wild, F. Schaich, A reduced complexity transmitter for UF-OFDM, in: Proc. IEEE 81st Vehicular Technology Conference (VTC Spring), 2015, May, pp. 1–6.
[12] T. Wild, F. Schaich, Y. Chen, 5G air interface design based on universal filtered (UF-)OFDM, in: 19th International Conference on Digital Signal Processing (DSP), 2014, Aug., pp. 699–704.
[13] Y.H. Yun, C. Kim, K. Kim, Z. Ho, B. Lee, J.Y. Seol, A new waveform enabling enhanced QAM-FBMC systems, in: Proc. IEEE SPAWC, 2015, June, pp. 116–120.
[14] A.A. Zaidi, R. Baldemair, H. Tullberg, H. Bjorkegren, L. Sundstrom, J. Medbo, C. Kilinc, I.D. Silva, Waveform and numerology to support 5G services and requirements, IEEE Communications Magazine (ISSN 0163-6804) 54 (11) (2016, November) 90–98, https://doi.org/10.1109/MCOM.2016.1600336CM.
[15] A.A. Zaidi, J. Luo, R. Gerzaguet, H. Wang, X. Chen, Y. Qi, N. Cassiau, A. Wolfgang, J. Vihriala, A. Kakkavas, T. Svensson, J. Mohammadi, R.J. Weiler, M. Dieudonne, H. Halbauer, V. Moles-Cases, H. Miao, Evaluation of waveforms for mobile radio communications above 6 GHz, in: 2016 IEEE Globecom Workshops (GC Wkshps), 2016, Dec., pp. 1–6.
[16] L. Zhang, A. Ijaz, P. Xiao, M.M. Molu, R. Tafazolli, Filtered OFDM systems, algorithms, and performance analysis for 5G and beyond, IEEE Transactions on Communications (ISSN 0090-6778) 66 (3) (2018, March) 1205–1218, https://doi.org/10.1109/TCOMM.2017.2771242.

CHAPTER

NR WAVEFORM

6

CP-OFDM waveform has been widely adopted for wireless and wireline communication as well as for digital audio and video broadcasting. The Asymmetric Digital Subscriber Line (ADSL) was the first wireline technology that used OFDM for high speed transmission. Digital Audio Broadcasting (DAB) and Digital Video Broadcasting (DVB-T) standards employ OFDM for broadcasting digital audio and video services. Various wireless IEEE and 3GPP standards have adopted OFDM, for example, IEEE 802.11a/WLAN, IEEE 802.16/WiMAX, 3GPP 4G LTE, and more recently 3GPP 5G NR. In this chapter, we discuss the OFDM design for 5G NR.

This chapter is organized as follows. Section 6.1 discusses suitability of CP-OFDM for NR. Section 6.2 describes a scalable OFDM design for NR, as adopted by the 3GPP. Section 6.3 provides a detailed discussion of how the scalable parameters of OFDM should be chosen considering various factors such as quality of service requirements, type of deployment, carrier frequency, user mobility, hardware impairments, and implementation aspects. Section 6.4 discusses the need for improving power efficiency of the NR waveform for coverage limited scenarios, along with a brief overview of the commonly used techniques. The effects of different synchronization errors in a CP-OFDM system are discussed in Section 6.5. Finally, some mitigation techniques for the hardware impairments (phase noise, carrier-frequency offset, and sampling frequency offset) are presented in Section 6.6.

6.1 SUITABILITY OF OFDM FOR NR

For the NR waveform, the design requirements vary depending on the carrier-frequency range and the link type, as discussed in Chapter 5. The 3GPP has selected CP-OFDM waveform for NR after a thorough investigation of several multicarrier and single-carrier waveforms considering requirements that are important for the NR design. In Chapter 5, we have also provided a comparison of state-of-the art waveforms. In the following, we discuss the suitability of OFDM for NR for various link types (uplink, downlink, sidelink,[1] vehicle-to-anything (V2X)[2] link, backhaul link) considering the waveform key performance indicators:

- ***Spectral efficiency*** OFDM is well known for its high spectral efficiency. The spectral efficiency is vital to meet extreme data rate requirements with NR. In general, the spectral efficiency is more crucial at lower carrier frequencies than at higher frequencies due to potentially much larger channel bandwidths at higher carrier frequencies. The spectral efficiency is important for both uplink and

[1] Sidelink refers to direct device-to-device (D2D) communication.

[2] Vehicle-to-anything (V2X) communication includes vehicle-to-vehicle (V2V), vehicle-to-infrastructure (V2I), vehicle-to-network (V2N), and vehicle-to-pedestrian (V2P).

5G Physical Layer. https://doi.org/10.1016/B978-0-12-814578-4.00011-4

the downlink. The requirements are even more stringent for the backhaul link due to the massive volumes of data transmission between the base stations. Vehicular communication can also create capacity bottlenecks in dense urban scenarios with large number of vehicles periodically broadcasting signals in an asynchronous fashion.

- ***MIMO compatibility*** OFDM enables a straightforward use of MIMO technology (see Chapter 7). With increase in the carrier frequency with NR, the number of antenna elements would increase in the base stations as well as in the devices. The use of various MIMO schemes is essential for enhancing spectral efficiency by enabling SU-MIMO/MU-MIMO and achieving greater coverage via beam-forming (see Chapter 7). Beam-forming is instrumental in overcoming high propagation losses at very high frequencies where coverage is limited.
- ***Peak-to-Average-Power-Ratio (PAPR)*** OFDM has a high PAPR like other multicarrier waveforms (see Section 5.5.2). A low PAPR is essential for power-efficient transmissions from the devices (for example, in the uplink and the sidelink). A low PAPR becomes even more important at very high frequencies where coverage can be limited. It is noteworthy that small sized low cost base stations are envisioned at high frequencies, therefore, low PAPR is also important for the downlink transmissions at high carrier frequencies. A high PAPR in OFDM can also be substantially reduced via various well-known PAPR reduction techniques with only minor compromise in performance [10] (see Section 6.4). For NR, OFDM with PAPR reduction is an attractive option for the uplink and the sidelink.
  LTE uses DFTS-OFDM for both uplink and sidelink due to its lower PAPR. However, DFTS-OFDM has certain drawbacks in comparison with OFDM such as lesser flexibility for scheduling and more complex MIMO receiver with degraded link level and system level performance [21]. For NR, DFTS-OFDM is optional for the uplink and can be used only for a single-stream (without MIMO) transmission. Since MIMO transmission is a key component of NR, DFTS-OFDM is not a preferred option for the uplink and the sidelink in general. The use of one waveform (i.e., OFDM) for all link types also makes transceiver designs and implementations symmetric for all transmissions.
- ***Robustness to channel time selectivity*** is vital in high mobility scenarios. High-speed UEs are relevant in large cell deployments. The large cell deployments are not expected at very high frequencies due to harsh propagation conditions (coverage limitation). At very high frequencies, the deployments are expected in the form of small cells where mobility is not a major concern. However, the V2V services may be enabled at very high frequencies, making robustness to the channel time selectivity a very important performance indicator at very high frequencies. Traditionally, the backhaul link is fixed and mobility is not a concern, however, for the envisioned mobile backhaul (e.g., access nodes on vehicles), robustness to channel time selectivity will become relevant. OFDM can be made robust to channel time selectivity by a proper choice of the subcarrier spacing.
- ***Robustness to channel frequency selectivity*** Channel frequency selectivity is always relevant to the transmission of large bandwidth signals over wireless multipath channels. Channel frequency selectivity depends on various factors such as the type of deployment, the beam-forming technique, and the signal bandwidth. OFDM is robust to frequency-selective channels and requires only single-tap frequency-domain equalization.
- ***Robustness against phase noise*** Typically, phase-noise increases with an increase in the oscillator frequency (see Section 4.2). An OFDM system can be made robust to phase-noise by a proper choice of its subcarrier spacing (see Section 6.3.1). Phase-noise robustness is crucial for all link types where a UE (transmitter/receiver) is involved. In particular, low phase-noise oscillators may be

too expensive and power consuming for devices (UEs). Phase-noise robustness is also important for future low cost base stations. Basically, any link that involves a device and/or a low cost base station puts a high requirement on phase-noise robustness of the waveform, especially if the communication takes place at high carrier frequencies.

- ***Transceiver baseband complexity*** The baseband complexity of an OFDM receiver is lowest among all 5G candidate waveforms as discussed in Chapter 5 (see Section 5.5.4 for a complexity comparison). The baseband complexity is always very important at the device (UE) side, especially from the receiver perspective. For NR, the baseband complexity is also a major consideration for the base stations, since an NR base station can be a small sized access node (especially at high frequencies) with a limited processing capability. At very high frequencies and large bandwidths, the receiver also has to cope with severe RF impairments.
- ***Time localization*** OFDM is very well localized in the time domain, which is important to efficiently enable (dynamic) TDD and support latency critical applications such as URLLC (see Section 2.7). Dynamic TDD is envisioned at high frequencies and provision of low latency is essential for all link types; especially backhaul and V2V links may impose very high requirements.
- ***Frequency localization*** OFDM is less localized in the frequency domain (see Section 5.5.1). Frequency localization can be relevant to support the coexistence of different services potentially enabled by mixing different waveform numerologies in the frequency domain on the same carrier (see Section 6.3.3). Frequency localization is also relevant if asynchronous access is allowed in uplink and sidelink. In general, frequency localization of a waveform may not be crucial at high frequencies where large channel bandwidths are available.
- ***Robustness to synchronization errors*** The provision of the cyclic prefix in OFDM makes it robust to timing synchronization errors (see Section 6.5.1). Robustness to the synchronization errors is relevant in scenarios where the synchronization is hard to achieve such as in the sidelink. It can also be important if asynchronous transmissions are allowed in the uplink.[3]
- ***Flexibility and scalability*** OFDM is a flexible waveform which can support diverse services in a wide range of frequencies by a proper choice of the subcarrier spacing and the cyclic prefix. (See Section 6.2 for a discussion on the OFDM numerology design that fulfills a wide range of requirements for NR.)

A summary of the waveform design requirements for different link types is shown in Fig. 6.1. A link requirement "High" for a waveform KPI tells that the given KPI is important for the given link type in general. Furthermore, a high-level assessment of OFDM is given in Table 6.1. The OFDM assessment "High" means that OFDM has good performance in general for the given KPI. We assess the D2D and V2V cases separately due to different levels of requirements. For example, V2V communication has higher requirements on the mobility and the system capacity, whereas there are lower requirements on the power efficiency when compared with UE-to-UE communication. Based on the overall assessment, it can be concluded that OFDM is an excellent choice for the NR air interface for all link types.

[3]We note that LTE only supports synchronous uplink transmission, which is realized via a timing advance mechanism at the UEs.

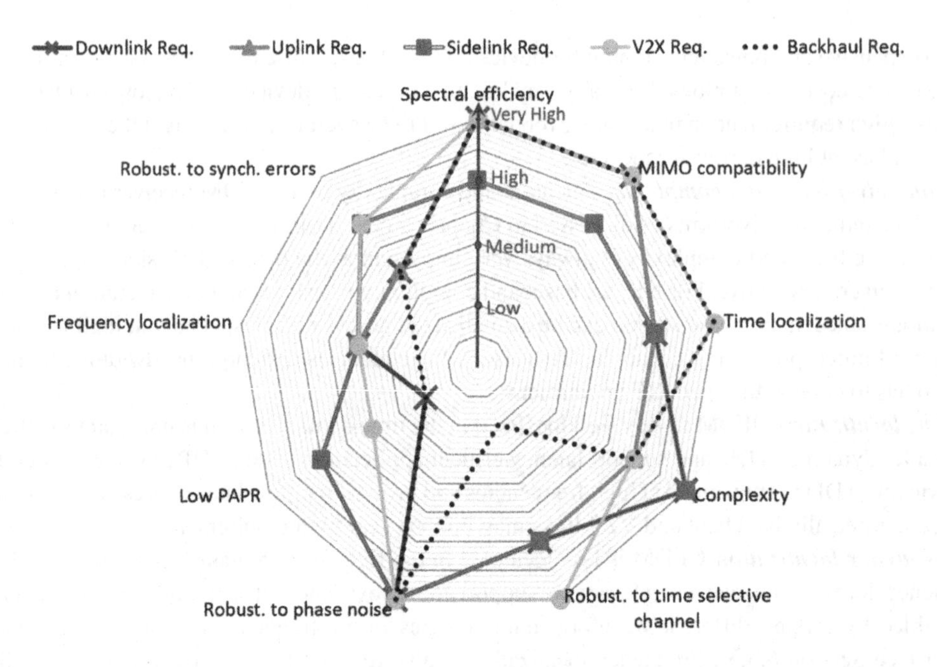

FIGURE 6.1

A high-level summary of the waveform design requirements for different link types: uplink, downlink, sidelink, V2V link, and backhaul link.

Table 6.1 A high-level assessment of OFDM

Performance indicators	OFDM assessment
Spectral efficiency	High
MIMO compatibility	High
Time localization	High
Transceiver baseband complexity	Low
Flexibility/Scalability	High
Robust. to freq.-selective chan.	High
Robust. to time-selective chan.	Medium
Medium Robust. to phase noise	Medium
Robust. to synch. errors	Medium
PAPR	High (can be improved)
Frequency localization	Low (can be improved)

6.2 SCALABLE OFDM FOR NR

CP-OFDM waveform has a few design parameters: subcarrier spacing, cyclic prefix, and the number of subcarriers. For a communication system, these parameters can be optimized based on a number of factors including carrier frequency, user mobility, phase noise, channel delay spread, quality of service

Table 6.2 Scalable OFDM numerology for 5G NR

OFDM subcarrier spacing	15 kHz	30 kHz	60 kHz	15 x 2^n kHz ($n \geq 3$)
OFDM symbol duration	66.67 μs	33.33 μs	16.67 μs	66.67/2^n μs
Cyclic-prefix duration	4.69 μs	2.34 μs	1.17 μs	4.69/2^n μs
OFDM symbol with CP	71.35 μs	35.68 μs	17.84 μs	71.35/2^n μs
OFDM symbols per slot	14	14	14	14
Slot duration	1000 μs	500 μs	250 μs	1000/2^n μs

Numerologies with $n = \{-2, -1\}$ are possible but not listed in this table.

requirements, signal bandwidth, and implementation complexity. For a given carrier frequency, phase noise and the Doppler effect set requirements on the minimum subcarrier spacing. The use of smaller subcarrier spacings either results in a high error vector magnitude (EVM) due to phase noise or in undesirable strict requirements on the local oscillator. Too narrow subcarrier spacings lead to performance degradations in high Doppler effect scenarios. The required cyclic-prefix overhead (and thus anticipated delay spread) sets an upper limit for the subcarrier spacing; selecting larger subcarriers would result in an undesirable high CP overhead (degrading spectral efficiency). The maximum FFT size of the OFDM modulator (or the number of total subcarriers) together with subcarrier spacing determines the channel bandwidth. Based on these relationships, the subcarrier spacing is typically chosen as small as possible while still being robust against phase noise and Doppler effect and providing the desired channel bandwidth. These design principles also apply to NR, as we will discuss in the sequel.

NR will operate from sub-1 GHz to 100 GHz using a wide range of deployment options (e.g., macro cells, micro cells, pico cells). Furthermore, NR will cover a very wide application range, including mobile broadband and several types of machine type communications. A single OFDM numerology cannot fulfill the desired frequency range and all envisioned deployment options and applications; therefore, the 3GPP has adopted a family of OFDM numerologies for NR. In the 3GPP context, OFDM numerology refers to the choice of OFDM subcarrier spacing, cyclic-prefix duration, and the number of OFDM symbols per transmission slot.[4] In particular, it has been agreed that the subcarrier spacing of OFDM can be chosen according to 15×2^n kHz, where n is an integer-valued parameter and the cyclic-prefix overhead is 7% as in LTE. The details related to NR OFDM numerologies (as agreed in the 3GPP) such as CP duration, symbol duration, and slot size, are given in Table 6.2. (We note that in 3GPP NR Release 15, the numerology is specified for up to 52.6 GHz carrier frequency with 120 kHz as the maximum subcarrier spacing as given in Table 2.1.[5]) The 15 kHz numerology is exactly the same as in LTE, with a minor exception that in LTE a slot contains seven OFDM symbols whereas a slot in NR has 14 OFDM symbols. For all numerologies, the cyclic prefix of the first OFDM symbol in every 0.5 ms interval is 16 T_s ($T_s = 32.6$ ns is the chip duration corresponding to an LTE sampling rate of 30.72 MHz) longer than the cyclic prefix of the remaining OFDM symbols in the interval. This implies that a slot length can slightly vary, depending on where it starts.

[4]See Section 2.3 for definition of the slot in NR.

[5]In 3GPP NR Release 15, the OFDM numerology is not specified for the frequency range 6 GHz to 24 GHz, since no spectrum has been identified for NR in this range. When the spectrum becomes available in the future, the numerology can be specified.

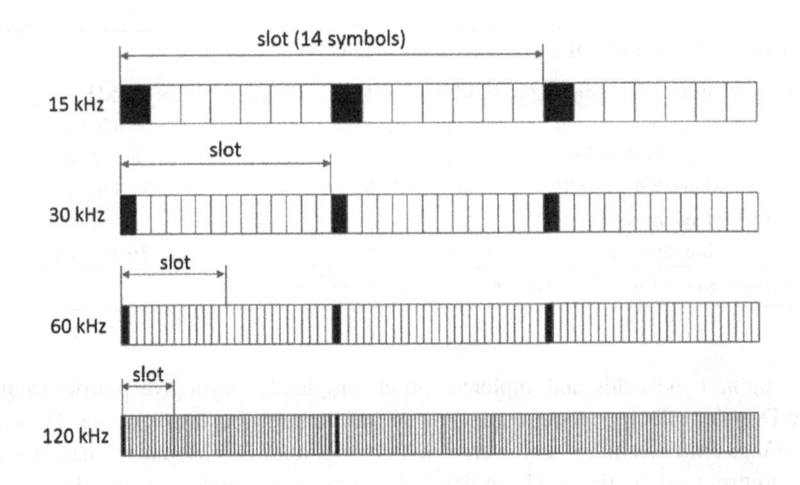

FIGURE 6.2

The NR numerology ensures symbol-wise and slot-wise time alignment. The time alignment is important for efficiently enabling TDD networks.

Fig. 6.2 illustrates OFDM symbols and slots for different numerologies (15 kHz, 30 kHz, 60 kHz, and 120 kHz), where symbols with darker colors represent OFDM symbols with longer CPs (every 0.5 ms). We observe that an integer number of slots of one numerology (with wider subcarrier spacing) fits into a slot of another numerology (with narrower subcarrier spacing). This time alignment of the slots is important for TDD networks to allow for time-aligned uplink and downlink transmission periods, which is discussed further in Section 6.2.2.

The 3GPP has introduced the concept of mini-slots in NR to support transmissions shorter than regular slot duration (see Section 2.3). A mini-slot can start at any OFDM symbol in a regular slot. Mini-slots can be useful in various scenarios—three important utilizations are shown in Fig. 6.3: i) low-latency transmissions, where the regular slot duration is too long and/or a transmission need to start immediately without waiting for the start of a slot boundary, ii) transmissions in an unlicensed spectrum, where it is beneficial to start transmission immediately after LBT, and iii) transmission in the millimeter-wave band, where the large channel bandwidths imply that the payload supported by one or two OFDM symbols can be sufficient for many of the packets. Furthermore, in the millimeter wave band the multiplexing of different users should primarily be done in the time domain to support analog/hybrid beam-forming, calling for a transmission duration shorted than a slot.

In principle, a mini-slot for a numerology can be as short as one OFDM symbol.[6] Some desirable characteristics of the mini-slot are: a mini-slot should be aligned with OFDM symbol boundaries in the regular slot; a mini-slot should end at the slot boundaries at latest; it should be possible to aggregate a mini-slot with a subsequent slot; and it may either contain control information at its beginning or at its end.

[6]In 3GPP NR Release 15, the downlink mini-slots are restricted to 2, 4, and 7 OFDM symbols.

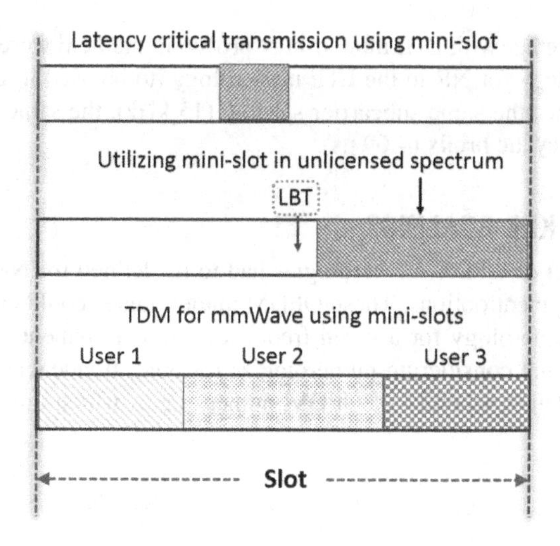

FIGURE 6.3

Mini-slots are useful in various situations, e.g., to achieve low latency, to efficiently transmit in an unlicensed spectrum based on listen-before-talk, and to schedule users at shorter time scales in the millimeter-wave band where large channel bandwidths are available.

6.2.1 WHY 15 KHZ AS BASELINE NUMEROLOGY?

The LTE OFDM numerology is 15 kHz subcarrier spacing with 7% cyclic-prefix overhead (4.69 μs). The numerology for LTE was specified after a thorough investigation in the 3GPP. For NR, it was straightforward for the 3GPP to aim for a similar OFDM numerology at LTE-like frequencies and deployments. Different subcarrier spacing options close to 15 kHz were therefore considered by the 3GPP as a baseline numerology for NR. The conclusion is to keep the LTE numerology as the baseline numerology for NR. There are two important reasons for keeping the LTE numerology as the base numerology:

- Narrow Band-IoT (NB-IoT) is a new radio access technology (already deployed since 2017) to support massive machine-type communications. NB-IoT uses the LTE numerology and provides different deployments, among others, an in-band deployment within an LTE carrier that is enabled by the selected LTE numerology. NB-IoT devices are designed to operate for 10 years or more on a single battery charge. Once such an NB-IoT device is deployed it is likely that within the device life time the embedding carrier (assuming in-band LTE deployment) gets refarmed to NR. The main reason for selecting the LTE-based numerology for NB-IoT was the option of the in-band deployment; in-band NB-IoT deployments after refarming LTE to NR would benefit from the LTE-based numerology.
- NR deployments can happen in the same band as LTE. With an adjacent LTE TDD carrier, NR must adopt the same uplink/downlink switching pattern as the LTE TDD does. Every numerology where (an integer multiple of) a subframe is 1 ms can be aligned with regular subframes in LTE. In LTE, the duplex switching happens in special subframes. To match the transmission direction in the special subframes, the same numerology as in LTE is needed.

The above arguments together with LTE numerology proven in the field were strong enough to set the baseline OFDM numerology for NR to the LTE numerology (to be used in LTE-like frequencies and deployments). This implies the same subcarrier spacing (15 kHz), the same OFDM symbol duration (66.67 μs), and the same cyclic prefix (4.69 μs).

6.2.2 WHY 15 × 2^n KHZ SCALING?

As discussed earlier, a set of OFDM numerologies had to be defined for NR to handle a wide range of frequencies and deployment options. These OFDM numerologies could either be unrelated to each other, i.e., the OFDM numerology for a given frequency and deployment is only based on this frequency and deployment, not considering numerologies for other frequencies and deployments at all. Another option was to define a family of OFDM numerologies which are related to each other via scaling, i.e.,

$$\Delta f_i = n_i \Delta f_{i-1}, \quad T_{cp,(i)} = \frac{T_{cp,(i-1)}}{n_i}, \tag{6.1}$$

where Δf_i and $T_{cp,(i)}$ denote the subcarrier spacing and the cyclic-prefix duration of the ith numerology and $n_i \in \mathbb{N}$ is a scaling factor. The duration of the OFDM symbol is the inverse of the subcarrier spacing. With this scaling approach, the sampling clock rates of different OFDM numerologies relate to each other via the scaling factors $\{n_i\}$, which simplifies the implementation. The 3GPP has adopted this scaling approach, i.e., the NR OFDM numerologies are derived from a baseline OFDM numerology (which is same as LTE numerology) via the scaling. In principle, the scaling factors $\{n_i\}$ could be selected independently of each other, however, it is desirable that the scaling factors follow a certain relationship to allow an efficient TDD based operation, which we discuss next.

For NR, the number of OFDM symbols per slot is equal for all numerologies, at-least up to the 120 kHz subcarrier spacing specified in 3GPP NR Release 15. Maintaining an equal number of OFDM symbols per slot for different numerologies simplifies scheduling and reference signal design. With the number of OFDM symbols per slot equal for all numerologies, the slot duration shrinks with the increase in the subcarrier spacing. This enables shorter latencies for wider subcarrier numerologies (to be used in high frequency small cell deployments where some of the low-latency applications are envisioned). Considering the equal numbers of OFDM symbols per slot for all numerologies, the following relationship holds for the respective slot durations between different numerologies:

$$T_{slot,(i)} = \frac{T_{slot,(i-1)}}{n_i} = \frac{T_{slot,(i-2)}}{n_i n_{i-1}} = \cdots = \frac{T_{slot,(1)}}{\prod_{k=1}^{i} n_k},$$

where $T_{slot,(i)}$ denotes the slot duration of the ith numerology. For adjacent TDD networks that are using different OFDM numerologies, it is desirable that an integer number of slots from one OFDM numerology fits into a slot of the other OFDM numerology to enable time-aligned downlink and uplink transmission periods. If the slot durations of different numerologies do not fulfill the above condition, then two neighboring TDD networks would require a guard time in the frame structure to enable synchronous operation, which will not be an efficient resource utilization. Therefore, the scaling factors are chosen such that a subcarrier spacing is integer divisible by all smaller subcarrier spacings, i.e.,

$$\Delta f_i = 2^{L(i)} \Delta f_1, \quad \forall i \in \{1, 2, ..., M\}, \tag{6.2}$$

where $L_{(i)} \in \mathbb{Z}$, M is the number of OFDM numerologies, and Δf_1 is subcarrier spacing of the base numerology. With this in mind, the 3GPP has specified the scaling factor as $n_i = 2^L$ in (6.1), where L is an integer.

6.3 OFDM NUMEROLOGY IMPLEMENTATION

The NR waveform is scalable—the subcarrier spacing of OFDM can be chosen according to 15×2^n kHz, where the integer-valued n is a design parameter and can be optimized for different scenarios. The 3GPP has agreed that NR should allow subcarrier spacing ranging from at least 3.75 kHz ($n = -2$) to 480 kHz ($n = 5$). A lower subcarrier spacing than 15 kHz (and a correspondingly longer cyclic prefix) is beneficial to support multicast-broadcast single-frequency network (MBSFN) transmission. For example, $n = -2$ results in a subcarrier spacing of 3.75 kHz, which is in the same range as the subcarrier spacings used in various digital broadcast standards, such as DVB, as well as being in line with some modes of NB-IoT. Hence, $n = -2$ is an option in the set of scaling factors to be considered. Choosing the value of n is not straightforward. It depends on various factors including type of deployments, service requirements, hardware impairments, mobility, performance, and implementation complexity. In the following, we provide a comprehensive discussion on these factors involved in selecting n for the NR waveform.

6.3.1 PHASE NOISE

Typically, phase noise increases with frequency of the local oscillator. It can be a major hardware impairment for NR deployments in high carrier frequencies, for example in the millimeter-wave band. Multicarrier waveforms are in general sensitive to phase noise. In an OFDM system, phase noise produces two types of degradations: a common phase error (CPE) and an intercarrier interference (ICI). We will discuss these in the following.

Consider a received baseband signal subject to phase noise $r[n] = x[n]e^{j\phi[n]}$, where $\phi[n]$ is a random process. Typically, in Phase-Locked-Loop (PLL)-based oscillators, $\phi[n]$ is a Wide-Sense Stationary (WSS) random process and $|\phi[n]| \ll 1$ [11], which leads to the following approximation:

$$r[n] = x[n]e^{j\phi[n]} \approx x[n]\left(1 + j\phi[n]\right). \tag{6.3}$$

Under this approximation, the demodulation of subcarrier $l \in \{1, 2, \ldots, N\}$ in an OFDM system with total N subcarriers is given by

$$\begin{aligned} R_l &= \frac{1}{N}\sum_{n=0}^{N-1} r[n]\mathrm{e}^{-j2\pi\frac{l}{N}n} \\ &\overset{(a)}{=} \frac{1}{N}\sum_{n=0}^{N-1}\sum_{i=0}^{N-1} X_i \mathrm{e}^{j2\pi\frac{i}{N}n}\mathrm{e}^{-j2\pi\frac{l}{N}n}(1 + j\phi[n]) \\ &= X_l + j\frac{X_l}{N}\sum_{n=0}^{N-1}\phi[n] + j\sum_{n=0}^{N-1}\sum_{i=0,i\neq l}^{N-1} X_i \mathrm{e}^{j2\pi\frac{i-l}{N}n}\phi[n] \end{aligned}$$

$$\stackrel{(b)}{=} X_l\,(1+\varphi) + N_l^{ICI}, \tag{6.4}$$

where (a) follows from $x[n] = \sum_{i=0}^{N-1} X_i e^{j2\pi \frac{i}{N} n}$ and (6.3) and (b) follows by defining $\varphi := j\sum_{n=0}^{N-1}\phi[n]$ and $N_l^{ICI} := jX_l \sum_{n=0}^{N-1}\sum_{i=1, i\neq l}^{N} X_i e^{j2\pi \frac{i-l}{N} n}\phi[n]$. According to (6.4), the transmitted symbol X_l is multiplied by the term $(1+\varphi)$. This is known as the common phase error (CPE) as it causes identical phase rotations in all subcarriers. The additive noise term N_l^{ICI} refers to the intercarrier interference, which is different for different subcarriers.

The CPE is an identical phase rotation in all subcarriers within an OFDM symbol, therefore it can easily be compensated for using pilot subcarriers (reference signals). (See Section 2.5 for an introduction to different reference signals specified for NR.) For a fast varying channel, the CPE can be compensated as part of a channel estimation process. However, for a slowly varying channel, the CPE needs to be tracked and compensated for more frequently. The 3GPP has introduced the PT-RS in NR mainly for the CPE compensation. In addition, the DM-RS whenever present, can also be used for the CPE compensation. Both the DM-RS and the PT-RS are discussed in Section 2.5.

The ICI is an additive noise (not always Gaussian) and often hard to compensate for, depending on how fast the phase noise variations are. Although the PT-RS can also be utilized for the ICI compensation, it typically requires denser pilot allocation which can degrade the system throughput.[7] An alternative is to chose the waveform numerology (the subcarrier spacing) such that there is sufficient robustness to the ICI. For a typical oscillator, the impact of the ICI decreases as the subcarrier spacing of the signal increases, as discussed in Section 5.5.5. Therefore, an OFDM system can be made robust against the ICI by choosing a larger subcarrier spacing.[8] A too low choice of the subcarrier spacing either puts strict requirements on the local oscillator or phase noise puts an upper limit on the achievable performance under the ICI. In Section 5.5.5, we have quantified the signal-to-interference ratio (SIR) due to the ICI in an OFDM system (cf. (5.24)). Assuming a perfect CPE compensation, the achievable signal-to-interference ratio (SIR) over the demodulated subcarrier l in an OFDM symbol (subject to phase noise) is given by

$$\mathrm{SIR}_l = \frac{1}{\left(\int_{-0.5}^{+0.5} S_\phi(\nu)\, W_l(\nu)\, d\nu\right)}, \tag{6.5}$$

where

$$W_l(\nu) := \sum_{i\in I_{ICI}} \frac{1}{N^2}\left|\frac{\sin\left(\pi\left(\nu - \frac{i-l}{N}\right)N\right)}{\sin\left(\pi\left(\nu - \frac{i-l}{N}\right)\right)}\right|^2. \tag{6.6}$$

In Fig. 6.4 we show the achievable SIRs for different subcarrier spacings based on a PLL based phase-noise model presented in Section 4.2.3. This phase-noise model was developed in mmMAGIC

[7] The density of the PT-RS is configurable in both time and frequency domains.

[8] It is noteworthy that the value of the CPE is constant only within one OFDM symbol. The value of CPE varies from one OFDM symbol to another OFDM symbol and its variance can increase with the subcarrier spacing (in contrast to the ICI). Therefore, an OFDM system becomes more robust against phase noise by employing the larger subcarrier spacings only if the CPE compensation is in place, which is the case for NR.

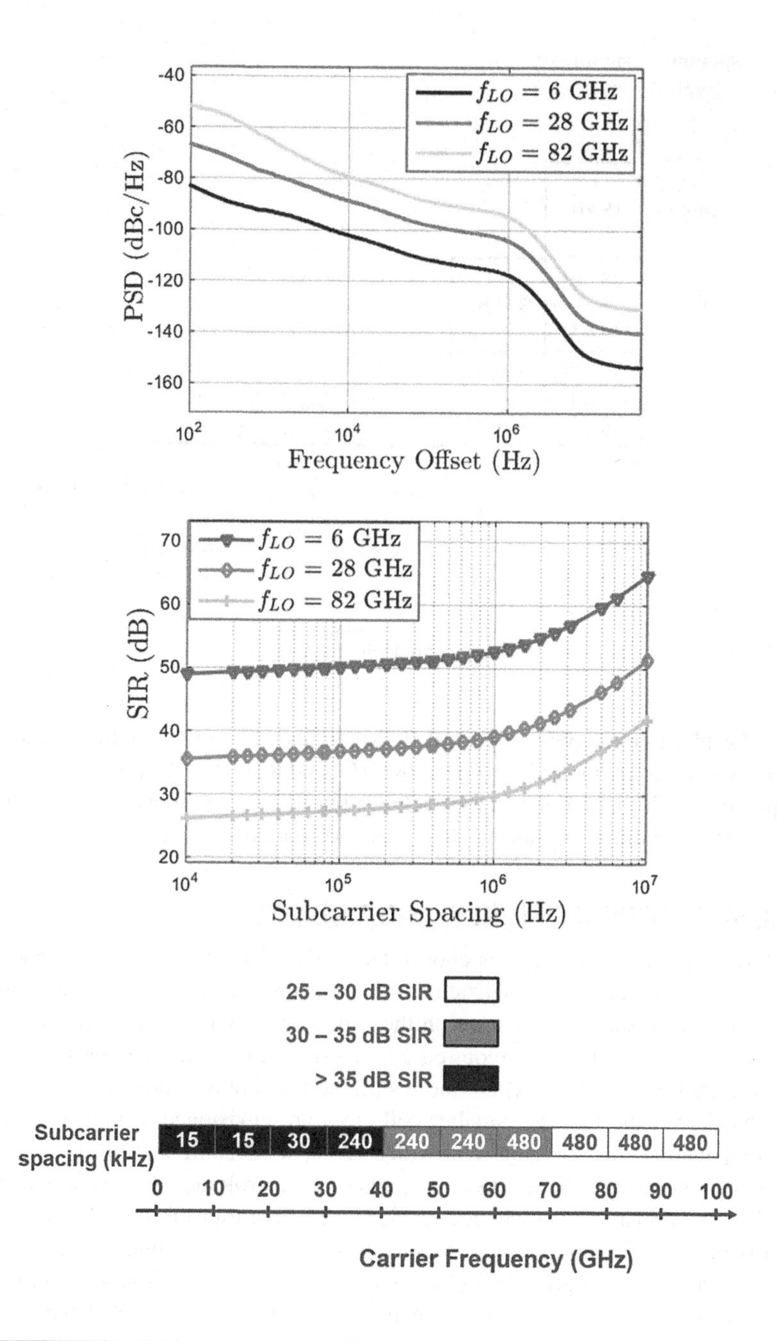

FIGURE 6.4

Achievable Signal-to-Interference Ratio (SIR) in a 100 MHz OFDM system subject to phase-noise for different subcarrier spacings. The CPE is assumed to be perfectly compensated for. (The phase-noise model is developed within mmMAGIC project.)

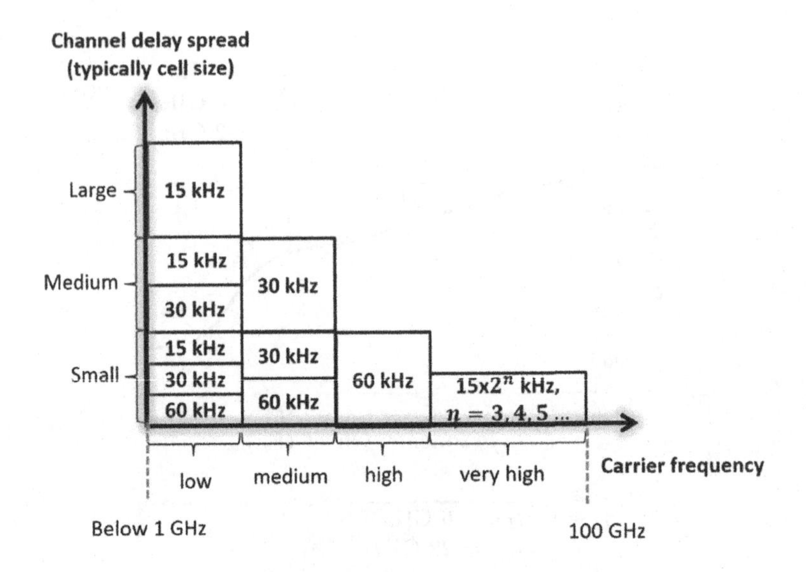

FIGURE 6.5

The NR numerology for wide range of frequencies and deployment types.

project[9] [7]. The phase-noise power spectral densities and the corresponding achievable SIRs are shown for three oscillator frequencies (6 GHz, 28 GHz, and 82 GHz). Towards the bottom of Fig. 6.4, we further specify which subcarrier spacings achieve the SIR values above 25 dB for carrier frequencies up to 100 GHz (assuming the mmMAGIC phase-noise model).

6.3.2 CELL SIZE, SERVICE LATENCY, AND MOBILITY

In an OFDM system, the cyclic prefix is chosen larger than the channel delay spread to avoid intersymbol interference and complex equalization. This means that for a given cyclic-prefix overhead, the channel delay spread sets a lower limit on the subcarrier spacing. Typically, the delay spread reduces with cell size and so does the required cyclic-prefix duration, meaning that wider subcarrier spacings (having shorter cyclic prefix) are more suitable for deployments with smaller cell size. This goes hand-in-hand with the fact that smaller cell sizes are envisioned at higher carrier frequencies due to harsh propagation characteristics and wider subcarrier spacing that make the system robust to phase noise (which increases with oscillator frequency). Numerologies with wider subcarrier spacings are also suitable for supporting low-latency services, since the transmission slot duration (as defined in Table 6.2) is inversely proportional to the subcarrier spacing. This implies that low latencies can be achieved in small cells using a numerology with wider subcarrier spacing. Fig. 6.5 and Fig. 6.6 illustrate the relationships between cell size, carrier frequency, and achievable latency for NR.

[9]mmMAGIC was an EU project (funded by the H2020 program) that developed radio interface concepts and solutions for above 6 GHz mobile radio communications. https://5g-mmmagic.eu/.

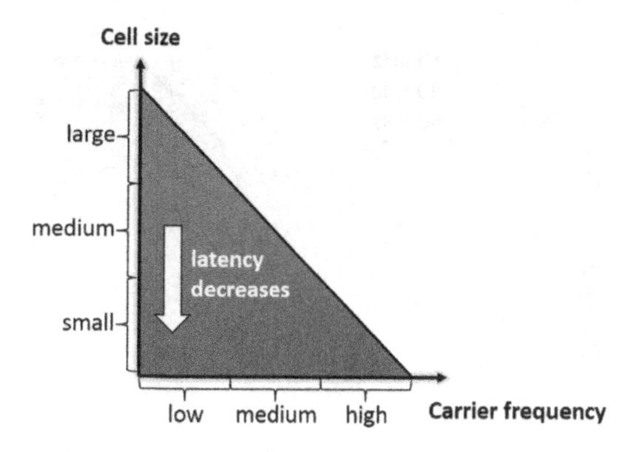

FIGURE 6.6

The relationship of achievable latency, carrier frequency, and deployment type for NR.

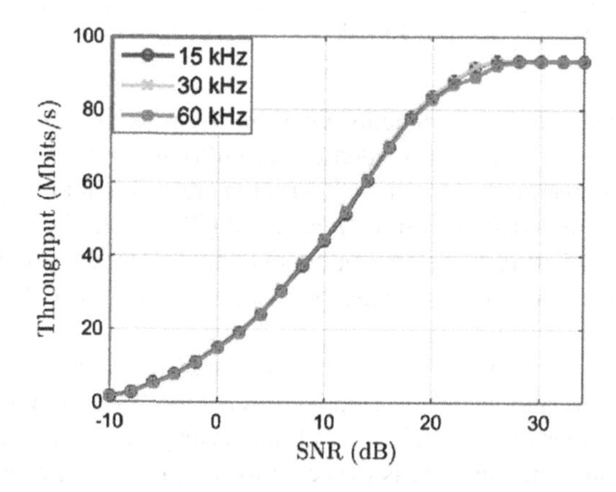

FIGURE 6.7

TDL-A channel model with 300 ns RMS delay spread.

Low latency can also be achieved in larger cells using wider subcarrier spacing, however, this can cost loss in performance either in terms of reliability or throughput. To understand this, we evaluate achievable throughput for three numerologies (15 kHz, 30 kHz, and 60 kHz) assuming the 7% CP overhead (as specified in 3GPP) subject to the Tapped Delay Line-A (TDL-A) channel model [14] with two different delay spread values (300 ns and 1000 ns). The results are shown in Fig. 6.7 and Fig. 6.8, respectively. In these simulations, we have assumed 1 ms TTI (i.e., 14 OFDM symbols for 15 kHz numerology, 28 OFDM symbols for 30 kHz numerology, and 56 OFDM symbols for 60 kHz numerology). Furthermore, we have considered link adaptation, with 64 QAM as the highest modulation order. In Fig. 6.7, we observe that all numerologies perform similarly, meaning that one can reduce the CP duration to four times smaller than what we have in LTE or one can also enable wider subcarriers to support latency critical services. Fig. 6.8 shows that the 60 kHz numerology should not

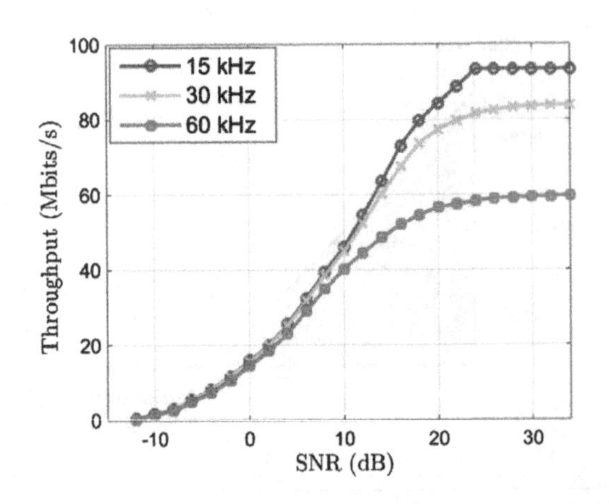

FIGURE 6.8

TDL-A channel model with 1000 ns RMS delay spread.

be used in delay spread intensive environments. One can use 60 kHz subcarrier spacing in the delay spread intensive environments with an extended cyclic prefix to support very low latency machine-type communications. However, extending the CP duration means higher overhead, and hence reduced throughput. The 3GPP has included an option of extended CP with 25% overhead for 60 kHz subcarrier spacing in NR Release 15. It is, however, important to note that the mini-slot based transmission is an alternative method to achieve very low latency without increasing the CP overhead (as discussed in Section 6.2).

In an OFDM system, the relative mobility between a user and an access node causes intercarrier interference (channel time variations or the Doppler effect leads to loss of subcarrier orthogonality). On the one hand, the Doppler effect increases with increasing carrier frequency for a given mobile scenario. On the other hand, the cell size is expected to reduce at higher carrier frequencies for 5G deployments, which implies UEs with low mobility (speeds) relative to the base station (access node). If small cells are realized at high carrier frequencies, mobility may not be a limiting factor in choosing numerology. However, for high-speed scenarios (e.g., V2V) in the millimeter-wave band, the Doppler effect can set a lower limit on the choice of the subcarrier spacings.

6.3.3 MULTIPLEXING SERVICES

Multiple services with different requirements (eMBB, mMTC, URLLC) can be efficiently supported on the same carrier by either using mini-slots or by multiplexing different OFDM numerologies in the frequency domain. For example, the mini-slot can be used to support a URLLC service within the regular slot that supports an eMBB service. Alternatively, a wideband subcarrier numerology and a narrowband subcarrier numerology can be multiplexed to support URLLC and MBB services, respectively. Fig. 6.9 illustrates these two options. Mini-slot is the preferred option due to the following drawbacks of the mixed numerology.

In an OFDM system with two (or more) different numerologies (subcarrier bandwidth and/or cyclic-prefix length) multiplexed in the frequency domain, only subcarriers within a numerology are

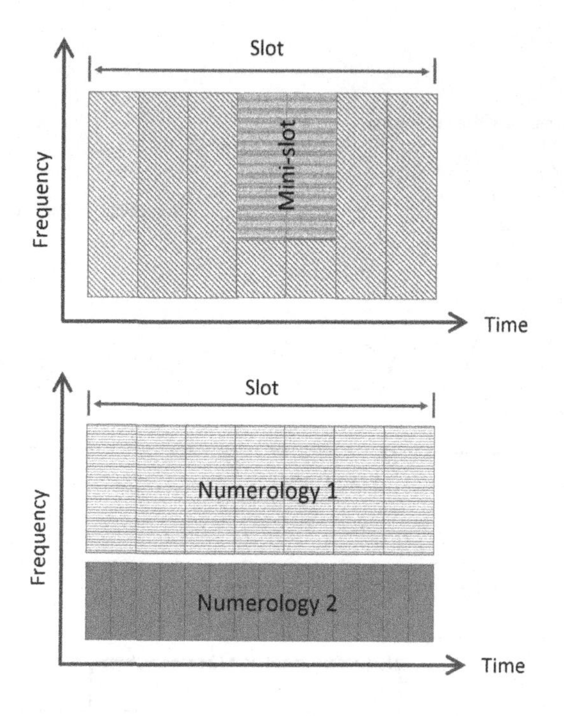

FIGURE 6.9

The mini-slot and the mixed numerology are two alternative tools for multiplexing services with different requirements on the same carrier.

mutually orthogonal. The subcarriers from one numerology interfere with the subcarriers of the other numerology, since energy leaks outside the subcarrier bandwidth and is picked up by subcarrier filters of the other numerology, as illustrated in Fig. 6.10. To reduce the internumerology interference, the transmit spectrum of each numerology must be better confined and a guard band is typically required. Moreover, an extended CP may be required with wider subcarrier spacings, depending on the channel delay spread.

6.3.4 SPECTRAL CONFINEMENT

Spectral confinement is important to reduce out-of-band (OOB) emissions. The 3GPP specifies the OOB emission requirements[10] for both base stations and devices. The better the spectrum roll-off of the signal, the easier it is to fulfill these requirements. The spectrum of the OFDM signal decays rather slowly (see Chapter 5) and the specified OOB emission requirements cannot be fulfilled without inserting guardband[11] and performing additional filtering/windowing operations. The spectrum roll-off of the OFDM signal becomes steeper as its subcarrier spacing reduces. As an example, PSDs of two

[10]The OOB emission requirements are specified in terms of spectrum emission mask and adjacent channel leakage ratio.

[11]LTE occupies 90% of the channel bandwidth. In 3GPP Release 15, the spectrum utilization for NR is above 94%.

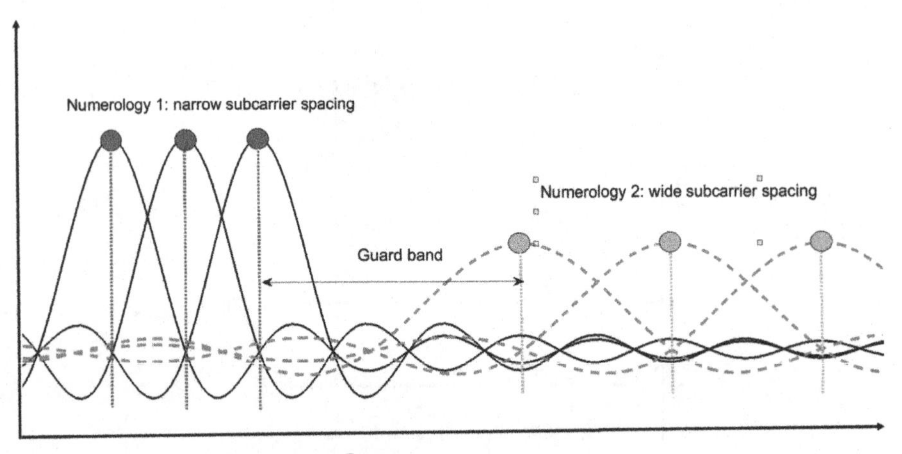

FIGURE 6.10

An illustration of the internumerology interference.

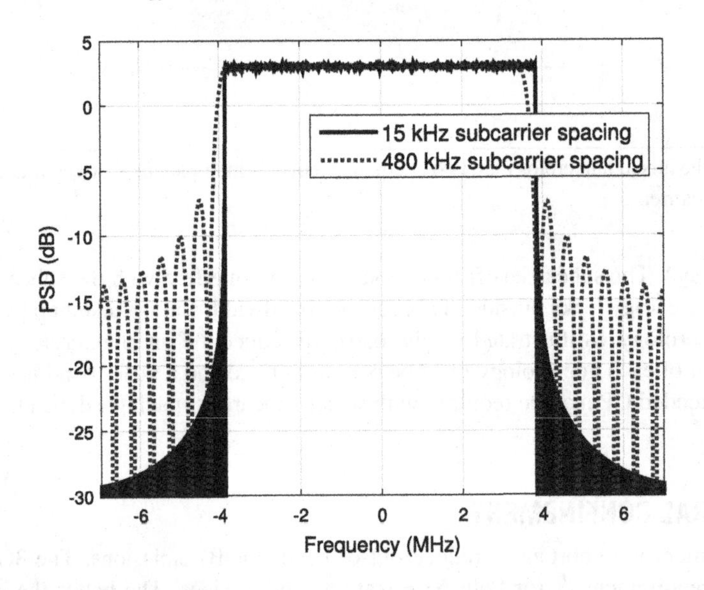

FIGURE 6.11

Comparison of OOB emissions of OFDM signals with 15 kHz and 480 kHz subcarrier spacing.

equal bandwidth OFDM signals with subcarrier spacing of 15 kHz and 480 kHz are shown in Fig. 6.11. This speaks in favor of avoiding too large subcarrier spacings in NR.

Spectral confinement is also important for reducing the internumerology interference. This can be understood with the help of Fig. 6.12 that shows two subbands with different numerologies multiplexed

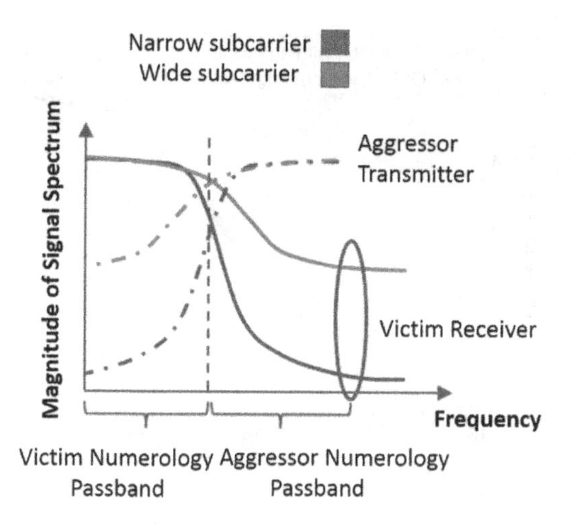

FIGURE 6.12

Both the transmitter and the receiver filtering functionality must be improved to reduce the internumerology interference efficiently.

in the frequency domain. The aggressor numerology (dash-dotted lines) must apply a spectrum emission confinement technique to reduce the energy transmitted in the passband of the victim numerology (blue[12] dash-dotted line). However, the emission control alone is not sufficient, since a victim receiver without steeper roll-off (solid red[13] curve) picks up high interference from the passband of the aggressor numerology. Only if the victim receiver (solid blue curve) and the aggressor transmitter (dashed blue curve) have improved filter functions, the internumerology interference is efficiently reduced.

Windowing and filtering are well-known techniques to confine spectrum of the OFDM signal [1,12]. Windowing has lower implementation complexity than filtering. A low complexity windowing operation can significantly improve spectrum confinement at the expense of consuming a small portion of the CP [20]. In Fig. 6.13, we compare throughput performance of OFDM without mixed numerology and W-OFDM[14] in a mixed numerology scenario where a 15 kHz numerology is multiplexed with a 60 kHz numerology. We assume that the 15 kHz numerology has 6 PRBs[15] and the 60 kHz numerology has 21 PRBs. Fig. 6.13 shows the throughput achieved by the 15 KHz numerology (i.e., considering the 60 kHz numerology as the aggressor and the 15 kHz numerology as the victim). We have used the TDL-A channel model with 1000 ns RMS delay spread (which is larger than what is typically experienced in most scenarios) and up to 64-QAM modulation for link adaptation. The two numerology subbands are separated by a guard band of 120 kHz (i.e., eight 15 kHz subcarriers). This result demonstrates that the throughput performance of W-OFDM with interference from the aggressor is close to CP-OFDM without any interference from the aggressor, even under extreme conditions with very large channel delay

[12]Dark gray in print version.

[13]Light gray in print version.

[14]W-OFDM is introduced in Section 5.1.2.2.

[15]A physical resource block contains 12 subcarriers in NR, like LTE.

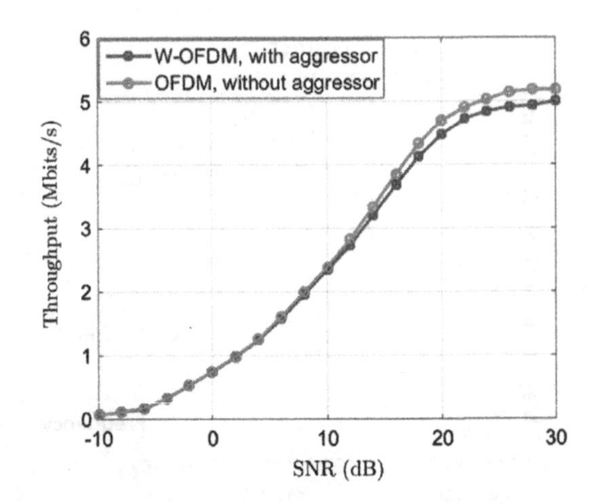

FIGURE 6.13

Comparison of W-OFDM with interference from the aggressor numerology and OFDM without interference from the aggressor numerology on TDL-A channel with 1000 ns RMS delay spread. The victim numerology is narrowband (15 kHz subcarrier spacing with 6 PRBs), the aggressor numerology is wideband (60 kHz subcarrier with 21 PRBs), and the guard band is 120 kHz.

spread, narrowband victim numerology, and wideband aggressor numerology. Hence, windowing on top of CP-OFDM is a low complexity viable technique to multiplex multiple numerologies in NR.

6.3.5 GUARD BAND CONSIDERATIONS

Guard tones can be inserted between numerologies to reduce internumerology interference and/or relax the degree of required spectrum confinement. Adding guard tones slightly increases the overhead. In a 20 MHz system with 1200 subcarriers, one guard tone corresponds to less than 0.1% overhead. It may not be worth the effort to reduce the guard band to an absolute minimum, since it increases the requirements on the spectrum confinement technique (both at transmitter and receiver) and also complicates other system design aspects, as discussed in the following.

Consider Fig. 6.14, where one narrowband subcarrier is inserted as a guard tone between numerology 1 (15 kHz subcarrier spacing) and numerology 2 (60 kHz subcarrier spacing). A physical resource block in NR contains 12 subcarriers for all numerologies. If the scheduling is done as indicated for numerology 2, then the subcarriers of numerology 2 are not on the 60 kHz resource grid (the first subcarrier of the blue (light gray in print version) resource block is on narrow subcarrier 41, which corresponds to the wide subcarrier 10.25, i.e., there is a fractional subcarrier shift).

To avoid such fractional subcarrier shift, subcarrier frequencies in each numerology should coincide with the natural grid of the numerology $n \times \Delta f$, with Δf the subcarrier spacing of the numerology. However, even with this requirement the wide resource blocks (numerology 2) are still not on their natural grid if compared to cell 2; see Fig. 6.15. Such a misaligned resource grid implies that all users of numerology 2 would have to be dynamically informed as regards this offset (since this offset depends on the scheduling decision). In another cell, a different offset may be present or another cell

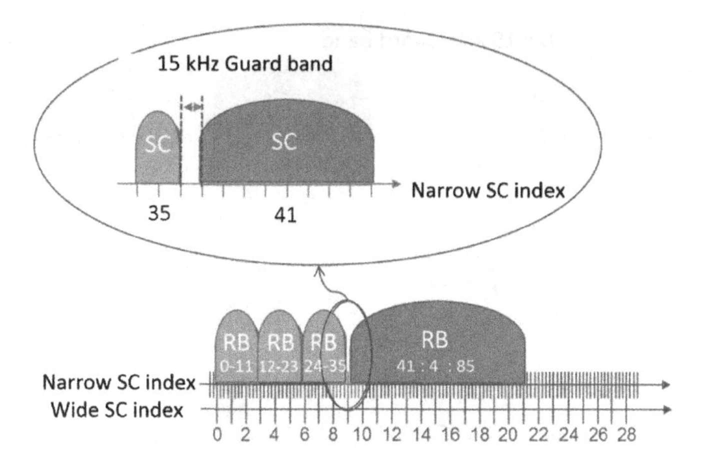

FIGURE 6.14

One narrowband subcarrier (SC) is inserted as a guard between numerology 1 (15 kHz) and numerology 2 (60 kHz). The first subcarrier of numerology 2 is located at 41 × 15 kHz, which corresponds to subcarrier 10.25 in a 60 kHz subcarrier grid.

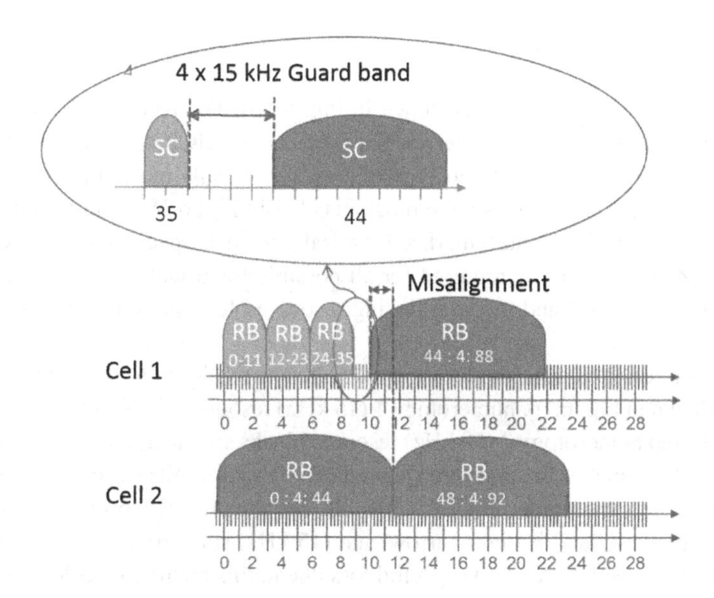

FIGURE 6.15

Four narrowband subcarriers are inserted as a guard between numerology 1 (15 kHz) and numerology 2 (60 kHz). The subcarriers of numerology 2 are located on its natural resource grid. However, numerology 2 resource blocks are still misaligned across cells.

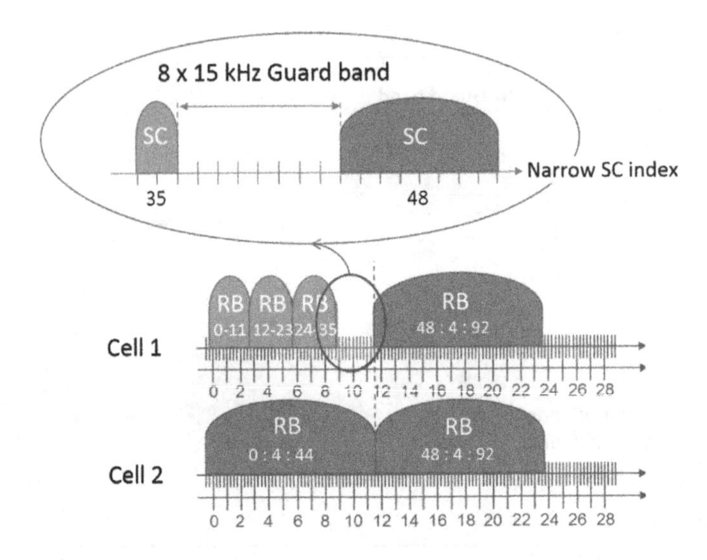

FIGURE 6.16

Eight narrowband subcarriers are inserted as guard between numerology 1 (15 kHz) and numerology 2 (60 kHz). The subcarriers of numerology 2 are located at its natural resource grid and numerology 2 resource blocks are aligned across cells.

may only operate with numerology 2 (as shown in Fig. 6.15). The resource blocks in different cells would not be aligned making intercell interference coordination (ICIC),[16] creation of orthogonal reference signals across cells, and interference prediction across cells more difficult. Alternatively, the first red (dark gray in print version) resource block in cell 1 in Fig. 6.15 could be a fractional resource block (corresponding to the bandwidth marked by misalignment). Special definitions of the reference signals and rate matching would be required for all possible fractional resource blocks. For the fractional resource block in cell 1 and the overlapping resource block in cell 2, the same disadvantages exist.

The cleanest solution is to limit the location of the resource blocks to their natural resource grids. For example, as shown in Fig. 6.16, numerology 1 (15 kHz) resource blocks always start at a frequency of $n \times 12 \times 15$ kHz and numerology 2 (60 kHz) resource blocks start at a frequency of $n \times 12 \times 60$ kHz. This simplifies the ICIC, makes the interference predication across cells easier, and enables orthogonal reference signals of the same numerology across cells. For the (15 kHz, 60 kHz) numerology combination, the resulting guard band is eight narrowband (15 kHz) subcarriers. In a 20 MHz system with around 1200 narrowband subcarriers, the spectral loss due to this guardband is less than 1%.

[16]The ICIC refers to the coordination between neighboring base stations to avoid severe interference situations. The overall system efficiency and user experience can be significantly improved with the ICIC, especially in heterogeneous network deployments with overlapping layers of base stations with large differences in the downlink transmission power.

6.3.6 IMPLEMENTATION ASPECTS

OFDM is implemented in baseband via the fast Fourier transform (FFT) operation (see Fig. 5.7). The FFT/IFFT size should be selected considering the subcarrier spacing, the channel bandwidth, and the affordable computational complexity. The asymptotic complexity of an N-point FFT operation is $\mathcal{O}(N \log N)$. For a given subcarrier spacing, the FFT size determines the maximum channel bandwidth (without carrier aggregation). The maximum channel bandwidth in LTE is 20 MHz, with 2048 FFT size, 30.72 MHz sampling rate, and 1200 is the number of active carriers (100 PRBs). In 3GPP NR Release 15, the maximum number of active subcarriers is 3300 and the largest currently defined channel bandwidth occupies 3276 subcarriers. This can be implemented using an FFT size of 4096. Note that a smaller FFT size may be used if the number of active subcarriers is less.

6.4 IMPROVING POWER EFFICIENCY OF NR WAVEFORM

A major drawback of OFDM waveform is its high PAPR. A high PAPR in the OFDM system causes the power amplifier to operate in a nonlinear region, which contributes to a spectral growth in the form of intermodulation between subcarriers and out-of-band emissions, as described in Section 4.1. Therefore, to keep the nonlinear effects within certain limits, the power back-off is commonly used in the power amplifier, which reduces the coverage. Another possibility is to increase the linear region of the power amplifier, which in turn results in larger amplifiers and hence, higher power consumption. A very common solution is to deploy DPD algorithms to cancel out the nonlinear distortion from the amplifier. This, however, is a high complexity solution, in particular for large antenna arrays.

The power efficiency of the waveform can be very important in the uplink and the sidelink, for coverage limited UEs. A high PAPR (with a high power back-off requirement) mainly affects the highest output transmission power level for the UE. Hence, a low PAPR is more relevant to the coverage-limited UEs. In LTE, the downlink uses OFDM, while the uplink and the sidelink use DFTS-OFDM due to its lower PAPR than OFDM. (See Section 5.2 for an introduction to DFTS-OFDM.) For NR, the 3GPP has specified DFTS-OFDM as an optional[17] waveform for the uplink, however, without spatial multiplexing (MIMO). A MIMO DFTS-OFDM transceiver has a significantly higher implementation complexity than OFDM and worse performance than OFDM. Furthermore, in the coverage limited scenarios spatial multiplexing is often not useful. Typically, UEs at the cell edge use only a small number of resource blocks (i.e., narrow bandwidth allocation) for transmission with low spectral efficiencies (low code rate and low order modulation formats, e.g., QPSK). Due to these reasons, NR supports DFTS-OFDM only for a single stream transmission in the uplink.

When a UE is in good cell coverage and has a wide bandwidth allocation, a high PAPR is often not critical. In this scenario, OFDM is the natural choice. Using OFDM instead of DFTS-OFDM in the uplink (and also in the sidelink) has several advantages:

- MIMO transmission is one of the key features of NR and it is expected that even uplink MIMO will be widely adopted for the enhanced mobile broadband usage (see Chapter 7). With MIMO

[17] In 3GPP NR Release 15, DFTS-OFDM is optional in the uplink and it is up to the base station to select which waveform (OFDM or DFTS-OFDM) should be used in the uplink. This implies that an NR capable UE need to implement both waveforms and a gNB (or base station) can choose to implement only OFDM.

transmission, OFDM is known to provide a significant superior link performance compared to DFTS-OFDM [8].

- DFTS-OFDM requires an additional DFT precoder in the transmitter and an extra IDFT decoder and equalizer in the receiver which increases the baseband complexity, significantly with the MIMO transmissions.
- OFDM is more flexible than DFTS-OFDM in terms of scheduling users' data, control signals, and reference signals, resulting in an enhanced system capacity.
- Having the same transmission scheme in both uplink and downlink makes the whole system design symmetric. A single waveform (i.e., OFDM) for all link types (uplink, downlink, sidelink, backhaul link) simplifies the design by avoiding the need for separate baseband implementations for different waveforms.

Although using OFDM in the uplink comes with major advantages, yet it suffers from the problem of the high PAPR which should be addressed if it is to be employed in all scenarios. There exist several PAPR reduction techniques for OFDM with their pros and cons [10]. While choosing an appropriate PAPR reduction technique, several factors need attention. For example, the PAPR reduction capability, implementation complexity, transparency between the transmitter and the receiver sides, and link performance degradation. Some of the techniques require implementation of multiple FFTs and/or transmission of side information to the receiver, which may be either undesirable or not possible.[18] In the following, we provide a brief overview of some well-known PAPR reduction schemes based on [10]. Broadly speaking, one can divide the PAPR reduction methods into two groups: i) techniques with distortion, and ii) distortion-less techniques.

6.4.1 TECHNIQUES WITH DISTORTION

The following schemes distort the OFDM signal while reducing its PAPR:

- **Clipping** The most simple and widely used technique is to limit the signal envelope to a predefined threshold by means of an amplitude clipper. The amplitude clipping function is given by

$$f(x[n]) = \left\{ \begin{array}{cc} x[n], & |x[n]| \leq A_{max}, \\ A_{max} \exp(j\phi(x[n])), & |x[n]| > A_{max}, \end{array} \right\} \tag{6.7}$$

  where A_{max} is the amplitude threshold and $\phi(x[n])$ is the phase of $x[n]$. Clipping is a nonlinear transformation of the signal. It causes both in-band and out-of-band distortions. The out-of-band distortion can be reduced by performing filtering after clipping. However, the filtering operation may increase the signal amplitude above the clipping threshold. To address this issue, repeated clipping and filtering operations may be employed at the cost of increased implementation complexity. The in-band distortion cannot be improved by filtering and therefore degrades the BER performance and the spectral efficiency.
- **Companding** Companding refers to a nonlinear transformation of a signal in which lower signal amplitude levels are increased and peak signal amplitudes remain unchanged. The average signal

[18] In 3GPP NR Release 15, any operation performed on CP-OFDM at the transmitter side has to be receiver agnostic.

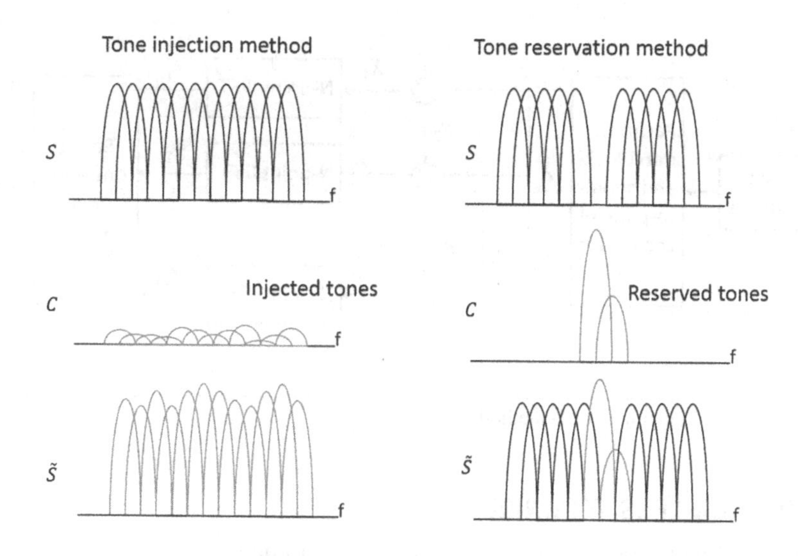

FIGURE 6.17

An illustration of Tone Injection (TI) and Tone Reservation (TR) methods.

power increases after the companding operation, which reduces PAPR. Several companding transformations exist in the literature, for example, exponential companding and polynomial companding are widely used. Companding is an invertible transformation since the companding function is a monotonically increasing function, unlike clipping. This means that a companded signal can be recovered at the receiver by applying an inverse transformation. Both the clipping and the companding methods have low implementation complexity, however, they give rise to relatively high spectral emissions.

- **Tone Injection (TI)** Tone injection refers to superimposing additional subcarriers with optimized amplitude and phases on the information bearing data subcarriers. An illustration is shown in Fig. 6.17 for the injection of the additional tones C into the data subcarriers S to get a PAPR reduced signal $\tilde{S} = S + C$. This is equivalent to expanding the constellation size of the signal, i.e., each point in the original constellation can be mapped in different points in the expanded constellation. As each data symbol can be mapped to one of the several equivalent constellation points, the transmitter can use this degree of freedom to reduce the PAPR. In this case, the receiver needs to know how to map the redundant constellations to the original constellation.
- **Active Constellation Extension (ACE)** ACE is similar to TI, but with the difference that in ACE only outer constellation points are dynamically extended away from the original constellation. Constellation extension gives additional degrees of freedom to reduce the PAPR by optimizing the constellation points. By extending the outer constellation, the spacing between constellation points also increases, which improves BER performance. A drawback of ACE is the increase in transmit power. The usefulness of ACE is restricted to the modulation schemes with large constellation size.

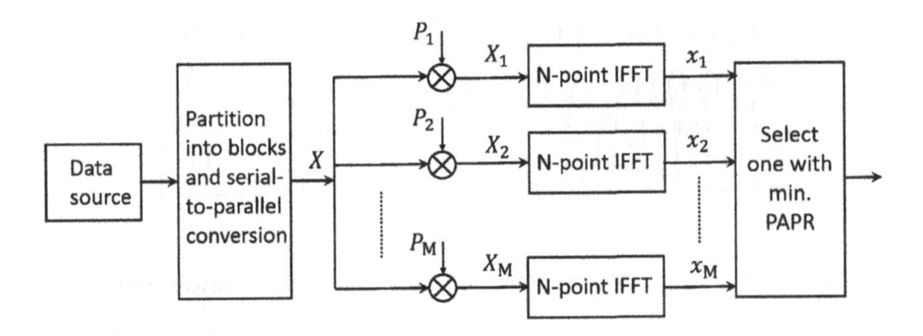

FIGURE 6.18

A schematic of Selective Mapping (SLM) method.

6.4.2 DISTORTION-LESS TECHNIQUES

The following PAPR reduction schemes do not introduce distortion:

- **Tone Reservation (TR)** In this technique, a set of subcarriers within an OFDM symbol are reserved solely for the PAPR reduction. These reserved subcarriers do not carry any data; they are modulated such that the summation of all subcarriers (the data subcarrier and the reserved subcarriers) generate a signal with a low PAPR. In Fig. 6.17, an illustration is given for the superposition of the reserved tones C with the data subcarriers S to get a PAPR reduced signal $\tilde{S} = S + C$. The procedure of finding and optimizing the reserved tones is typically of moderate complexity. The effectiveness of this method depends on the number of reserved tones.
- **Selective Mapping (SLM)** In this method, a number of data blocks carrying same information are generated and the data block that has lowest PAPR is selected for transmission, as shown in Fig. 6.18. Different data blocks are generated by multiplying modulated symbols with different phase vectors before IFFT. The information as regards phase vectors has to be transmitted to the receiver. Therefore, this scheme is not transparent to the receiver and there is a loss of spectral efficiency due to the transmission of side information. The PAPR reduction ability of this method is directly related to the implementation complexity; the larger the set of phase vectors and the number of IFFTs, the higher will be the PAPR reduction capability.
- **Partial Transmit Sequence (PTS)** In this method, the input data block is divided into a set of disjoint subblocks. Each subblock is padded with zeros, IFFT operation is performed on each subblock, which is then multiplied with a phase vector, as shown in Fig. 6.19. The subblocks are then combined such that PAPR of the time-domain OFDM signal is minimized.
- **Suitable Coding** The idea is to select the codewords that reduce the PAPR of the transmitted signal. Peak power is achieved in the OFDM signal when N subcarriers with same phase values are superimposed. The peak power is N times the average power. However, not all codewords result in a high PAPR. The codewords can be selected such that likelihood of occurrence of the same phase for N subcarriers can be minimized. Various coding methods have been shown effective for reducing the PAPR. A drawback of the coding-based PAPR reduction methods is the potential loss in the coding rate (i.e., lower spectral efficiency).

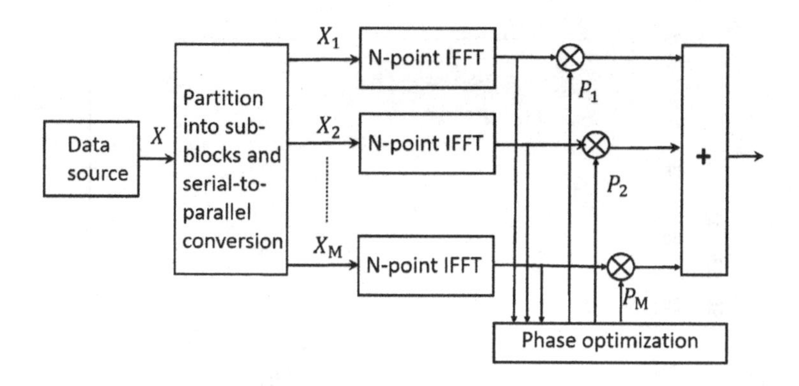

FIGURE 6.19

A schematic of Partial Transmit Sequence (PTS) method.

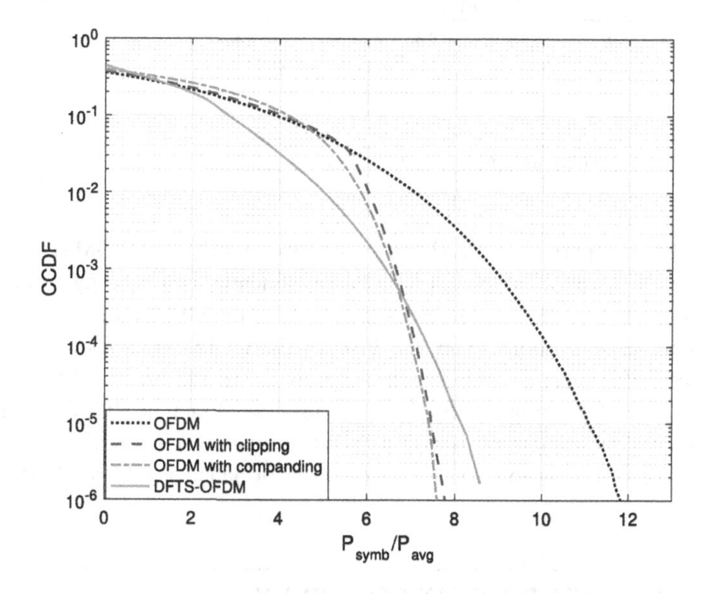

FIGURE 6.20

CCDF of PAPR (P_{symb}/P_{avg}).

Next, we look at the effectiveness of two low complexity PAPR reduction techniques—the clipping and the companding methods. We assume the following OFDM parameters: 2048 FFT size, 60 kHz subcarrier spacing, and 16 QAM modulation. The exponential companding degree is set to 1 and the clipping ratio (i.e., the ratio of the amplitude threshold to the square root of the average power of the signal) of the amplitude clipping scheme is set to 1.7, in order to have similar level of PAPR reduction with the two schemes. Fig. 6.20 shows the CCDF of the ratio of the instantaneous power of an OFDM symbol (P_{symb}) to the average power (P_{avg}) of the OFDM system with and without different PAPR

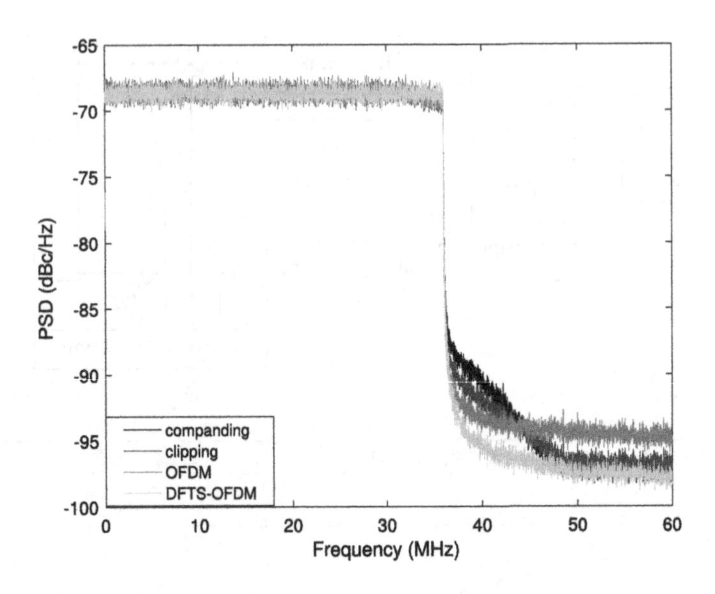

FIGURE 6.21

PSD of OFDM signals with and without different PAPR reduction schemes.

reduction schemes. For comparison, the PAPR performance of a DFTS-OFDM signal is also plotted in the same figure. The effectiveness of the clipping and exponential companding schemes in the OFDM system is clearly demonstrated. Fig. 6.21 shows the PSDs of the OFDM signals with and without the PAPR reduction schemes. For comparison, the PSD of the DFTS-OFDM signal is also included in the figure. We observe that there is some spectral regrowth when these PAPR reduction schemes are employed (which is one of the drawbacks). Nevertheless, we conclude that the high PAPR issue of OFDM can be addressed by implementation of a PAPR mitigation scheme which is not complex from the implementation point of view and achieves similar performance as DFTS-OFDM.

6.5 EFFECTS OF SYNCHRONIZATION ERRORS

In general, an OFDM system is robust to timing synchronization errors due to the presence of the cyclic prefix. However, it is sensitive to frequency synchronization errors due to its overlapping orthogonal narrowband subcarriers that can lose orthogonality when subject to frequency errors. In this section, we will briefly discuss the effects of timing and frequency synchronization errors in an OFDM system.

6.5.1 EFFECT OF TIMING OFFSET

Consider a discrete time OFDM signal with its mth OFDM symbol denoted as $x_m[n] = \sum_{i=0}^{N-1} X_{m,i} \times e^{j2\pi \frac{i}{N}n}$, having N subcarriers (generated by an N point IFFT) and N_{cp} cyclic-prefix samples. To

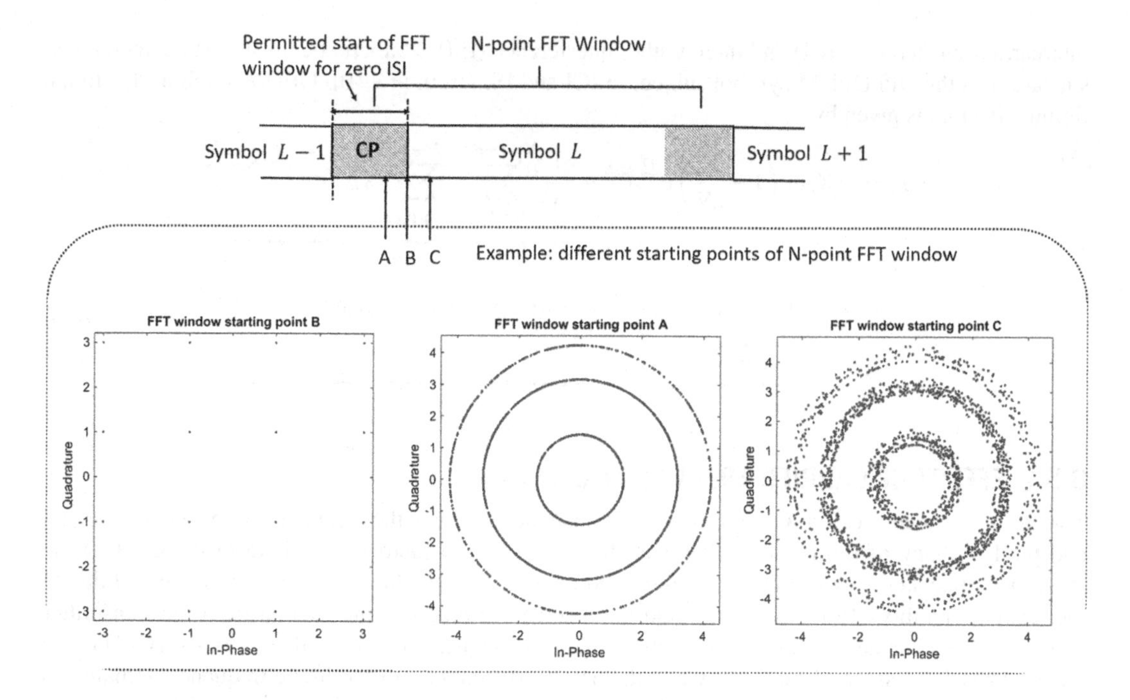

FIGURE 6.22

Effects of timing offset in an OFDM system.

demodulate this signal, an OFDM receiver must place an FFT window (perform an FFT operation) on N consecutive samples of each OFDM symbol. The ideal FFT window starting point is the first sample after the removal of the cyclic prefix, which leads to a perfect reconstruction of the transmitted modulation symbols. This is illustrated in Fig. 6.22 as the starting point B, along with the effects of starting an FFT window within the CP duration (the starting point A) and outside the CP duration, capturing portion of another symbol (the starting point C). For the starting point A (FFT window within the CP duration), the demodulated QAM symbols experience phase rotation (a frequency error). Mathematically, the demodulated subcarriers in this case are given by

$$\hat{X}_{m,k} = X_{m,k} e^{\frac{-j2\pi n_o k}{N}}, \quad k = 1, 2, \ldots, N, \tag{6.8}$$

where n_o is the timing offset. According to (6.8), the kth subcarrier experiences a phase shift $e^{\frac{-j2\pi n_o k}{N}}$. This phase shift (or the timing offset n_o) can be estimated (via pilot subcarriers) and easily compensated for in the frequency domain (post FFT). Hence, a timing offset of $1 \le n_O \le N_{cp}$ is not problematic. In practice, when a signal is received over a multipath channel with maximum delay spread L, a timing offset of $1 \le n_O \le N_{cp} - L$ is tolerable.

The timing offset become problematic when an FFT window captures samples of an adjacent OFDM symbol (i.e., the FFT window starts after the first OFDM sample in a CP removed symbol), illustrated by the starting point C in Fig. 6.22. In this case the demodulated subcarriers suffer from intercarrier interference (ICI) and intersymbol interference (ISI). Mathematically, the kth demodulated subcarrier in the mth OFDM symbol subject to ICI and ISI (from $m+1$th OFDM symbol) due to the timing offset n_o is given by

$$\begin{aligned} R_{m,k} = & X_{m,k}\left(1-\frac{n_o}{N}\right)e^{\frac{j2\pi n_o k}{N}} + \underbrace{\frac{1}{N}\sum_{n=0}^{N-1-n_o}\sum_{\bar{k}=0;\bar{k}\neq k}^{N-1} X_{m,k}e^{\frac{j2\pi(n+n_o)\bar{k}}{N}}}_{\text{ICI}} \\ & + \underbrace{\frac{1}{N}\sum_{n=N-n_o}^{N-1}\sum_{\bar{k}=0}^{N-1} X_{m+1,k}e^{\frac{j2\pi(n-N-N_{cp}+n_o)\bar{k}}{N}}e^{\frac{-j2\pi n_o k}{N}}}_{\text{ISI}}. \end{aligned} \tag{6.9}$$

6.5.2 EFFECT OF CARRIER FREQUENCY OFFSET

The carrier-frequency offset (CFO) refers to the mismatch between the frequency of the received signal and the frequency of the local oscillator at the receiver. Two factors contribute to the CFO: i) the frequency mismatch between the transmitter and the receiver oscillators; ii) the Doppler effect due to the relative mobility of the transmitter and the receiver. In practice, the oscillators at the transmitter and the receiver can never oscillate at identical frequencies; therefore, there always exists a CFO in the received baseband signal. Due to the CFO, the baseband signal is shifted in the frequency domain. As we go higher in carrier frequencies, the CFO is more pronounced due to both the oscillator frequency mismatch and the Doppler effect.

Multicarrier waveforms are more sensitive to the CFO than single-carrier waveform, since a subcarrier bandwidth is typically much smaller than the overall bandwidth in multicarrier waveforms. A small CFO can cause significant degradation in the symbol error rate performance. In an OFDM system, the CFO produces two effects: i) a common phase error (CPE), ii) an intercarrier interference (ICI). The CPE refers to a common phase rotation in all subcarriers and the ICI refers to an interference between the subcarriers due to the loss of subcarrier orthogonality. Let us define the CPE and the ICI, mathematically. Consider a received baseband signal subject to the CFO: $r[n] = x[n]e^{j2\pi\frac{\epsilon}{N}n}$, where ϵ is a normalized fractional CFO. The demodulation of the subcarrier l is then given by

$$\begin{aligned} R_l &= \frac{1}{N}\sum_{n=0}^{N-1} r[n]e^{-j2\pi\frac{l}{N}n} \\ &\overset{(a)}{=} \frac{1}{N}\sum_{n=0}^{N-1}\sum_{i=0}^{N-1} X_i e^{j2\pi\frac{i}{N}n}e^{-j2\pi\frac{l}{N}n}e^{j2\pi\frac{\epsilon}{N}n} \\ &= X_l\frac{1}{N}\sum_{n=0}^{N-1} e^{j2\pi\frac{\epsilon}{N}n} + \sum_{n=0}^{N-1}\sum_{i=0,i\neq l}^{N-1} X_i e^{j2\pi\frac{i-l+\epsilon}{N}n} \\ &\overset{(b)}{=} \underbrace{\alpha X_l e^{j\pi\epsilon\frac{N-1}{N}}}_{\text{CPE}} + \underbrace{\sum_{n=0}^{N-1}\sum_{i=0,i\neq l}^{N-1} X_i e^{j2\pi\frac{i-l+\epsilon}{N}n}}_{\text{ICI}_l}, \end{aligned} \tag{6.10}$$

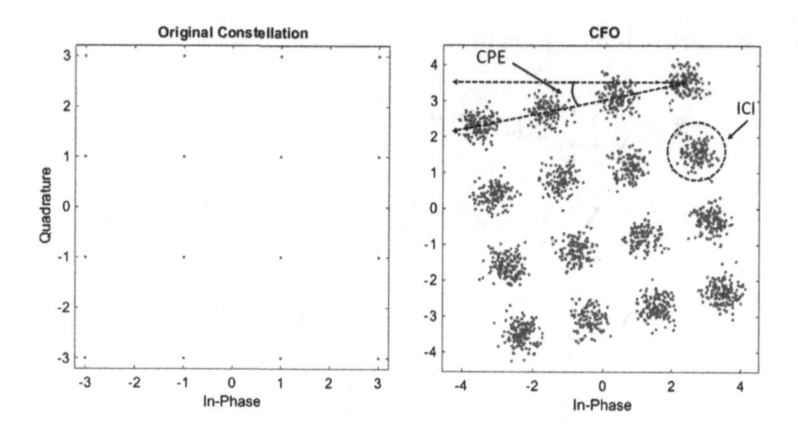

FIGURE 6.23

Effects of carrier-frequency offset in an OFDM system.

where α is an attenuation factor common to all subcarriers, which is given by

$$\alpha = \frac{sinc\,(\epsilon)}{sinc\,(\epsilon/N)}. \tag{6.11}$$

As an example, the two effects produced by CFO in an OFDM system (CPE and ICI) are illustrated in Fig. 6.23 for a 16-QAM constellation. CPE causes an identical phase rotation in all subcarriers within an OFDM symbol and ICI acts as an additive noise in the demodulated subcarrier. CPE can easily be compensated for as part of the channel equalization process. Assuming perfect CPE compensation, the achievable signal-to-interference ratio (SIR) due to ICI in the demodulated subcarrier l in the OFDM symbol subject to CFO is given by

$$\mathrm{SIR}_l = \frac{\alpha^2 E\left[|X_l|^2\right]}{E\left[|\mathrm{ICI}_l|^2\right]}, \tag{6.12}$$

where $E[\cdot]$ is the expectation operator.

6.5.3 SAMPLING FREQUENCY OFFSET

The sampling frequency offset (SFO) refers to a mismatch between the oscillator of the transmitter and the oscillator of the receiver. In practice, the sampling clocks of the transmitter and the receiver are offset by a few parts per million (ppms), which can cause major degradation in an OFDM system if not properly compensated for. In the following, we briefly discuss the effect of the SFO in an OFDM system.

Consider an OFDM transmitter with the sampling time T_T and an associated OFDM receiver with the sampling time T_R (i.e., assuming that the sampling rates are not identical). Then the received baseband OFDM signal can be expressed as $\sum_{i=0}^{N-1} X_i e^{j2\pi i n \frac{T_R}{T_T}}$. Let us define $\eta := \frac{T_R - T_T}{T_T}$. The demodulation of a subcarrier l is then given by

$$
\begin{aligned}
R_l &= \frac{1}{N}\sum_{n=0}^{N-1} r[n]e^{-j2\pi \frac{l}{N}n} \\
&= \frac{1}{N}\sum_{n=0}^{N-1}\sum_{i=0}^{N-1} X_i e^{j2\pi i n \frac{T_R}{T_T}} e^{-j2\pi l \frac{n}{N}} \\
&= X_l \sum_{n=0}^{N-1} e^{-j2\pi l n \frac{(T_R - T_T)}{T_T N}} + \sum_{n=0}^{N-1}\sum_{i=0, i\neq l}^{N-1} X_i e^{j2\pi n \frac{iT_R - lT_T}{T_T N}} \\
&= \underbrace{\alpha_l X_l \sum_{n=0}^{N-1} e^{j\pi l \eta \frac{N-1}{N}}}_{\text{linear phase drift}} + \underbrace{\sum_{n=0}^{N-1}\sum_{i=0, i\neq l}^{N-1} X_i e^{j2\pi \frac{i-l+\epsilon}{N} n}}_{\text{ICI}},
\end{aligned}
\tag{6.13}
$$

where α_l is given by

$$
\alpha_l = \frac{sinc\,(l\eta)}{sinc\,(l\eta/N)}. \tag{6.14}
$$

From (6.13), we observe that the SFO induces a linear phase drift across the subcarriers in contrast to CFO that produces a common phase rotation (CPE) in all subcarriers. Moreover, different subcarriers are also attenuated differently when subject to the SFO, unlike the CFO. Typically, the ICI term due to the SFO is very small and can be neglected.

6.6 IMPAIRMENT MITIGATION

We have seen that an OFDM system is sensitive to the phase-noise, the carrier-frequency offset, and the sampling frequency offset. NR is expected to be deployed at frequencies as high as 100 GHz, where these impairments would be a major bottleneck. In this section, we discuss mitigation schemes for these impairments assuming a slow time varying frequency-selective channel. In a fast varying channel, these impairments are typically corrected as part of the channel estimation and equalization process.

6.6.1 A PHASE-NOISE MITIGATION SCHEME

In order to focus on the PN impairment, we assume a perfect time-synchronization and a quasi-static multipath channel (with a channel length of L samples). The PN impaired signal at the receiver at the nth time sample can be expressed as

$$
y[n] = \exp(j\phi[n]) \sum_{l=0}^{L-1} h_l x[n-l] + w[n], \tag{6.15}
$$

where $x[n]$ represents the transmitted OFDM signal including the CP, h_l ($l = 0, \cdots, L$-1) is the lth tap of the channel impulse response (CIR), $w[n]$ denotes the additive white Gaussian noise (AWGN), and $\phi[n]$ denotes the PN.

Let $\mathbf{y} = \left[y[0]\ y[1]\ \cdots\ y[N-1]\right]^T$ be the received time-domain OFDM symbol (after CP removal), where N is the number of subcarriers. The PN is mitigated by the operation $\mathbf{\Phi y}$, where

$$\mathbf{\Phi} = \text{diag}\left\{\left[\exp\left(-j\hat{\phi}[0]\right)\ \cdots\ \exp\left(-j\hat{\phi}[N-1]\right)\right]^T\right\}, \tag{6.16}$$

with $\hat{\phi}$ denoting the estimate of ϕ. Hence, the task of the PN mitigation scheme is to essentially estimate the PN samples.

Let $\mathbf{H}_\text{p}$ be an $N_\text{p} \times N_\text{p}$ diagonal matrix consisting of the channel transfer functions (CTFs) at the N_p pilot subcarriers ($N_\text{p} \leq N$), $\mathbf{s}_\text{p}$ be a vector consisting of the N_p subcarriers, $\mathbf{D}$ be an $N_\text{p} \times N$ submatrix of the $N \times N$ discrete Fourier transform (DFT) matrix $\mathbf{F}$ (whose elements are given by $\exp(-j2\pi nk/N)/\sqrt{N}$ with $k, n \in \{0, \cdots, N-1\}$) corresponding to the N_p pilot subcarriers, and $\tilde{\mathbf{w}}$ be an $N \times 1$ vector consisting of the time-domain AWGNs. We have

$$\mathbf{D\Phi y} = \mathbf{H}_\text{p}\mathbf{s}_\text{p} + \mathbf{D\Phi}\tilde{\mathbf{w}}. \tag{6.17}$$

Let $\mathbf{Y} = \text{diag}(\mathbf{y})$ and $\mathbf{T}$ be an $N \times q$ transformation matrix, such that $\mathbf{\Phi} \approx \text{diag}(\mathbf{T}\boldsymbol{\alpha})$, where $\boldsymbol{\alpha}$ consists of q unknowns or anchors ($q \leq N_\text{p}$), and it is given by

$$\boldsymbol{\alpha} = (\mathbf{DYT})^\dagger\, \mathbf{H}_\text{p}\mathbf{s}_\text{p} + (\mathbf{DYT})^\dagger\, \tilde{\mathbf{w}}. \tag{6.18}$$

The first term in the right hand side (RHS) of (6.18) is the least-square (LS) estimator of $\boldsymbol{\alpha}$, whereas the second term in the RHS of (6.18) is additive noise.

The physical meaning of $\boldsymbol{\alpha}$ depends on the type of the transformation matrix. If $\mathbf{T}$ is a linear interpolation matrix [5], the elements in $\boldsymbol{\alpha}$ are estimates of the inverse carrier PN $\exp(-j\phi)$ at the q anchors (time samples). These anchors are usually evenly distributed in the time-domain OFDM symbol.

Typically, the phase-noise power is concentrated in low frequencies. As an example, the power spectral density (PSD) of mmMAGIC phase-noise model at 82 GHz is shown in Fig. 6.24 [7]. As phase noise is dominant in lower frequencies, $\mathbf{T}$ can be a $N \times q$ submatrix of $\mathbf{F}$ corresponding to the q lowest spectral components. In this case, the elements in $\boldsymbol{\alpha}$ are the spectral components of $\exp(-j\phi)$.

The computational complexity of the PN mitigation scheme mainly depends on the pseudoinverse of an $N_\text{p} \times q$ matrix in (6.18), whose complexity increases linearly with N_p, yet cubically with q [2].

For convenience of the analysis, we study the performance of the PN mitigation/estimation in the preamble (where $N_\text{p} = N$). (The findings hold for the payload as well, where $N_\text{p} < N$.)

The LS estimator of the PN, i.e., the first term in the RHS of (6.18), contains the modeling error $\mathbf{\Phi} - \text{diag}(\mathbf{T}\boldsymbol{\alpha})$. Assuming perfect estimation of $\boldsymbol{\alpha}$ (by setting the second term in the RHS of (6.18), i.e., the additive noise, to zero), it is self-evident that the modeling error reduces to zero as q increases to N.

Now we examine the effect of the additive noise in (6.18) w.r.t. q. Let $\breve{\mathbf{w}} = \mathbf{F\Phi}\tilde{\mathbf{w}}$. Since $\mathbf{F\Phi}$ is a unitary matrix, $\breve{\mathbf{w}}$ and $\tilde{\mathbf{w}}$ have the same statistics. Thus, the additive noise in (6.18) can be equivalently written as

$$\mathbf{z} = (\mathbf{FYT})^\dagger\, \breve{\mathbf{w}} = (\mathbf{T})^\dagger\, (\mathbf{Y})^{-1}\, (\mathbf{F})^H\, \breve{\mathbf{w}}. \tag{6.19}$$

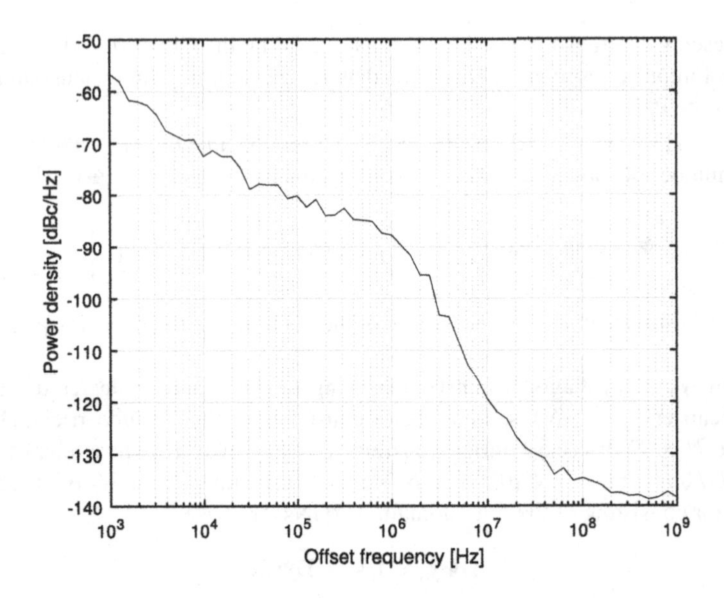

FIGURE 6.24

PSD of phase-noise.

For simplicity, we assume that **T** is the DFT transformation matrix. The power of the additive noise is given as

$$
\begin{aligned}
\mathrm{E}\left[\mathbf{z}^H\mathbf{z}\right] &= \sigma_w^2 \mathrm{E}\left\{\mathrm{Tr}\left[(\mathbf{T})^\dagger (\mathbf{Y})^{-1}\left((\mathbf{Y})^{-1}\right)^* \left((\mathbf{T})^\dagger\right)^H\right]\right\} \\
&= \sigma_w^2 \mathrm{E}\left\{\mathrm{Tr}\left[\left(\mathbf{T}\mathbf{T}^H\right)^\dagger (\mathbf{Y})^{-1}\left((\mathbf{Y})^{-1}\right)^*\right]\right\} \\
&\geq \frac{q}{N}\frac{\sigma_w^2}{\sigma_h^2\sigma_s^2+\sigma_w^2} = \frac{q}{N}\frac{1}{\gamma_0+1},
\end{aligned}
\tag{6.20}
$$

where the last inequality follows from the Jensen's inequality and by denoting σ_h^2, σ_s^2, and σ_w^2 as variances of the CIR, subcarrier symbol, and AWGN, respectively; and $\gamma_0 = \sigma_h^2\sigma_s^2/\sigma_w^2$ is the signal-to-noise ratio (SNR). As can be seen, increasing q will increase the power of the additive noise in (6.18) for fixed N and SNR. This is because there are more anchors to the estimate as q increases. All in all, for the PN estimation, increasing q reduces the modeling error, yet increases the additive noises. As mentioned before, the computational complexity of the PN estimation increases cubically with increasing q. Therefore, q should not be unnecessarily large. For example, it has been shown that $q = 7$ is sufficient for mitigating the PN [4].

For simulations, we assume a DFT size of 512, including 32 scattered pilots. The remaining active subcarriers are loaded with 16- or 64-QAM. The CP length is set to be larger than the channel length. The sampling frequency is set so that the subcarrier spacing is 240 or 480 kHz. The carrier frequency

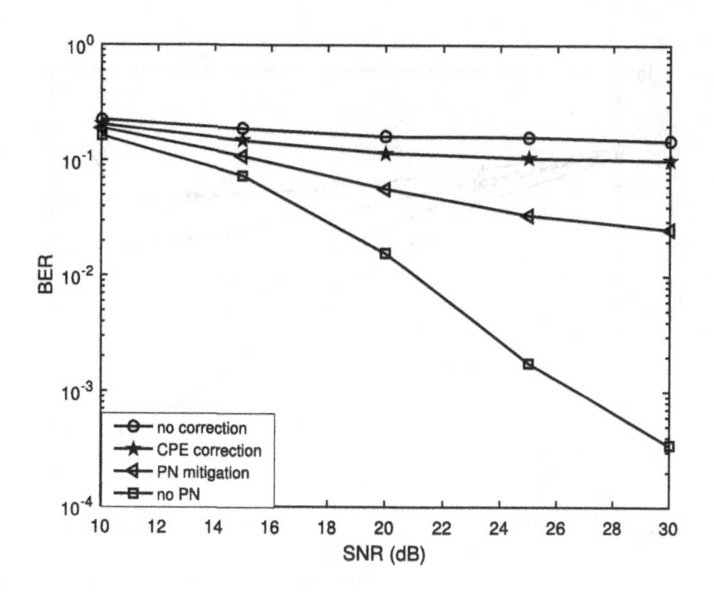

FIGURE 6.25

SER performance with 240 kHz subcarrier spacing and 64-QAM.

is set to fc = 82 GHz. The QuaDRiGa channel model [9] is used for channel emulation. Both the CPE correction [17] and the PN mitigation scheme with $q = 7$ are evaluated. Figs. 6.25–6.27 show the BER performances with/without phase-noise compensations with 16- and 64-QAM, and 240 and 480 kHz subcarrier spacing. It can be seen that the phase-noise mitigation scheme clearly outperforms the CPE correction (at the expense of increased complexity); and the corresponding BER performance improves as the subcarrier spacing increases from 240 to 480 kHz and as the modulation order decreases from 64-QAM to 16-QAM.

6.6.2 CFO AND SFO MITIGATION

In this section, we discuss estimation (and correction) of the CFO and the SFO. A scheme for mitigation of the SFO and the CFO (jointly with PN) is presented. This is an extension of the PN mitigation scheme presented in Section 6.6.1.

There exists an extensive literature on mitigation of the synchronization errors in OFDM systems. Thanks to the seminal work on the CFO estimation [16], [15], the (coarse) frequency synchronization can be readily achieved and our focus will be on the residual carrier-frequency offset estimation. The SFO mitigation is studied in [3], [19]. A pilot-based SFO estimator is proposed in [3], which estimates the SFO using scattered pilots in each OFDM symbol. The SFO can also be tracked using a delay-locked loop (DLL) [18]. A blind SFO estimator is proposed in [19], which estimates the SFO using current and previously received OFDM symbols. In general, the pilot-based SFO estimator outperforms the blind estimator and because the scattered pilots are needed for the PN estimation, the pilot-based SFO estimator is more popular in practical systems. To be robust to perturbations from an imperfect

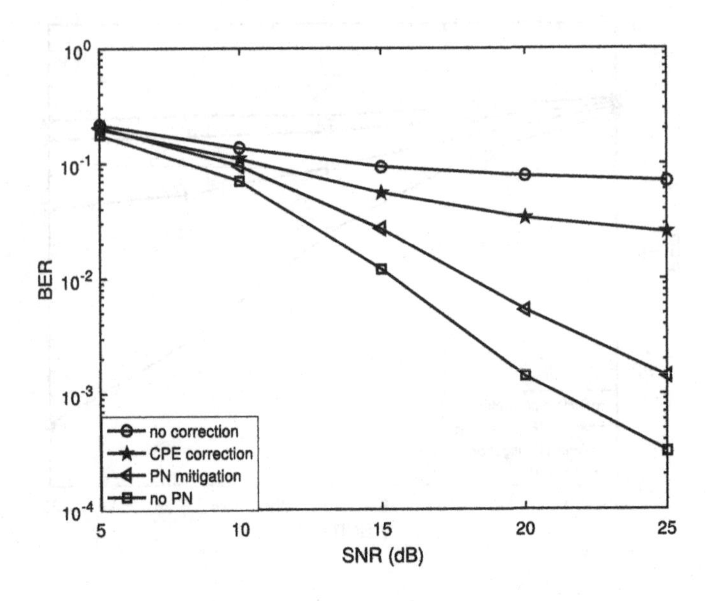

FIGURE 6.26

SER performance with 480 kHz subcarrier spacing and 64-QAM.

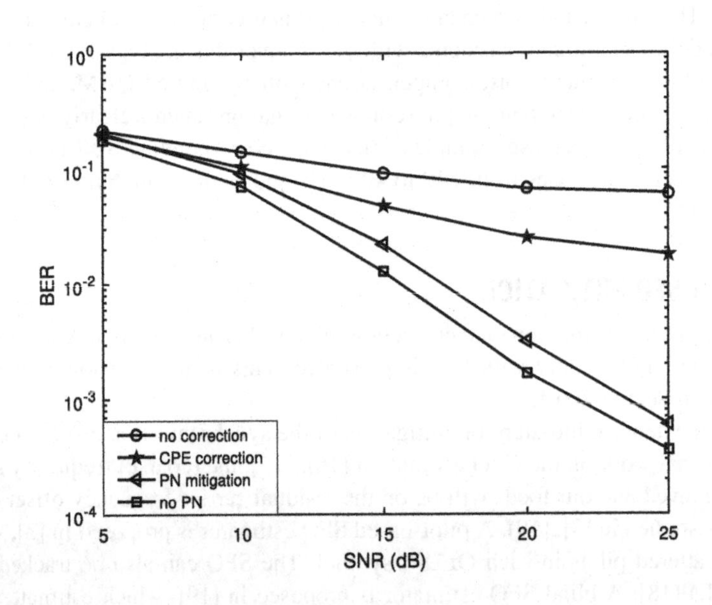

FIGURE 6.27

SER performance with 480 kHz subcarrier spacing and 16-QAM.

PN mitigation, ICI, and AWGN, the pilot-based SFO estimator can be combined with the Kalman filter to track the SFO. Unlike the DLL, the Kalman filter can also be used for channel prediction, which is necessary for the PN mitigation (prior to the SFO correction).

In the following, we present a mitigation scheme for the CFO and the SFO, assuming that there is a single clock and a single oscillator in the transmitter/receiver. For the case of multiple oscillators per transceiver, the interested reader may refer to [13], [6]. For simplicity and without loss of generality, we assume that coarse frequency synchronization has been achieved using the classical CFO estimation ([16,15]) with a residual CFO (RCFO) of ε. The residual CFO (RCFO) is normalized to the sampling frequency. Let T_{T} and T_{R} be the sampling intervals at the transmitter and at the receiver, respectively. The SFO can then be expressed as

$$\eta = \frac{T_{\mathrm{R}} - T_{\mathrm{T}}}{T_{\mathrm{T}}}. \tag{6.21}$$

The SFO is introduced to the transmit signal by sequentially sending the OFDM symbols to a buffer (which is longer than the duration of one OFDM symbol with CP), and sampling the buffered signal using the windowed sinc interpolation function with a sampling interval of $T_{\mathrm{T}}(1-\eta)$.

Consider the PN mitigation scheme presented in Section 6.6.1 with $\hat{\phi}[n]$ now denoting the estimate of the PN plus the RCFO at nth time sample,

$$\hat{\phi}[n] \approx \phi[n] + 2\pi\varepsilon n. \tag{6.22}$$

Once $\hat{\phi}[n]$ is estimated, the RCFO can be approximated by

$$\hat{\varepsilon} \approx \frac{1}{2\pi N}\sum_{n=0}^{N-2}\left(\hat{\phi}[n+1] - \hat{\phi}[n]\right). \tag{6.23}$$

The above expression assumes that the time derivative of the PN has zero mean, which holds for practical oscillators. After the PN mitigation, the RCFO becomes $\tilde{\varepsilon} \approx \varepsilon - \hat{\varepsilon}$.

The SFO causes a phase rotation (which varies linearly across subcarriers) and ICI. SFO is typically just a few ppm in practical systems; thus, the resulting ICI is negligible. For notational convenience, we drop the ICI term caused by the SFO in the following expressions. The received signal of the mth OFDM symbol at the kth subcarrier is given by

$$r_m[k] = J_m[0]\exp\left(j\frac{k\psi_m}{N}\right)H_m[k]s_m[k] + \sum_{l=0,l\neq k}^{N-1} J_m[k-l]\exp\left(j\frac{l\psi_m}{N}\right)H_m[l]s_m[l] + w_m[k], \tag{6.24}$$

where $s_m[k]$ and $r_m[k]$ are transmitted and received signals, $J_m[k]$ is the spectral component of the phase noise together with the RCFO (after PN mitigation), ψ_m is the rotation angle caused by the SFO together with the RCFO, $H_m[k]$ is the channel transfer function, and $w_m[k]$ is the AWGN. We have

$$J_m[k] = \frac{1}{N}\sum_{n=0}^{N-1}\exp\left(j\left(\tilde{\phi}_m[n] + 2\pi\tilde{\varepsilon}n\right)\right)\exp\left(-\frac{2\pi kn}{N}\right), \tag{6.25}$$

$$\psi_m = 2\pi m(N + N_g)(N\tilde{\varepsilon}/k + \eta), \tag{6.26}$$

where $\tilde{\phi}_m[n]$ is the residual PN after the PN mitigation at the nth time sample of the mth OFDM symbol, and N_g denotes the CP length of the OFDM symbol. As can be seen, μ_m varies linearly over the OFDM symbol time. After an effective PN mitigation, $J_m(0)$ is close to one and the ICI term (i.e., the second term in (6.24)) is close to zero. Let $\tilde{H}_m[k] = J(0)H_m[k]$ and $\tilde{w}_m[k]$ be the sum of the ICI and AWGN terms, Eq. (6.24) can then be rewritten as

$$r_m[k] = \exp\left(j\frac{k\psi_m}{N}\right)\tilde{H}_m[k]s_m[k] + \tilde{w}_m[k]. \tag{6.27}$$

The (frequency-domain) phase rotation ψ_m must be corrected prior to the data detection. To be robust to perturbations of the residual PN and the AWGN, we apply the Kalman filter to track ψ_m. Let δ_m be the time derivative of ψ_m. The state-space model is given as

$$\mathbf{s}_{m+1} = \mathbf{M}\mathbf{s}_m + \mathbf{n}, \tag{6.28}$$

where $\mathbf{s}_m = [\psi_m\ \delta_m]^T$, $\mathbf{n}$ denotes the 2×1 state model noise vector, and

$$\mathbf{M} = \begin{bmatrix} 1 & 1 \\ 0 & 1 \end{bmatrix}.$$

The covariance matrix of $\mathbf{n}$ is denoted $\mathbf{Q} = E[\mathbf{n}\mathbf{n}^H]$, where E is the expectation operator. The observation model is

$$\mu_m = \psi_m + \zeta, \tag{6.29}$$

where ζ denotes the observation noise with a variance of $\kappa = E[\zeta\zeta^*]$. By denoting the observation vector as $\mathbf{o} = [1\ 0]^T$; we can write

$$\mu_m = \mathbf{o}^T\mathbf{s}_m + \zeta. \tag{6.30}$$

The Kalman filter algorithm is now given by:

Initialize:

$$\begin{aligned} \mathbf{s}_{0|0} &= \begin{bmatrix} 0 & 0 \end{bmatrix}^T \\ \mathbf{P}_{0|0} &= \mathbf{I}_2. \end{aligned} \tag{6.31}$$

Iterate:

predict:

$$\begin{aligned} \mathbf{s}_{m|m-1} &= \mathbf{M}\mathbf{s}_{m-1|m-1} \\ \mathbf{P}_{m|m-1} &= \mathbf{M}\mathbf{P}_{m-1|m-1}\mathbf{M}^T + \mathbf{Q} \end{aligned} \tag{6.32}$$

update:

$$\begin{aligned}
\alpha_m &= \mu_m - \mathbf{o}^T \mathbf{s}_{m|m-1} \\
\mathbf{k}_m &= \frac{\mathbf{P}_{m|m-1}\mathbf{o}}{\mathbf{o}^T \mathbf{P}_{m|m-1}\mathbf{o}} \\
\mathbf{s}_{m|m} &= \mathbf{s}_{m|m-1} + \mathbf{k}_m \alpha_m \\
\mathbf{P}_{m|m} &= (\mathbf{I}_2 - \alpha_m \mathbf{k}_m)\,\mathbf{P}_{m|m-1}.
\end{aligned} \tag{6.33}$$

At the mth OFDM symbol time, the phase rotation has to be estimated based on the scattered pilots in the mth OFDM symbol. Let k_p $(p = 1, \ldots, N_\text{p})$ be the indices of the N_p pilots in the mth OFDM symbol, and denote

$$\vartheta_m[k] = \angle\left(\frac{\left(\tilde{H}_m[k]s_m[k]\right)^* r_m[k]}{\left(\tilde{H}_m[k]s_m[k]\right)^* \left(\tilde{H}_m[k]s_m[k]\right)}\right), \tag{6.34}$$

where $\angle(\cdot)$ denotes the angle of its argument, and the superscript * denotes the complex conjugate operation. The phase rotation can be estimated as

$$\mu_m = \frac{N}{N_\text{p} - 1} \sum_{p=1}^{N_\text{p}-1} \frac{\vartheta_m[k_p] - \vartheta_m[k_{p+1}]}{k_p - k_{p+1}}. \tag{6.35}$$

When $|\mu_m|$ is larger than 2π, the synchronization module shifts the signal by one sample accordingly; otherwise, the SFO is corrected by $\exp(-jk\mu_m/N)\,\mathbf{y}_m[k]$. Note that the SFO correction also compensates for the RCFO $\tilde{\varepsilon}$. The PN mitigation and the SFO correction are applied to both the training signal and the payload. In the payload, the estimated channel is used together with the Kalman filter for the channel prediction. Specifically, we predict the phase rotation for the next OFDM symbol:

$$\mu_{m+1}[k] = \mathbf{o}^T \mathbf{M}\mathbf{s}_{m|m}. \tag{6.36}$$

The channel prediction for the next OFDM symbol is

$$\hat{H}_{m+1}[k] = \exp\left(\frac{jk\mu_{m+1}}{N}\right)\hat{H}_m[k], \tag{6.37}$$

which is used for the PN mitigation in the next OFDM symbol. We assume that the OFDM payload has 512 subcarriers, including 354 active subcarriers, 32 pilot subcarriers, and 158 guard band subcarriers (79 null subcarriers in the beginning and 79 at the end of each OFDM symbol). The guard band helps to reduce the out-of-band emission and eliminates the distortions in the interpolation and the decimation process. The CP length is set to be larger than the channel length. We further assume that the system is operating at 60 GHz with 256-QAM and sampling frequency 100 MHz. For tracking the frequency-domain phase rotation (due to the SFO and the RCFO) using the Kalman filter, we assume the covariance matrix of the state model noises to be $\mathbf{Q} = 10^{-15}\mathbf{M}\mathbf{M}^T$ and the variance of

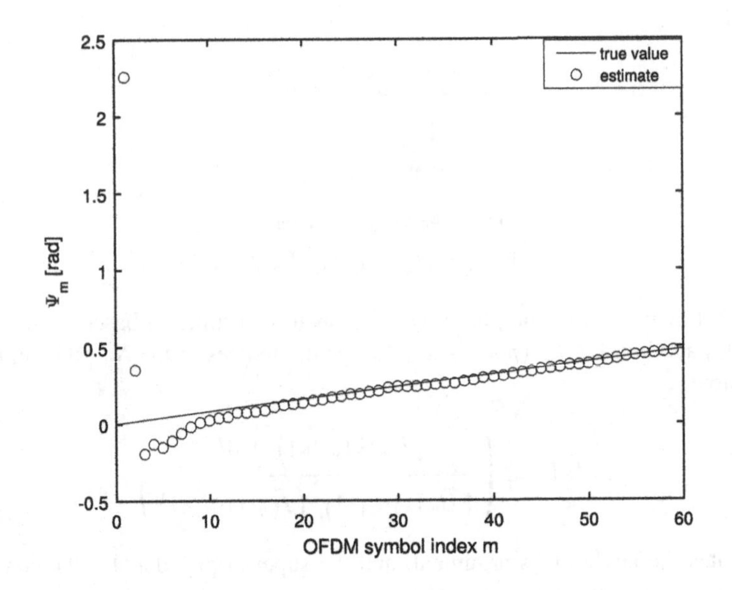

FIGURE 6.28

Example of the SFO phase rotation and its estimate as a function of the OFDM symbol index m.

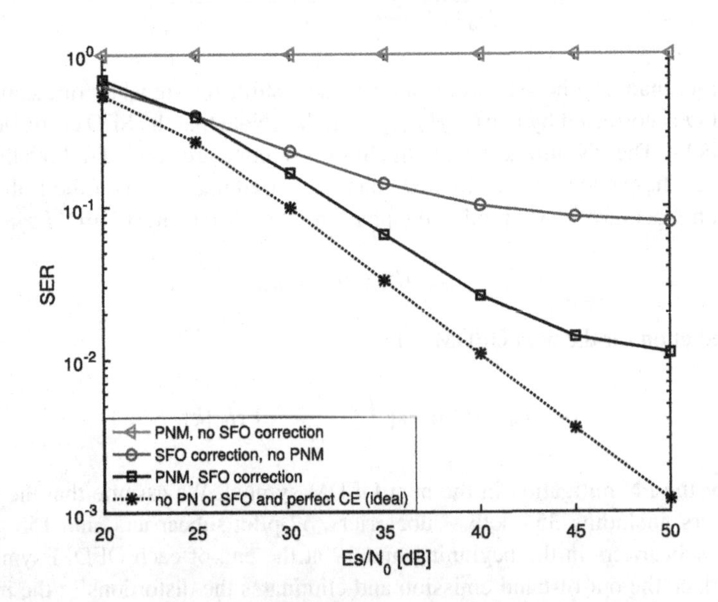

FIGURE 6.29

SER performances of the uncoded MIMO-OFDM system with/without the PN mitigation or the SFO correction in an urban microcell scenario at 60 GHz.

the observation noise to be $\kappa = 0.01$. Note that the small covariance matrix $\mathbf{Q}$ is justified by the time-invariant state model. The observation noise ζ is actually the estimation error of the phase rotation estimator, whose mean square error (MSE) is nontrivial to derive. However, it is found that the error rate performance of the system is insensitive to the value of κ (i.e., the variance of ζ) ranging from 0.001 to 0.1. Fig. 6.28 shows an example of the estimate of the phase rotation ψ_m and its true values as functions of the OFDM symbol index m. As can be seen, the estimates converge after the first eight OFDM symbols. Note that, apart from being robust to perturbations, e.g., of the PN estimation error, the Kalman filter is also used for the channel prediction for the PN mitigation.

For evaluating the error rate performance of the MIMO-OFDM link in a multipath fading channel, we resort to the QuaDRiGa channel model [9]. In total 200 drops are generated. For each channel drop, 10 preamble symbols and 50 payload symbols are transmitted. Fig. 6.29 shows the symbol error rate (SER) performance of the uncoded MIMO-OFDM system in the presence of the PN, the SFO, and the RCFO with/without the PN mitigation (PNM) and the SFO correction. For comparison, we also plot SERs for the ideal case without the PN, the SFO, the RCFO, and with a perfect channel estimation. As can be seen, the PN and SFO must be jointly compensated for in order to have an acceptable SER performance.

REFERENCES

[1] E. Bala, J. Li, R. Yang, Shaping spectral leakage: a novel low-complexity transceiver architecture for cognitive radio, IEEE Vehicular Technology Magazine 8 (3) (2013, Sept.) 38–46.

[2] P. Bürgisser, M. Clausen, A. Shokrollahi, Algebraic Complexity Theory, Springer-Verlag, Berlin/Heidelberg, 1997.

[3] D.C. Chang, Effect and compensation of symbol timing offset in ofdm systems with channel interpolation, IEEE Transactions on Broadcasting (ISSN 0018-9316) 54 (4) (2008, Dec.) 761–770, https://doi.org/10.1109/TBC.2008.2002339.

[4] X. Chen, OFDM based multi-node transmission in the presence of phase noises for small cell backhaul, IEEE Communications Letters (ISSN 1089-7798) 21 (5) (2017, May) 1207–1210, https://doi.org/10.1109/LCOMM.2017.2655509.

[5] X. Chen, A. Wolfgang, Phase noise mitigation in ofdm-based backhaul in the presence of channel estimation and synchronization errors, in: 2016 IEEE 83rd Vehicular Technology Conference (VTC Spring), 2016, May, pp. 1–5.

[6] X. Chen, C. Fang, Y. Zou, A. Wolfgang, T. Svensson, Beamforming mimo-ofdm systems in the presence of phase noises at millimeter-wave frequencies, in: 2017 IEEE Wireless Communications and Networking Conference Workshops (WCNCW), 2017, March, pp. 1–6.

[7] D5.1, Initial multi-node and antenna transmitter and receiver architectures and schemes, Millimetre-Wave Based Mobile Radio Access Network for Fifth Generation Integrated Communications (mmMAGIC), 2016.

[8] G. Huang, A. Nix, S. Armour, Impact of radio resource allocation and pulse shaping on PAPR of SC-FDMA signals, in: Proc. IEEE 18th International Symposium on Personal, Indoor and Mobile Radio Communications (PIMRC), 2007, Sept., pp. 1–5.

[9] S. Jaeckel, L. Raschkowski, K. Börner, L. Thiele, Quadriga: a 3-d multi-cell channel model with time evolution for enabling virtual field trials, IEEE Transactions on Antennas and Propagation (ISSN 0018-926X) 62 (6) (2014, June) 3242–3256, https://doi.org/10.1109/TAP.2014.2310220.

[10] D.W. Lim, S.J. Heo, J.S. No, An overview of peak-to-average power ratio reduction, Journal of Communications and Networks 11 (3) (2009, June) 229–239.

[11] A. Mehrotra, Noise analysis of phase-locked loops, in: IEEE/ACM International Conference on Computer Aided Design. ICCAD – 2000. IEEE/ACM Digest of Technical Papers (Cat. No.00CH37140), 2000, Nov., pp. 277–282.

[12] S.H. Muller-Weinfurtner, Optimum Nyquist windowing in OFDM receivers, IEEE Transactions on Communications 49 (3) (2001, Mar.) 417–420.

[13] A. Puglielli, G. LaCaille, A.M. Niknejad, G. Wright, B. Nikolić, E. Alon, Phase noise scaling and tracking in OFDM multi-user beamforming arrays, in: 2016, IEEE International Conference on Communications (ICC), 2016, May, pp. 1–6.

[14] R1-162960, Summary of email discussion on link level channel model, 3GPP TSG RAN WG1 Meeting 84b, 2016, April.

[15] T.M. Schmidl, D.C. Cox, Robust frequency and timing synchronization for ofdm, IEEE Transactions on Communications (ISSN 0090-6778) 45 (12) (1997, Dec.) 1613–1621, https://doi.org/10.1109/26.650240.

[16] J.J. van de Beek, M. Sandell, P.O. Borjesson, Ml estimation of time and frequency offset in OFDM systems, IEEE Transactions on Signal Processing (ISSN 1053-587X) 45 (7) (1997, July) 1800–1805, https://doi.org/10.1109/78.599949.

[17] S. Wu, Y. Bar-Ness, OFDM systems in the presence of phase noise: consequences and solutions, IEEE Transactions on Communications (ISSN 0090-6778) 52 (11) (2004, Nov.) 1988–1996, https://doi.org/10.1109/TCOMM.2004.836441.

[18] B. Yang, K.B. Letaief, R.S. Cheng, Z. Cao, Timing recovery for OFDM transmission, IEEE Journal on Selected Areas in Communications (ISSN 0733-8716) 18 (11) (2000, Nov.) 2278–2291, https://doi.org/10.1109/49.895033.

[19] Y.H. You, S.T. Kim, K.T. Lee, H.K. Song, An improved sampling frequency offset estimator for ofdm-based digital radio mondiale systems, IEEE Transactions on Broadcasting (ISSN 0018-9316) 54 (2) (2008, June) 283–286, https://doi.org/10.1109/TBC.2008.915763.

[20] A.A. Zaidi, R. Baldemair, H. Tullberg, H. Bjorkegren, L. Sundstrom, J. Medbo, C. Kilinc, I.D. Silva, Waveform and numerology to support 5G services and requirements, IEEE Communications Magazine (ISSN 0163-6804) 54 (11) (2016, November) 90–98, https://doi.org/10.1109/MCOM.2016.1600336CM.

[21] J. Zhang, C. Huang, G. Liu, P. Zhang, Comparison of the link level performance between OFDMA and SC-FDMA, in: Communications and Networking in China (ChinaCom), 2006, Oct., pp. 1–6.

CHAPTER

MULTIANTENNA TECHNIQUES 7

Multiantenna techniques can be defined by the use of multiple antennas at the transmitter and/or receiver in combination with signal processing. A communication system with multiple antennas at both the transmitter and the receiver is often referred to as a MIMO system. These techniques can be used to improve the system performance in terms of capacity, coverage, data rates, and link reliability. Fundamentally, multiple antennas can provide:

- Diversity. Diversity gives robustness against a fading radio channel and improves the link reliability. Multiple antennas can be used in different ways to provide diversity, e.g., space, pattern, polarization, and delay diversity.
- Array gain. By coherently combining the signals from/to multiple antennas, so that the desired signal adds constructively, a spatial processing gain can be achieved that increases the SNR. In a LoS channel, array gain is obtained by directivity but array gain can also be obtained in rich scattering channels by proper antenna combining.
- Interference suppression. Multiple antennas cannot only be used to enhance the desired signal but they can also be combined so that undesired signals are suppressed, thereby also improving the signal-to-interference-ratio (SIR).
- Spatial multiplexing. By transmitting multiple data streams over several antennas using the same time-frequency radio resource, spectral efficiency can be improved. The multiple data streams can be transmitted to a single UE, often called single-user MIMO (SU-MIMO), or multiple UEs, often called MU-MIMO.

Multiple antennas have been used in global wireless communication systems such as GSM, WCDMA/HSPA, and LTE. Initially, only simple antenna diversity techniques were used. More advanced multiantenna techniques such as MIMO were introduced in HSPA Release 7 and developed further in LTE. While multiantenna techniques have been useful for improving performance in current and previous generations of the standards, in 5G NR they have a more fundamental role to play in the system design.

This chapter gives an overview of multiantenna techniques in cellular wireless communication and their particular use in NR. The NR-specific features and functionalities described in this chapter are according to the first NR release (Release 15) which was finalized in 2018. NR is continuously evolving and new features will be added and existing features will be enhanced in future releases of the specifications.

The chapter is organized as follows. In Section 7.1 we discuss the role of multiantenna techniques in NR, for both low and high frequency bands. The fundamental theory of the multiantenna techniques relevant for NR is provided in Section 7.2 in order to give a better understanding of the particular features adopted in the NR specifications. The NR-specific multiantenna techniques and features are then described in Section 7.3. Finally, the effectiveness of the discussed techniques is illustrated by some experimental examples in Section 7.4.

5G Physical Layer. https://doi.org/10.1016/B978-0-12-814578-4.00012-6

7.1 THE ROLE OF MULTIANTENNA TECHNIQUES IN NR

This section gives a brief overview of the role of multiantenna techniques in NR. More details on specific features are provided in later sections. NR has been designed for millimeter-wave spectrum in addition to traditional cellular frequency bands at lower frequencies. The motivation for having multiple antennas and the techniques used are different in the low and high frequency bands. Some of these aspects are discussed in the following. More attention is paid to the high frequency bands, since this is where multiantenna techniques have a more fundamental impact on the system design and where most new features have been developed. Clearly, there is no sharp border between low and high frequencies in general. However, 3GPP NR has defined two frequency ranges, FR1 and FR2, where FR1 is between 450 MHz and 6 GHz and FR2 is between 24.25 GHz and 52.6 GHz[1] [5]. Therefore, in this chapter, low frequencies will mean carrier frequencies ≤ 6 GHz and high frequencies will mean carrier frequencies ≥ 24.25 GHz. High frequencies will sometimes also be referred to as millimeter-wave frequencies

7.1.1 LOW FREQUENCIES

In the low frequency bands, spectrum is congested. To meet the never-ceasing quest for higher data rates, a higher spectral efficiency is needed. This can be achieved by advanced multiantenna techniques such as spatial multiplexing and interference suppression. The former is attained by reusing radio resources in an efficient manner and the latter by ultimately enabling a multicell, multiuser system to be limited by thermal noise and not by interference.

At low frequencies the physical size may limit the number of antenna elements in an array that is practical to have since the antenna element area is proportional to the wavelength squared; see Chapter 3. However, advances in active array antenna technology have made it feasible to have digital control over more of the antenna elements in the array. This can be used to exploit more details in the spatial domain in order to increase performance.

For low frequencies, multiantenna techniques for NR are mainly refinements and evolution of multiantenna techniques that have been used in LTE. Some of the enhancements include improved support for reciprocity-based operation and more detailed feedback of CSI to achieve a higher spectral efficiency with MU-MIMO transmission.

7.1.2 HIGH FREQUENCIES

At high frequencies, the spectral efficiency is less crucial since there is plenty of spectrum available. Instead, obtaining coverage is the main challenge as substantially higher transmission losses may occur due to smaller antenna apertures and in some cases also higher attenuations, as explained in Chapter 3. A large bandwidth can exacerbate this further due to increased thermal noise power in the receivers. Compensating this with higher transmission power might not be possible due to limitations in millimeter-wave hardware design and to current regulations on transmitted power being stricter at frequencies above 6 GHz [11]. On the other hand, increasing the carrier frequency also means that, for a given physical size of an antenna, it becomes more directive. This may, depending on the directional

[1] The gap between 6 GHz and 24.25 GHz is due to the fact that no spectrum allocations have been identified for NR in this range.

properties of the channel, compensate for the increase of transmission loss with frequency and even be turned into a gain, as explained in Chapter 3. To make use of this directivity, dynamic and user-specific beam-forming is needed, since the directions to the users in a wireless access network are not known a priori and are dynamically changing.

In free-space propagation, the transmission loss increases with frequency if the antenna gain is assumed to be constant at both ends of the link; see Chapter 3. This is an effect of the fact that the antenna effective area is proportional to the wavelength squared, as explained in Chapter 3. If instead the antenna area is assumed to be constant at one end of the link, the free-space transmission loss is frequency independent. If the antenna area is assumed to be constant at both ends of the link, the free-space transmission loss will actually decrease with frequency since the antenna gain increases with frequency if the antenna area is constant. However, to make use of the increased antenna gain the transmit (Tx) and receive (Rx) beams must be aligned. For a point-to-point radio link this can be achieved by mechanically aligning the Tx and Rx antennas. However, in a mobile communication system with moving users this solution is not practically feasible. Instead, beam-forming and beam tracking using array antennas are needed to dynamically adjust the directions of the Tx and Rx beams. If the frequency is increased, more antenna elements can be accommodated by an antenna array with a given physical size, since the individual antenna elements become smaller. Therefore the potential beam-forming gain increases with frequency for a given physical size of the array.

Clearly, a mobile communication system cannot rely only on free-space propagation since many users have NLoS and/or are located indoors. Therefore, other propagation characteristics such as diffraction, reflection, scattering, and penetration are important to consider. These different propagation mechanisms have a varying degree of frequency dependence; see Chapter 3 for more details. However, in many NLoS scenarios there is a substantial increase of transmission loss with frequency making it more challenging to maintain an adequate link budget at high frequencies. To what extent beam-forming can compensate for the increased propagation loss depends on the scenario and also on how the beam-forming is implemented. Both theoretical and experimental examples of this issue are given later in this chapter.

The adverse propagation conditions and current hardware technology at millimeter-wave frequencies have a fundamental impact on the NR system design. To ensure sufficient coverage, in the millimeter-wave spectrum NR has a beam-centric design in which not only data transmissions can be beam-formed but also control and broadcast signals. This is different from previous generations of cellular systems in which typically only data transmissions are beam-formed.[2] Furthermore, support for beam-forming also in the UE has been introduced in NR in order to increase the potential beam-forming gain even further. UE beam-forming is possible in millimeter-wave bands since more antenna elements can be fit into the limited form factor of a UE. Due to hardware constraints, analog beam-forming will be a common implementation in millimeter-wave bands, particularly for hand-held devices. Therefore, support for analog beam-forming procedures has been included in the NR specifications.

Another difference between low and high frequency bands is that spectrum allocations in millimeter-wave bands are foreseen to mainly be unpaired. This has impact on the duplexing method

[2]Exceptions exist, e.g., the enhanced physical downlink control channel (EPDCCH) introduced in LTE Release 11 is a control channel that can be beam-formed.

used and thereby also on how different multiantenna techniques can operate. Frequency division duplex (FDD) operation is used in paired frequency bands where different frequency ranges are assigned for downlink and uplink, while TDD is used in unpaired bands where a single frequency range is shared between downlink and uplink. This has significant impact on multiantenna techniques, since the propagation channel can be assumed to be reciprocal under TDD operation, i.e., the downlink channel state is identical to the uplink channel state. In FDD operation, the downlink and uplink will typically experience independent fast fading due to the frequency difference between the uplink and downlink carriers. Advanced multiantenna transmission techniques often rely on detailed channel state information at the transmitter (CSIT). If reciprocity holds, this can be obtained from uplink measurements while extensive pilot and feedback signaling may be needed in the case that it does not. Therefore, reciprocity-based multiantenna transmission techniques can benefit from TDD operation, especially for antenna arrays with many elements where the signaling overhead may become prohibitive. Note that not only the propagation channel needs to be reciprocal, but also the multiantenna transceivers, which may require calibration of the transceivers [50].

7.2 MULTIANTENNA FUNDAMENTALS

In this section we provide some fundamental theory of multiantenna techniques relevant for NR. We try to keep the presentation general, thus being agnostic to any particular wireless communication standard. However, when deemed relevant, we at times point out relations to current LTE and NR specifications. Details on particular techniques adopted in the NR specifications are deferred to Section 7.3.

7.2.1 BEAM-FORMING, PRECODING, AND DIVERSITY

Beam-forming, precoding,[3] and diversity are techniques to coherently combine multiple antenna elements in an antenna array; at the transmitter, receiver or both.[4] By doing so, two types of gain can be achieved:

- Array gain. Array gain is the increase in the average SNR obtained by combining multiple antenna elements compared to a single element.
- Diversity gain. Diversity techniques are used to reduce the impact of fading by combining antenna elements that experience different fading. Diversity performance can be characterized by diversity order, which is the number of independently fading antenna elements.

The difference between array gain and diversity gain is that array gain gives an increase in the average SNR, while diversity gain makes the probability density function of the instantaneous SNR more concentrated around its average value. In some cases, array and diversity gain can be achieved simultaneously, while in other cases only one of the gains is achieved. This is described later in this section.

[3] Precoding is a transmission technique. For lack of better terminology we simply call it antenna combining when performed in the receiver.

[4] Diversity also includes other techniques than antenna combining such as antenna selection, frequency diversity, etc.

One of the most simple, intuitive and often used multiantenna techniques is beam-forming. Beam-forming is a technique to focus the transmitted or received radio energy in a particular direction using an array of antenna elements. This is achieved by applying a progressive time delay to the antenna elements so that the signals from the different elements add constructively in a desired direction. By controlling the progressive time delay, a beam can be steered in a desired direction. To steer a beam in a certain direction the time delays should compensate the propagation delay between the antenna elements of a plane wave impinging upon the array from that direction. For a narrowband system, a time delay can be approximated by a phase shift. Beam-forming is implemented by multiplying the signals on the antenna elements by complex beam-forming weights. The phase of the beam-forming weights determine the beam direction and the amplitude can be used to control beam width and sidelobe level.

The notion of beam-forming is coupled to free-space propagation of a single plane wave.[5] Under such conditions, beam-forming is the optimal transmission/reception scheme in the sense that it maximizes the SNR if the noise is spatially white. An antenna array with N elements can achieve an array gain that is equal to N. In wireless communication, beam-forming is a suitable technique for scenarios in which there is one dominating propagation path, e.g., when there is LoS or a strong specular reflection. The transmitter can apply Tx beam-forming to focus the transmitted energy in the dominating angle of departure and the receiver can apply Rx beam-forming to obtain a focused reception in the dominating angle of arrival.

The Tx and Rx array gains in a "pure" LoS channel, i.e., a fully correlated channel with no multipath propagation, are multiplicative so that the composite link array gain for a transmitter with N_T elements and a receiver with N_R elements is $N_T N_R$. Hence, full Tx and Rx array gain can be obtained simultaneously in a fully correlated channel. Under these circumstances there is no fast or frequency-selective fading. Furthermore, in most scenarios, the LoS direction varies slowly. Therefore, in this case, the beam-forming weights can be updated on a slow time basis and the same weights can be used over the entire bandwidth.[6]

However, most cellular deployments are characterized by multipath propagation. This means that the communication between a transmitter and receiver is not conveyed by a single plane wave, rather a superposition of multiple plane waves, or channel rays. This superposition is the cause of fast fading in wireless communications; see Chapter 3. The rays have different angles of arrival/departure as well as different amplitudes and phases. This means that it is no longer optimal to transmit/receive in a single direction by applying a simple progressive phase shift. Instead, the optimal approach is to utilize the different propagation paths in the channel. By proper coherent combining of the antenna elements, energy can actually be focused in space (within a limited resolution) rather than in direction. In effect, the different propagation paths are then aligned so that they add constructively at the receiver. Precoding can also be applied to antenna elements having different polarizations to match the polarization properties of the channel.

---

[5] The plane wave assumption implies that the receiver is in the far field of the Tx antenna. One rule of thumb for the far field is a minimum distance of $2L^2/\lambda$, where L is the largest array dimension and λ is the carrier wavelength [30]. This is fulfilled in a mobile communication network.

[6] If not too large to violate the narrowband assumption. This depends both on the relative bandwidth and the size of the antenna array in wavelengths. Different rules of thumb can be derived depending on the criterion. One example is that the maximum relative bandwidth in per cent is equal to the array beam width in degrees [30].

This more general amplitude and phase combining of antenna elements is commonly referred to as precoding when applied at the transmitter, or antenna combining when applied at the receiver. Also precoding is implemented by applying complex weights to the antenna elements. Although there is no strict distinction between beam-forming and precoding, beam-forming may be viewed as a special case of precoding for correlated channels. In a rich scattering channel, the fading correlation between different antenna elements will be low.[7] Precoding can therefore also provide diversity gain in addition to array gain. This requires more detailed channel knowledge, and the precoding weights need to be updated more frequently to follow the fast fading. Furthermore, fast fading is usually frequency selective so that different precoding weights may be needed in different parts of the scheduled bandwidth.

To design an optimal precoder, the complex channel coefficients between the transmitter and receiver antenna elements need to be known. How to acquire this information is a challenging task, which is described in more detail in Section 7.2.6. For now, assume that the complex channels between all Tx/Rx antenna pairs are known to both the transmitter and the receiver. If the transmitter emits a symbol, s, precoded with an $N_T \times 1$ complex weight vector $\mathbf{w}_T$ using an array with N_T elements, the signal at a receiver having an array with N_R elements can be modeled by

$$\mathbf{y} = \sqrt{P}\mathbf{H}\mathbf{w}_T s + \mathbf{n}, \tag{7.1}$$

where $\mathbf{y}$ is an $N_R \times 1$ vector containing the signal samples from the Rx antennas, P is the transmitted power per antenna, $\mathbf{H}$ the $N_R \times N_T$ complex channel matrix, and $\mathbf{n}$ is an $N_R \times 1$ vector modeling additive noise, which is assumed to be spatially white. The receiver can apply a $1 \times N_R$ antenna combining weight vector $\mathbf{w}_R$ to produce the complex scalar output

$$z = \mathbf{w}_R\mathbf{y} = \sqrt{P}\mathbf{w}_R\mathbf{H}\mathbf{w}_T s + \mathbf{w}_R\mathbf{n}. \tag{7.2}$$

First, assume that the transmitter has a single antenna element and the receiver has N_R antenna elements so that $\mathbf{H} = \mathbf{h}$, where $\mathbf{h}$ is the $N_R \times 1$ channel vector. The Rx combining vector that maximizes the SNR after the combiner when the noise is spatially white is easily shown[8] to be

$$\mathbf{w}_R = \frac{\mathbf{h}^H}{\|\mathbf{h}\|_F} \tag{7.3}$$

when $\mathbf{w}_R$ is constrained to fulfill $\|\mathbf{w}_R\|_F = 1$. Here, $\mathbf{h}^H$ denotes the complex conjugate transpose of $\mathbf{h}$ and $\|\cdot\|_F$ denotes the Frobenius norm. This solution is commonly referred to as maximum ratio combining (MRC). It is also easy to show that MRC achieves full array gain equal to N_R, regardless of the channel correlation.

Second, assume that the receiver has a single antenna and the transmitter has N_T antennas. The precoding vector that maximizes the SNR at the receiver is given by

$$\mathbf{w}_T = \frac{\mathbf{h}^H}{\|\mathbf{h}\|_F} \tag{7.4}$$

[7] The fading correlation is determined by the distance between antenna elements and the channel angular spread.
[8] This follows from the Cauchy–Schwarz inequality.

when $\mathbf{w}_T$ is constrained to fulfill $\|\mathbf{w}_T\|_F = 1$ and where $\mathbf{h}$ is the channel vector, presently a $1 \times N_T$ vector. The precoder in (7.4) is often called an MRT precoder. The MRT precoder achieves full array gain, N_T, regardless of the channel correlation. A physical explanation of this is that the precoder makes the signals scattered by the channel arrive in phase and add coherently at the receiver. It is important to realize that the power constraint assumed in the derivation is on the total transmitted power summed over all antenna elements, not on the power per antenna element. In general, the MRT solution gives a precoder with non-constant modulus weights, i.e., the weights for different antenna elements have different amplitudes. An active array typically has a PA per antenna element so that power cannot be shared between antenna elements. Therefore, a per-antenna power constraint would be a more realistic assumption for an active array antenna. Maximizing the SNR under a per-antenna power constraint is achieved by simply keeping the phase of the MRT precoder and setting all amplitudes equal [53]; sometimes this is called an equal gain transmission (EGT) precoder. The average SNR loss for an EGT precoder compared to MRT assuming an IID Rayleigh fading channel is $N_T/(1+(N_T-1)\pi/4)$, which converges to $4/\pi$ or roughly 1 dB for large N_T [44]. Note that this loss is under the assumption of the same total Tx power for the two precoders. If an MRT precoder with non-constant modulus weights is applied to an array with a PA per antenna element, not all PAs would transmit with full power. This would lead to a reduction in the total transmitted power and a corresponding reduction in SNR for the MRT precoder.

To illustrate the difference between precoding in a LoS and NLoS scenario, Fig. 7.1 shows an example of azimuth radiation patterns for these two cases. The left plot shows the radiation pattern of a uniform linear array (ULA) with 16 antenna elements when the optimal MRT precoding weights have been applied to the array for a scenario in which there is LoS to the UE. The right plot shows a corresponding pattern when the UE is in NLoS, based on one realization of the ITU urban macro channel model in [22]. In the LoS case, there is a single, narrow beam pointing in the direction to the UE. In the NLoS case, energy is instead transmitted in several different directions in order to utilize different propagation paths in the environment. The radiation pattern of the MRT precoder matches the azimuth power spectrum of the channel. In the LoS case, MRT precoding is equivalent to beam-forming according to our previous "definition". In the NLoS case, we prefer to use the term precoding since it is hard to distinguish a well-defined main beam from such a radiation pattern.[9] Note that the maximum directivity of the radiation pattern is lower in the NLoS case due to the multipath propagation. Nevertheless, the array gain is the same for both cases since the "multiple beams" in the NLoS case are combined coherently at the receiver.

Third, and finally, assume that the transmitter has N_T antenna elements and the receiver has N_R antenna elements. The precoder and combiner that maximize the SNR at the receiver output under a sum-power constraint is given by the principal right- and left-singular vector of $\mathbf{H}$, respectively [32], i.e.,

$$\mathbf{w}_T = \mathbf{v}_1, \ \mathbf{w}_R = \mathbf{u}_1^H \tag{7.5}$$

[9]According to IEEE, a beam is defined as "The major lobe of the radiation pattern of an antenna" and a major lobe is defined as "The radiation lobe containing the direction of maximum radiation" [21].

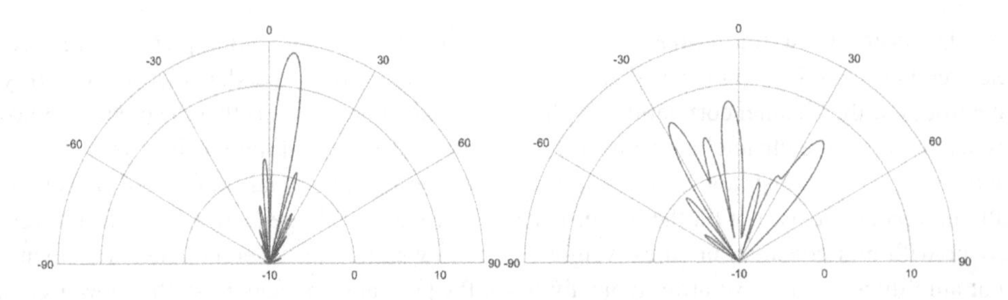

FIGURE 7.1

Radiation patterns for MRT precoding. Sample radiation pattern for a ULA with 16 elements performing MRT precoding. Left: There is LoS to the UE. Right: The UE is in NLoS.

where $\mathbf{v}_1$ is the first[10] column of $\mathbf{V}$ and $\mathbf{u}_1$ is the first column of $\mathbf{U}$ in the singular value decomposition (SVD) of $\mathbf{H}$

$$\mathbf{H} = \mathbf{U}\boldsymbol{\Sigma}\mathbf{V}^H. \tag{7.6}$$

The composite link array gain of this scheme, sometimes referred to as dominant eigenmode transmission, is equal to the expectation of the maximum eigenvalue of $\mathbf{H}\mathbf{H}^H$ [32]. In this case, the composite link array gain depends on the channel correlation. In a channel with full Tx and Rx correlation, e.g., a pure LoS channel, the composite link array gain is $N_R N_T$ so that full Rx and Tx gain is achieved simultaneously. In a scattering channel, however, the composite link array gain will be lower, since a single precoder cannot maximize the power at all Rx antenna elements simultaneously. It can be shown that, for an uncorrelated channel, the composite link array gain is upper bounded by $(\sqrt{N_T} + \sqrt{N_R})^2$ for large arrays [8]. Hence, full Tx array gain *or* full Rx array gain can be obtained in a scattering channel, but not the two simultaneously. This can only be achieved in a fully correlated channel. Similar to the previous case of a single-antenna receiver, all amplitudes in $\mathbf{w}_T$ can be set to the same value in order to fully utilize all PAs also when the receiver has multiple antennas.[11] It has been shown that the incurred SNR loss by doing this is not more than 1 dB for an IID Rayleigh channel, i.e., not more than when we have a single Rx antenna [45].

MRC and MRT were derived under the assumption of spatially white noise. This is a reasonable assumption for thermal noise but in many cases the dominating impairment is caused by interference. Interference is typically not spatially white, since it usually comes from a particular direction. The antenna elements can then be combined with weights so that the array beam pattern has nulls in the interference directions. Theoretically, $N - 1$ interference directions can be suppressed with an array having N antenna elements. Interference can be suppressed in the receiver, see Section 7.2.2.3, or by the transmitter, see Section 7.2.2.2.

Besides array gain, precoding and antenna combining also gives diversity gain in a scattering channel since the signals transmitted/received on spatially separated antennas or antennas with different

[10] Assuming the singular values in $\boldsymbol{\Sigma}$ have been ordered in descending order.

[11] The Rx combining vector should then be modified to $\mathbf{w}_R = \tilde{\mathbf{h}}/\|\tilde{\mathbf{h}}\|_F$, where $\tilde{\mathbf{h}} = \mathbf{H}\mathbf{w}_{\mathrm{EGT}}$, and $\mathbf{w}_{\mathrm{EGT}}$ is the EGT precoding vector.

polarizations will experience different fading. By combining a number of independently fading antenna elements the probability that all antenna elements are in a fade is reduced, thus creating a more stable communication link. Diversity performance can be characterized by diversity order, which is the effective number of independently fading antenna elements. The diversity order determines the slope of the symbol error rate curve as a function of SNR in a log-log scale. In the case of IID Rayleigh fading, MRT and MRC gives diversity order N_T and N_R, respectively, while dominant eigenmode transmission gives diversity order $N_T N_R$. Note that even in a scattering channel, closely spaced co-polarized antenna elements will have some fading correlation, making the IID assumption invalid.

So far it has been assumed that the channel is known to the transmitter and receiver. Without any channel knowledge, array gain cannot be obtained but diversity gain is still possible to achieve. A simple Tx diversity technique for two Tx antennas that can achieve full diversity order without any channel knowledge at the transmitter is the Alamouti scheme [6]. In this scheme, a complex symbol s_1 is transmitted on a first antenna and another complex symbol s_2 on a second antenna in a first symbol period. In the next symbol period, $-s_2^*$ is transmitted on the first antenna and s_1^* on the second antenna. Assume that the receiver has a single Rx antenna and arranges two consecutive received signal samples in a vector, $\mathbf{y} = [y_1\ y_2^*]^T$. If the channel is constant over the two symbol periods, the received signal vector is given by

$$\mathbf{y} = \begin{bmatrix} y_1 \\ y_2^* \end{bmatrix} = \sqrt{P/2}\tilde{\mathbf{H}}\mathbf{s} + \mathbf{n}, \tag{7.7}$$

where

$$\tilde{\mathbf{H}} = \begin{bmatrix} h_1 & h_2 \\ h_2^* & -h_1^* \end{bmatrix} \tag{7.8}$$

is the effective channel and

$$\mathbf{s} = \begin{bmatrix} s_1 \\ s_2 \end{bmatrix}. \tag{7.9}$$

Since $\tilde{\mathbf{H}}$ is an orthogonal matrix, the receiver can retrieve the transmitted symbols by simply multiplying by $\tilde{\mathbf{H}}^H$ according to

$$\mathbf{z} = \tilde{\mathbf{H}}^H\mathbf{y} = \sqrt{P/2}\left\|\tilde{\mathbf{H}}\right\|_F^2 \mathbf{s} + \tilde{\mathbf{H}}\mathbf{n} \tag{7.10}$$

In this way, the full Tx diversity order of two is obtained without any knowledge of the channel at the transmitter. However, no Tx array gain is achieved due to the absence of channel knowledge at the transmitter. The Alamouti scheme has rate one, i.e., two symbols are conveyed over two symbol periods. It is a special case of orthogonal space–time block codes (OSTBCs) for two antennas. OSTBCs exist also for more than two Tx antennas and provide full diversity order with simple linear processing in the receiver. However, for complex signal constellations, an orthogonal code with rate one exists only for two Tx antennas [43]. In more general terms, Tx diversity can be achieved by mapping the modulation symbols in the space–time or space–frequency domains, the so-called space–time transmit diversity (STTD) and space frequency transmit diversity (SFTD), respectively.

Antenna diversity can be combined with frequency diversity if the channel is frequency selective. Otherwise, a frequency-selective channel can be created artificially from a spatially dispersed channel using the so-called delay diversity. This is accomplished by transmitting delayed copies of the same signal on different antennas, thus creating time dispersion or, equivalently, frequency selectivity. If the antennas have uncorrelated fading, the signal at the receiver will appear as a signal that has passed through a channel having multiple taps with uncorrelated fading, i.e., a frequency-selective channel. Delay diversity is transparent to the UE since it only sees the effective frequency-selective channel. Hence, delay diversity can be implemented without any specification support. A particular type of delay diversity that is suitable for OFDM systems is cyclic delay diversity (CDD) [20]. In CDD a cyclic shift instead of a linear delay is applied on the antennas. This is equivalent to applying a frequency dependent phase shift prior to the OFDM modulation.

In LTE, downlink Tx diversity schemes for up to four antennas are supported in the specifications. For two antennas, Tx diversity is based on the so-called space frequency block coding (SFBC), which is equivalent to Alamouti coding in the frequency domain. With four Tx antennas, SFBC is used in combination with the so-called frequency-switched transmit diversity (FSTD) [13]. There is also a downlink transmission mode that combines precoded multilayer transmission with a CDD scheme called large-delay CDD. Uplink Tx diversity using two antennas was introduced for the control channel in LTE Release 10 and uses so-called spatial orthogonal-resource transmit diversity (SORTD) [13]. With SORTD, a signal is transmitted on the different antennas using orthogonal resources in frequency, time, and/or code domain. In NR, however, Tx diversity is currently not explicitly supported and one has to rely on specification-transparent methods.

7.2.2 SPATIAL MULTIPLEXING

As described in the previous section, multiple antennas at the transmitter and receiver can give array gain by precoding and Rx antenna combining. This will increase the SNR which in turn will increase the data rate. This is efficient when the data rate is power limited rather than bandwidth limited. From basic information theory, the achievable data rate grows approximately linearly with the SNR when the SNR is low [46]. However, at high SNR, the achievable rate starts to saturate, since it grows only logarithmically with SNR. In this regime, it would be a more efficient use of the available bandwidth if one could "split" the SNR over several weaker links that could communicate in parallel. Indeed, this is possible by utilizing multiple antennas at the transmitter and receiver and it is the basic principle of spatial multiplexing. To realize this, perform an SVD of a channel matrix $\mathbf{H}$ according to

$$\mathbf{H} = \mathbf{U}\mathbf{\Sigma}\mathbf{V}^{H}. \tag{7.11}$$

If the transmitter applies a precoder matrix $\mathbf{V}$ and the receiver multiplies the received signal vector by $\mathbf{U}^H$ the effective channel matrix $\tilde{\mathbf{H}}$ becomes

$$\tilde{\mathbf{H}} = \mathbf{U}^{H}\mathbf{U}\mathbf{\Sigma}\mathbf{V}^{H}\mathbf{V} = \mathbf{\Sigma}. \tag{7.12}$$

Since $\mathbf{\Sigma}$ is a diagonal matrix, the effective channel is composed of multiple parallel subchannels without any crosstalk, where the gain of each subchannel is determined by the corresponding singular value. Independent data streams, or layers, can then be transmitted over the subchannels without any mutual interference. In this way, the achievable data rate can increase linearly with the number of antennas,[12] thus circumventing the data rate saturation at high SNR. This approach is deceptively simple but involves several issues that need to be solved in a practical system. For example, it requires that the channel matrix is known to both the transmitter and the receiver. In practice, this cannot be known exactly and there will be some interference between the transmitted data layers. More advanced receivers can then be used to suppress this interference; see Section 7.2.2.3.

The performance of spatial multiplexing depends to a large extent on the channel properties. If the channel has a high correlation, some singular values will be small, leading to weak subchannels that will not give any significant contribution to the overall data rate. The number of non-zero singular values of the channel matrix is called the channel rank.[13] In a practical system it is important to dynamically adapt the number of used subchannels to the current channel and data traffic conditions in order to optimize the overall data rate, so-called rank adaptation. The number of used subchannels is often referred to as the transmission rank. The transmission rank can depend on the channel rank, but it can also depend on other parameters such as the SINR. To make use of many subchannels, i.e., a high transmission rank, a high channel rank may not be sufficient. A high SINR is usually also required. To benefit from spatial multiplexing, a high "basic" SINR is needed, since the transmission power has to be shared between the transmitted layers and in practice there will be some interference between the layers.

Precoding and spatial multiplexing can be combined. One example was given above where premultiplication of $\mathbf{V}$ orthogonalizes the channel on the Tx side which makes the processing at the Rx side easier. If the number of transmitted layers is lower than the number of Tx antenna elements, precoding can also increase the SNR per layer by providing array gain. In order to determine a suitable precoder matrix, CSIT is required. For example, the channel matrix can be estimated from uplink SRSs and the precoder matrix can be determined from an SVD of the estimated channel matrix. However, this requires TDD operation and calibration between uplink and downlink RF branches. For FDD, a common approach is to let the UE estimate the channel based on downlink reference signals and feed back a proposed precoder matrix. To save feedback overhead, a limited number of predefined precoder matrices can be collected in a so-called codebook. The UE then needs only to signal an index to the preferred matrix in the codebook, rather than the complexed-valued coefficients of the precoder matrix itself. The codebook can contain precoder matrices for different numbers of transmission ranks. The codebook needs to be known by both the BS and the UE and thus requires specification support. This scheme is referred to as codebook-based precoding and has been adopted in both the LTE and the NR specifications. More details on how these codebooks are constructed are given in Section 7.2.6.

Spatial multiplexing of several layers to/from a single UE is often referred to as SU-MIMO. The layers may also be multiplexed to/from different UEs. This is called MU-MIMO or space division multiple access (SDMA). SU-MIMO and MU-MIMO can be combined so that the spatially multiplexed UEs can have multiple layers each. Residual interference between the transmitted layers can be suppressed by the receiver as described in Section 7.2.2.3. However, this is more difficult in downlink MU-MIMO, since Rx antennas from different UEs cannot be processed coherently and the number of antenna elements per UE might not be sufficient to suppress all interference. Downlink MU-MIMO

[12] More precisely, under certain conditions the achievable rate grows as $\min\{N_R, N_T\}$.

[13] In practice, no singular values will be exactly zero so some kind of threshold could be used.

transmission may therefore require some interference suppression at the Tx side. In a cellular system the channel conditions for different UEs will vary rapidly with time. To optimize system performance it is therefore important to dynamically switch between SU- and MU-MIMO operation depending on channel conditions and traffic load.

7.2.2.1 SU-MIMO Precoding

The previous discussion of precoding concerned maximizing the SNR for a single layer. As alluded to previously, precoding and spatial multiplexing can be combined. We can then find the precoder that maximizes the sum rate of all layers. Assuming that the channel is known to both the transmitter and the receiver, the capacity-optimal precoder under a sum-power constraint is given by [32]

$$\mathbf{W}_T = \mathbf{V}\mathbf{P}^{1/2} \tag{7.13}$$

where $\mathbf{V}$ is obtained from the SVD of $\mathbf{H}$ according to (7.11) and $\mathbf{P} = \mathrm{diag}\{p_1, \ldots, p_{N_T}\}$ is a diagonal matrix containing the power allocated to each layer. The optimal power allocation is obtained from the well-known waterfilling algorithm; see [46]. This algorithm will allocate high power to strong subchannels. The capacity-optimal transmission scheme under a per-antenna power constraint is a more difficult problem and one may need to resort to numerical techniques to find the optimal solution [41]. A closed-form solution for the general problem seems to be unknown, but for the case when the channel matrix has full column rank and the input covariance matrix has full rank, the optimal input covariance is given by [48]

$$\mathbf{R}_{\mathrm{opt}} = \mathbf{I} + \mathrm{diag}\left[\left(\mathbf{H}^H\mathbf{H}\right)^{-1}\right] - \left(\mathbf{H}^H\mathbf{H}\right)^{-1} \tag{7.14}$$

where $\mathbf{I}$ denotes the identity matrix. The optimal precoder matrix is then obtained from

$$\mathbf{W}_T = \mathbf{E}\mathbf{\Lambda}^{1/2} \tag{7.15}$$

where $\mathbf{E}$ and $\mathbf{\Lambda}$ are obtained from the eigendecomposition $\mathbf{R}_{\mathrm{opt}} = \mathbf{E}\mathbf{\Lambda}\mathbf{E}^H$.

Obtaining knowledge of the instantaneous channel matrix at the transmitter may be difficult in a practical system, e.g., if the channel is varying rapidly due to UE movement. A precoder design can then be based on channel statistics, e.g., the Tx channel covariance matrix, $\mathbf{R}_T = \mathrm{E}[\mathbf{H}^H\mathbf{H}]$, where $\mathrm{E}[\cdot]$ denotes expectation, instead of the instantaneous channel. Assuming that the Tx covariance matrix is known to the transmitter, the optimal[14] precoding matrix under a sum-power constraint is given by [23,47]

$$\mathbf{W}_T = \mathbf{E}\mathbf{P}^{1/2} \tag{7.16}$$

where $\mathbf{E}$ is now obtained from the eigendecomposition of $\mathbf{R}_T = \mathbf{E}\mathbf{\Lambda}\mathbf{E}^H$ and $\mathbf{P}$ is the diagonal matrix that allocates the power over the eigenvectors of $\mathbf{R}_T$. An iterative method to find $\mathbf{P}$ has been proposed in [47]. An approximate solution is to apply "statistical waterfilling", i.e., to use the eigenvalues of $\mathrm{E}[\mathbf{H}^H\mathbf{H}]$ instead of $\mathbf{H}^H\mathbf{H}$ as is used in the conventional waterfilling algorithm.

[14]Optimal with respect to ergodic capacity.

7.2.2.2 MU-MIMO Precoding

In this section, we discuss precoding for MU-MIMO transmission. There are several aspects that make MU-MIMO different from SU-MIMO [32]:

- SU-MIMO performance can be characterized by a link capacity while MU-MIMO performance is characterized in terms of a capacity region, i.e., the set of simultaneously achievable rates for all UEs.
- In SU-MIMO, only the sum rate of all layers is of interest since all layers are transmitted to the same user. In MU-MIMO, the layers are transmitted to different UEs and fairness between UEs needs also to be taken into account.
- In SU-MIMO, the transmission loss for all Tx–Rx antenna pairs is usually similar, while there can be a large difference in transmission loss to different UEs in MU-MIMO.
- In SU-MIMO, the Rx antenna elements can be combined coherently to optimize performance, e.g., by suppressing interlayer interference. Rx antennas in different UEs cannot be combined coherently in MU-MIMO.
- MU-MIMO requires more accurate channel knowledge at the transmitter than SU-MIMO. One reason for this is that interlayer interference needs to be suppressed at the transmitter if the number of transmitted layers is larger than the number of Rx antennas in the UEs and interference suppression requires accurate channel knowledge.

For simplicity, we assume that all UEs have a single antenna element each. Furthermore, we assume that the channel is known to both the transmitter and the receivers. The signal received by the kth out of K co-scheduled UEs served by a BS with N_T Tx antenna elements can be modeled by

$$y_k = \mathbf{h}_k \sum_{i=1}^{K} \mathbf{w}_i s_i + n_k \tag{7.17}$$

where $\mathbf{h}_k$ is the $1 \times N_T$ channel vector to UE k, $\mathbf{w}_i$ is the $N_T \times 1$ precoding vector to UE i, s_i the signal transmitted to UE i, and n_k is additive receiver noise with power N_0. We assume that $\|\mathbf{w}_i\|_F^2 = P_i$ where P_i is the power allocated to UE i and $\mathrm{E}[|s_i|^2] = 1$. The SINR of UE k is thus given by

$$\mathrm{SINR}_k = \frac{|\mathbf{h}_k \mathbf{w}_k|^2}{\sum_{i \neq k} |\mathbf{h}_k \mathbf{w}_i|^2 + N_0}. \tag{7.18}$$

Finding the optimal linear precoder may be posed as maximizing some utility function[15] $f(\mathrm{SINR}_1, \ldots, \mathrm{SINR}_K)$ subject to a constraint on the total transmitted power according to $\sum_{i=1}^{K} P_i \leq P$. In general, this is a very difficult problem. However, the solution has a simple general structure according to [9]:

$$\mathbf{w}_{k,opt} = c\sqrt{P_k}\left(\mathbf{I} + \sum_{i=1}^{K} \frac{\lambda_i}{N_0} \mathbf{h}_i^H \mathbf{h}_i\right)^{-1} \mathbf{h}_k^H = c\sqrt{P_k}\left(\mathbf{I} + \frac{1}{N_0} \mathbf{H}^H \mathbf{\Lambda} \mathbf{H}\right)^{-1} \mathbf{h}_k^H \tag{7.19}$$

[15]One example of a utility function is the sum rate $f(\mathrm{SINR}_1, \ldots, \mathrm{SINR}_K) = \sum_{i=1}^{K} \log_2(1 + \mathrm{SINR}_i)$.

for some positive parameters $\lambda_1, \ldots, \lambda_K$ such that $\sum_{i=1}^{K} \lambda_i = P$. Here, $\mathbf{w}_{k,opt}$ is the optimal precoding vector for UE k, c is a normalization that makes $\|\mathbf{w}_{k,opt}\|^2 = P_k$, $\mathbf{H} = [\mathbf{h}_1^T \cdots \mathbf{h}_K^T]^T$, and $\mathbf{\Lambda} = \text{diag}\{\lambda_1, \ldots, \lambda_K\}$. In general it is difficult to find the optimal λ_i in closed form. However, several well-known precoders correspond to specific choices of λ_i such as the regularized ZF [33], minimum mean square error (MMSE), transmit Wiener filter [24], and signal-to-leakage-and-interference ratio (SLNR) [39] precoders. For example, letting $\lambda_i = P/K$ for all UEs leads to the MMSE solution

$$\mathbf{w}_k = c\sqrt{P_k}\left(\mathbf{I} + \frac{P}{KN_0}\mathbf{H}^H\mathbf{H}\right)^{-1}\mathbf{h}_k^H. \tag{7.20}$$

Other well-known precoders can also be obtained from (7.19) asymptotically for low and high SNR. At low SNR ($N_0 \to \infty$), (7.19) reduces to

$$\mathbf{w}_k = c\sqrt{P_k}\mathbf{h}_k^H, \tag{7.21}$$

i.e., the MRT precoder in (7.4). The MRT MU-MIMO precoder "beam-forms" a layer to its intended UE, while ignoring interference to co-schedules UEs. At high SNR ($N_0 \to 0$), we obtain the ZF precoder

$$\mathbf{w}_k = c\sqrt{P_k}\mathbf{h}_k^\dagger \tag{7.22}$$

where $\mathbf{h}_k^\dagger$ denotes the kth column of $\mathbf{H}^\dagger$, and $\mathbf{H}^\dagger = \mathbf{H}^H(\mathbf{H}\mathbf{H}^H)^{-1}$ is the pseudo-inverse of $\mathbf{H}$. The ZF precoder beam-forms a layer in the direction to the desired UE while placing nulls in the directions to co-schedules UEs.[16] Since this precoder places nulls at co-scheduled UEs, the UEs will receive no inter-user interference. However, if the channel vectors for two different UEs are close to parallel, there will be a gain reduction that leads to a loss in SNR. The regularization of the inverse in (7.19) and (7.20) mitigates this problem and provides a balance between gain reduction to the desired UE and interference suppression, which leads to good performance over a wide SNR range.

Linear precoding techniques suffer from performance degradation when the channels to co-scheduled UEs are highly correlated. This can be improved by using nonlinear precoding techniques. The optimal approach is the so-called dirty paper coding (DPC) which achieves the maximum sum rate [51]. The idea with DPC is to precancel interference at the transmitter by using perfect knowledge about the channel and transmitted signals. It rests on the fundamental result in [12] that interference does not decrease the channel capacity if the transmitter knows the interference at the receiver, even if the receiver has no knowledge regarding the interference.

DPC-based precoding can be used in MU-MIMO to eliminate interference between UEs through coding and interference presubtraction [10]. To realize the viability of such an approach, let $\mathbf{H} = \mathbf{RQ}$ be the QR decomposition of the channel matrix. Here, $\mathbf{R}$ is a $K \times K$ lower triangular matrix and $\mathbf{Q}$ is a $K \times N_T$ unitary matrix, i.e. $\mathbf{Q}\mathbf{Q}^H = \mathbf{I}$. By using the precoder

$$\mathbf{W}_T = \mathbf{Q}^H \tag{7.23}$$

[16]This is a LoS description, but it holds also for multipath channels if "direction" is interpreted in a wider sense.

the symbols received at the UEs are given by

$$\mathbf{y} = \sqrt{P}\mathbf{R}\mathbf{s} + \mathbf{n}. \tag{7.24}$$

Since $\mathbf{R}$ is lower triangular, the received symbol for the kth UE is, ignoring the additive noise term,

$$y_k = \sum_{i=1}^{k} r_{i,k} s_i, \; k = 1, \ldots, K, \tag{7.25}$$

where $r_{i,k}$ denotes the (i, k)th element of $\mathbf{R}$. Hence, the first UE receives no interference, the second UE receives interference only from the first UE, and so on. Interference presubtraction can then be achieved by replacing the transmitted symbols $s_k, \; k = 1, \ldots, K$, by

$$s'_k = s_k - \frac{1}{r_{k,k}} \sum_{i=1}^{k-1} r_{k,i} s'_i, \tag{7.26}$$

so that the UEs receive the symbols

$$y_k = r_{k,k} s_k, \; k = 1, \ldots, K. \tag{7.27}$$

Hence, the interference presubtraction has completely removed the interference between UEs. This scheme can be seen as the Tx equivalent of the successive interference cancellation (SIC) receiver described in Section 7.2.2.3. An advantage with performing the interference subtraction on the Tx side is that it does not suffer from error propagation since the transmitted signals are known.

DPC precoding is, however, complex and may be difficult to implement in practice. Numerous suboptimal, nonlinear precoding techniques with lower complexity have therefore been developed, e.g. the Tomlinson–Harashima method [14], and Vector Perturbation [18] precoding.

The MU-MIMO precoders described so far have been developed under the assumptions of a single-cell system with perfect channel knowledge at the transmitting BS. In a practical system the performance of such precoders is impaired by channel estimation errors and intercell interference. To mitigate this, the MMSE precoder can be generalized to take channel estimation errors and intercell interference into account [25,27,28]. Such a multicell MMSE precoder has roughly the general structure (see the cited references for details that have been omitted here for the sake of brevity)

$$\mathbf{w}_k \sim \left(\mathbf{I} + \alpha \hat{\mathbf{H}}_{\text{intra}}^H \hat{\mathbf{H}}_{\text{intra}} + \beta \hat{\mathbf{H}}_{\text{inter}}^H \hat{\mathbf{H}}_{\text{inter}} + \gamma \mathbf{C}\right)^{-1} \hat{\mathbf{h}}_k^H \tag{7.28}$$

where $\hat{\mathbf{H}}$ denotes estimated channels and $\mathbf{C}$ a channel estimation error covariance matrix. The different terms inside the inverse account for intracell interference, intercell interference, and channel estimation errors, respectively. The multicell MMSE precoder suppresses interference to co-scheduled UEs within the served cell as well as the interference to UEs in other cells, while taking the channel estimation quality into account.

7.2.2.3 MIMO Receivers

This section gives a brief overview of different MIMO receivers that can be used to suppress interference. They are presented in the context of interlayer interference suppression in spatial multiplexing, but the same principles apply for canceling other types of interference, e.g., intercell interference.

The optimal approach to MIMO receiver design is the maximum likelihood (ML) principle. The ML receiver searches over all possible transmitted signal vectors to find the most likely one. This is a nonlinear receiver which usually is too complex to implement. A simpler nonlinear receiver is the so-called SIC receiver [49]. The idea with SIC is to successively demodulate and decode the different layers, and to reencode and subtract their contributions to the received signal, layer by layer. Performance may be impaired by error propagation, which occurs if a layer is not decoded correctly. This can be mitigated by ordering the layers so that the layer with highest SINR is decoded in each stage, so-called ordered SIC.

To reduce complexity further, linear receivers can be used. A linear receiver applies a linear filter to separate the transmitted layers and then decodes each layer independently. The output of a linear receiver can be expressed as

$$\mathbf{z} = \mathbf{W}_R \mathbf{y} \tag{7.29}$$

where $\mathbf{z}$ is the $N_T \times 1$ vector output of the receiver, $\mathbf{W}_R$ an $N_T \times N_R$ Rx weight matrix, and $\mathbf{y}$ is the $N_R \times 1$ received signal vector before the receiver. The received signal vector can be modeled as

$$\mathbf{y} = \sqrt{P}\mathbf{H}\mathbf{s} + \mathbf{n} \tag{7.30}$$

where $\mathbf{H}$ is the $N_R \times N_T$ channel matrix,[17] $\mathbf{s}$ is the $N_T \times 1$ transmitted signal vector, and $\mathbf{n}$ is a noise vector, which is assumed to be spatially white with covariance matrix $N_0\mathbf{I}$.

A ZF receiver removes interlayer interference by inverting the channel according to

$$\mathbf{W}_R = \frac{1}{\sqrt{P}}\mathbf{H}^{\dagger} \tag{7.31}$$

where $\mathbf{H}^{\dagger} = (\mathbf{H}^H\mathbf{H})^{-1}\mathbf{H}^H$ is the pseudo-inverse of $\mathbf{H}$. The output of the ZF receiver is given by

$$\mathbf{z} = \mathbf{s} + \frac{1}{\sqrt{P}}\mathbf{H}^{\dagger}\mathbf{n}. \tag{7.32}$$

Hence, the ZF receiver eliminates the interference completely and decouples the matrix channel into N_T parallel scalar channels. However, it also increases the noise level at the output of the receiver since the noise vector $\mathbf{n}$ is premultiplied by $\mathbf{H}^{\dagger}$. If $\mathbf{H}$ is close to singular, $\|\mathbf{H}^{\dagger}\|_F$ will be large and the noise increase will be large. Furthermore, $\mathbf{W}_R$ makes the noise spatially colored at the output of the receiver.

[17] We ignore any potential precoding at the transmitter. In the case of precoding, the precoding matrix could be absorbed in $\mathbf{H}$ to represent an effective channel matrix. N_T is then the number of transmitted layers, which could be lower than the number of Tx antenna elements.

The SINR of the kth layer on the output of the ZF receiver is given by

$$\mathrm{SINR}_k = \frac{P}{N_0} \frac{1}{\left[\left(\mathbf{H}^H\mathbf{H}\right)^{-1}\right]_{k,k}} \tag{7.33}$$

where $[\cdot]_{k,k}$ denotes the kth diagonal element of a matrix. It can be shown that the diversity order and array gain for each layer is proportional to $N_R - N_T + 1$ [32].

The noise enhancement of the ZF receiver can be mitigated by suppressing the interference down to the noise level instead of canceling it completely. This is accomplished by the MMSE receiver which is derived by minimizing the squared error between the received and transmitted signal vector. The solution is

$$\mathbf{W}_R = \frac{1}{\sqrt{P}} \left(\mathbf{H}^H\mathbf{H} + \frac{N_0}{P}\mathbf{I}\right)^{-1} \mathbf{H}^H. \tag{7.34}$$

The MMSE receiver balances noise enhancement and interference suppression to minimize the total power of interference and noise. From (7.34) it can be seen that the MMSE solution converges to ZF for high SNR and to MRC for low SNR. The SINR of the kth layer on the output of the MMSE receiver is given by [32]

$$\mathrm{SINR}_k = \frac{1}{\left[\left(\frac{P}{N_0}\mathbf{H}^H\mathbf{H} + \mathbf{I}\right)^{-1}\right]_{k,k}} - 1. \tag{7.35}$$

7.2.3 ANTENNA ARRAY ARCHITECTURES

As the name suggests, multiantenna techniques require multiple antennas. Although not necessary, these are often collocated in a single enclosing structure. Multiantenna techniques can also be applied across several such enclosures of antenna elements, placed in different locations. Herein, we refer to a collection of collocated antenna elements as an antenna array. This section gives a high-level description of some different architectures for antenna arrays.

NR is expected to operate in a large span of carrier frequencies: from below 1 GHz up to 100 GHz. As with many other technology components also the antenna system designs will be different in different parts of this vast frequency span. This is partly due to building practices and hardware implementation issues and partly due to system level and propagation aspects. Without going into implementation details, this section discusses different antenna array architectures that are viable candidates in different parts of this frequency range. Although a 3GPP technical specification rarely dictates a particular hardware implementation or architecture, it has in several cases been developed to be suited for relevant implementations. Therefore, some parts of the specifications are suitable for certain antenna array architectures, while others are not.

7.2.3.1 Digital Arrays

In a digital array architecture each antenna element is equipped with its own RF chain and data converters (ADC and DAC). Fig. 7.2 shows a schematic illustration of the transmitting part of a digital array architecture. The receiving part would look the same if the DACs are replaced with ADCs. The antenna array is depicted as a 1-D linear array but it could have any topology, e.g., a 2-D planar array. It is common practice that each antenna element position is populated with two radiating elements having orthogonal polarization,[18] each polarization having its own RF chain and DAC. However, for ease of exposition, this is omitted here. The illustration should be seen as functional rather than an implementation description. Digital arrays designed for OFDM systems also have an FFT/IFFT for each antenna element, enabling frequency-selective precoding. In principle, different precoding weights can thus be applied for each subcarrier. In practice, however, this also requires CSI for every subcarrier, which may not always be available due to, e.g., constraints on the frequency granularity in the CSI feedback in an FDD system. Therefore, frequency-selective precoding is often used with a subband granularity, where a subband contains several consecutive subcarriers. Besides being limited by CSI, the granularity of frequency-selective precoding can also be limited by the signal processing capacity, since calculating precoding weights per subcarrier can be computationally demanding.

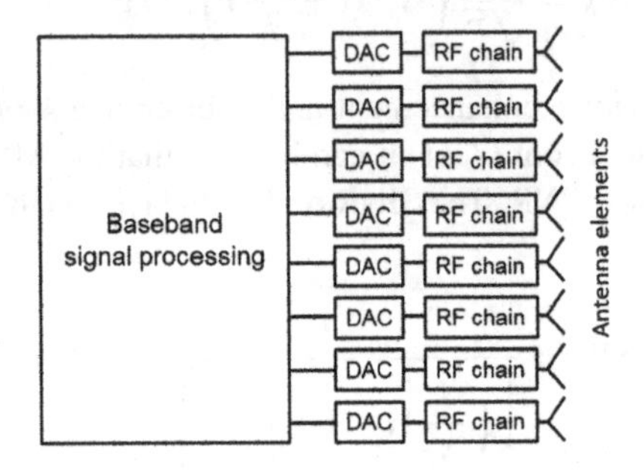

FIGURE 7.2

Digital array architecture. Schematic illustration of the digital array architecture.

Advances in active array antenna technology have made it possible to produce digital arrays with a large number of elements. Having an RF chain and data converter for each antenna element provides the highest performance and flexibility. Multiantenna techniques such as spatial multiplexing and interference suppression can reach its full potential when used in a fully digital array. However, a fully digital array can also be expensive in terms of cost and power consumption. These aspects are particularly pronounced at millimeter-wave frequencies, since the number of antenna elements is expected to be large in order to populate a sufficiently large physical antenna area that can achieve a required link budget. Furthermore, the large bandwidth foreseen to be used at these frequencies requires the data converters to operate at high sampling rates, leading to high power consumption and heat generation. A large bandwidth coupled with many digitized antenna elements is also challenging from a shear data shuffling perspective, putting high demands on data interfaces between the antenna array and signal processing units. This also leads to high demands on signal processing capacity. Therefore, a fully digital array is currently a likely implementation only in the low frequency bands. In millimeter-wave bands, analog and hybrid array architectures will be prevalent, at least in the near future. These architectures are described in the sequel.

[18] Physically, it may be a single element with different excitation points.

7.2.3.2 Analog Arrays

A schematic illustration of the transmitting part of an analog array architecture is shown in Fig. 7.3. With an analog array, analog beam-forming can be performed by applying a linear phase progression over the array by means of phase shifters[19] in order to steer a beam in the desired direction. If the array has some form of gain control per antenna element, amplitude tapering can be applied to reduce the sidelobes in order to mitigate interference to/from other UEs.

With an analog array, beam-forming is usually limited to wideband beam-forming, i.e., the same beam-forming weights are used over the entire bandwidth. With wideband beam-forming it is therefore not possible to adapt to the frequency selectivity of the channel. However, the spatial characteristics of the channel, such as the main direction of energy is typically not frequency dependent. Analog beam-forming is therefore suited for scenarios where the channel energy has a dominant direction, e.g., scenarios with LoS, a strong specular reflection or a dominating cluster with low angular spread.

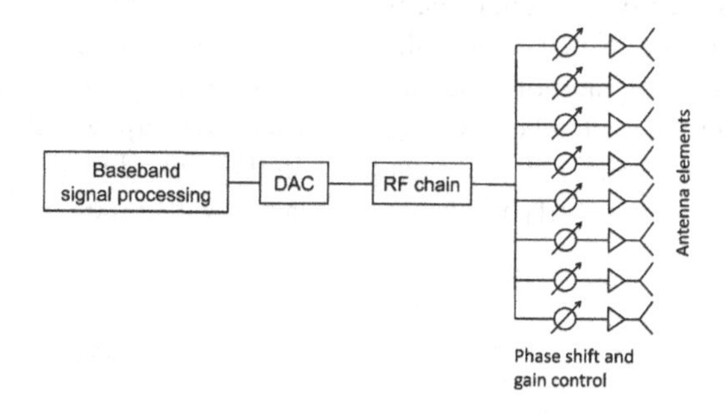

FIGURE 7.3

Analog array architecture. Schematic illustration of the analog array architecture.

A digital array has higher performance potential than an analog array since it provides more degrees of freedom in the spatial signal processing. However, at millimeter-wave frequencies many degrees of freedom may not be crucial to the system performance. It is likely that millimeter-wave systems will be deployed in small cells, at least initially. Since small cells have a high probability of LoS, the benefits with frequency-selective digital precoding over analog beam-forming becomes small since the angle and delay spread is small. In a LoS channel with no angle or delay spread, the optimal precoder is simply a wideband beam-former if interference is ignored. This can be implemented by an analog array. Efficient interference nulling is difficult to implement with analog arrays, but amplitude tapering can be performed in the analog beam-forming to reduce sidelobe levels and hence interference.

An important limitation of the analog array architecture from a system design point of view is that an analog beam-former can only transmit or receive in one direction at a time. Indeed, this has had a profound impact on the development of the NR specifications. In order to provide support for analog beam-forming of both data and control channels, a set of procedures called *beam management* has been developed. This is described in Section 7.3.4. Also the initial access procedures have been designed to support analog beam-forming.

[19]For systems with large relative bandwidth, phase shifts should be replaced by true time delays.

7.2.3.3 Hybrid Arrays

The fully digital and analog array architectures presented in previous sections represent two extreme cases, each having their advantages and disadvantages with regard to performance, cost, and complexity. In a hybrid array architecture, more flexibility in these trade-offs is provided by combining the digital and analog array architectures. Two examples of a hybrid array architecture are shown in Fig. 7.4, referred to as fully and partially connected, respectively. In the fully connected architecture, multiple (in this case two) arrays of phase shifters and gain controllers have the same antenna elements. In this example, two beams with full beam-forming gain can be generated independently of each other, since each analog chain has its own phase shifters and gain controllers and they are connected to all antenna elements. These two beams can then be combined digitally in the baseband signal processing, for example to perform spatial multiplexing or to increase the array gain in a multipath environment.

The partially connected architecture is a less complex hybrid architecture in which each analog chain is connected to a subarray of antenna elements. A drawback with this architecture compared to the fully connected architecture is that each analog beam-former cannot achieve the full beam-forming gain, only the gain provided by an individual subarray. However, full array gain may be achieved by digital beam-forming over the analog beams created by each subarray, so-called hybrid beam-forming. The analog and digital beam-former should then point in the same direction, otherwise grating lobes will appear and the gain is reduced. Hybrid beam-forming may also require some calibration between the analog subarrays.

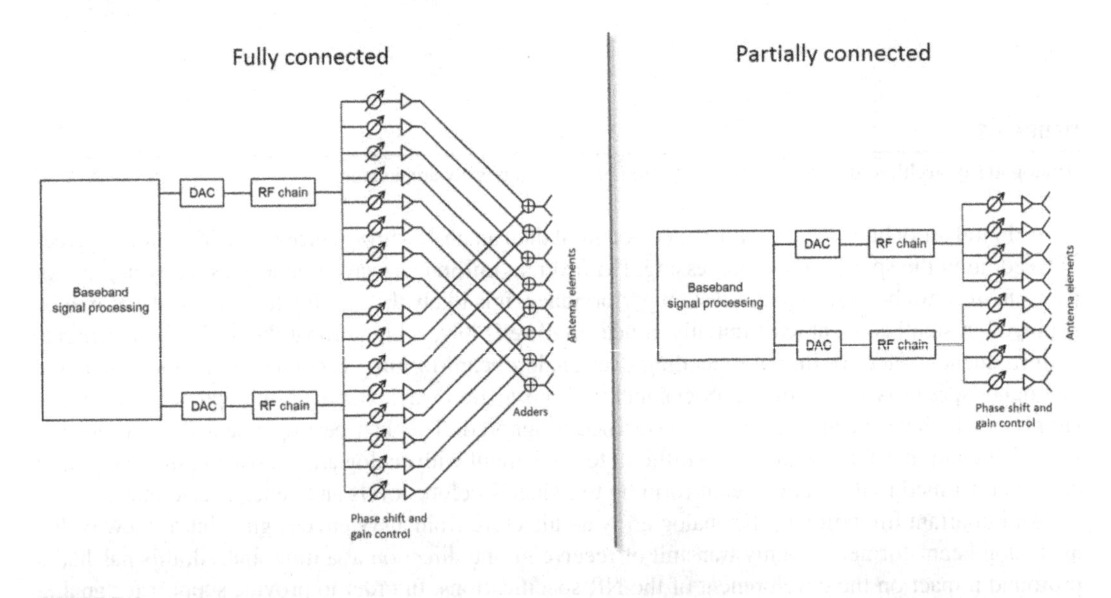

FIGURE 7.4

Hybrid array architecture. Schematic illustration of hybrid array architectures.

Drawbacks with the fully connected architecture compared to the partially connected architecture are higher complexity and losses in the adders and dividers (or combiners in an Rx array). Clearly, the fully connected array has more components due to the adders and the replication of phase shifters. The adders also introduce RF losses that scale linearly with the number of added signals. Furthermore, the losses in the dividers between the RF chains and phase shifters also scale linearly with the number of phase shifters, which is larger in the fully connected architecture [16].

With a hybrid array architecture, the channel at antenna element level cannot be observed due to the analog beam-forming. A challenge is then how to co-design the analog and digital beam-formers. For the highest performance they should be designed jointly, but this may be too complex. A suboptimal but simpler approach is to design the beam-formers independently; see Section 7.3.4.3.

A partially connected hybrid array architecture with rectangular subarrays that has been evaluated extensively in the development of the NR specifications is the so-called panel array [1,3]. In 3GPP, the subarrays are referred to as panels and each panel is a uniform planar array (UPA) with single- or dual-polarized elements. An example of panel array with 2×2 panels, each with 4×4 dual-polarized antenna elements is illustrated in Fig. 7.5. A possible implementation is that analog beam-forming is performed per polarization in each panel and that each panel has two RF chains, one per beam/polarization. The panels can be used in different ways. For example, they can be used independently to serve different UEs or be combined coherently to serve a single UE.

FIGURE 7.5

Panel array. An example a panel array with 2×2 panels, each with 4×4 dual-polarized antenna elements.

7.2.3.4 A Millimeter-Wave Antenna Array System Prototype

An example of a compact millimeter-wave active antenna array system is shown in Fig. 7.6. The upper left picture shows the top view of an active array antenna module for 28 GHz carrier frequency designed by Ericsson and IBM Watson Research Center [35,40,15]. One module has 64 dual-polarized antenna elements[20] and can generate one beam per polarization using analog beam-forming. Each antenna element and polarization has its own front-end radio with phase shifter and gain control, i.e., 128 front-end radio chains in one module. The size of one antenna array is 70 × 70 mm. The lower left picture shows the bottom view of a module with four radio frequency integrated circuits (RFICs). Each RFIC has 32 Tx and Rx branches containing mixer, phase shifter, attenuator, PA, low-noise amplifier (LNA), and TDD switches. The right picture shows a radio unit prototype designed by Ericsson that consists of two such modules.

[20] As can be seen in the picture, a module actually has 100 elements but the edge elements are just dummy elements to reduce border effects.

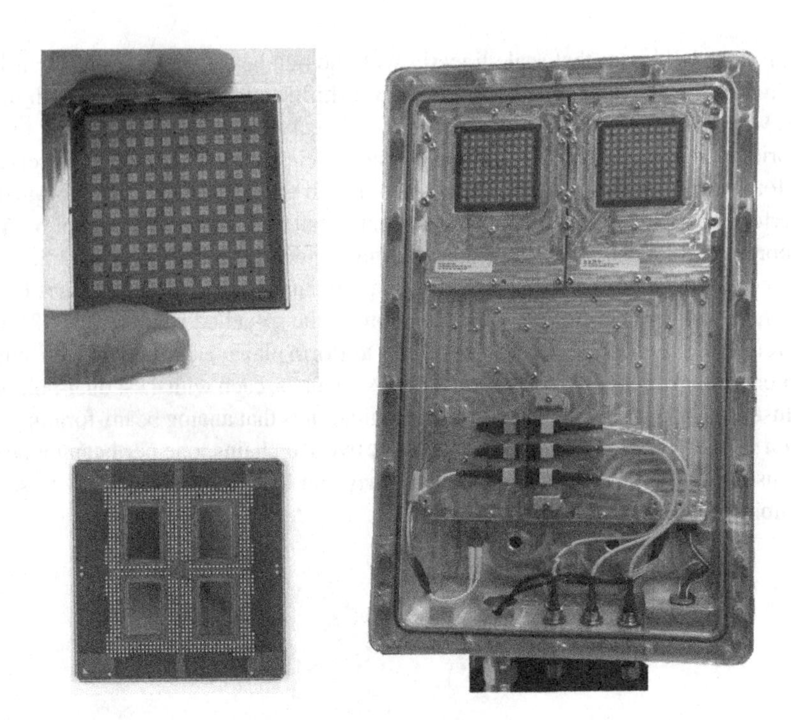

FIGURE 7.6

28 GHz active antenna array. Top left: Front view of one module with 8×8 dual-polarized active antennas. Low left: Back view of one module with four RFICs. Right: A prototype radio unit with two modules.

A high-level architecture of the Tx branches of a radio unit and a connected baseband unit is shown in Fig. 7.7. The receiver architecture is similar but with ADCs instead of DACs and FFTs instead of IFFTs. A radio unit is capable of generating four beams independently, one beam per module and polarization, each having its own ADC/DAC and digital radio. Frequency domain data are transmitted between the radio and baseband unit over a common public radio interface (CPRI). The bandwidth of the prototype is 800 MHz and it is capable of delivering four data streams with a total of 15 Gbps peak data rate. A beam from a module has 23 dBi gain and a half-power beam width of 12° in both azimuth and elevation. Measured azimuth beam patterns for five different beams from one module are shown in Fig. 7.8.

An Ericsson 5G Testbed system including the 28 GHz radio unit has been used in a 5G trial network in Pyeongchang, Republic of Korea, in a cooperation between KT Corporation, Ericsson and several other technology partners [34].

7.2.4 UE ANTENNAS

Antennas of hand-held devices are usually designed to have as close to omni-directional coverage as possible since the incident waves may come from any direction. At traditional cellular frequencies, the size of a typical hand-held device is of the same order as the carrier wavelength. At millimeter-wave

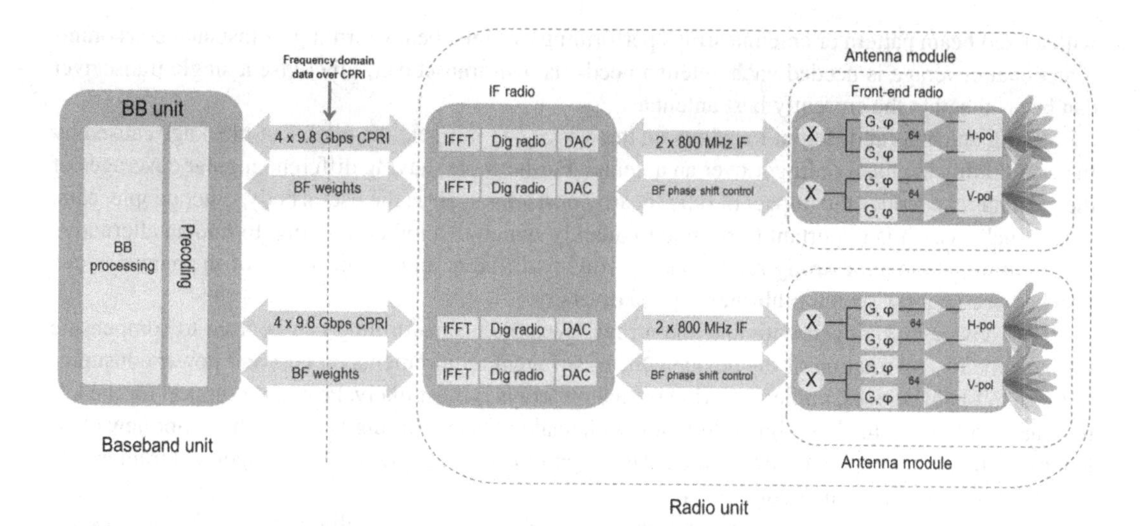

FIGURE 7.7

Architecture. High-level architecture of radio and baseband unit.

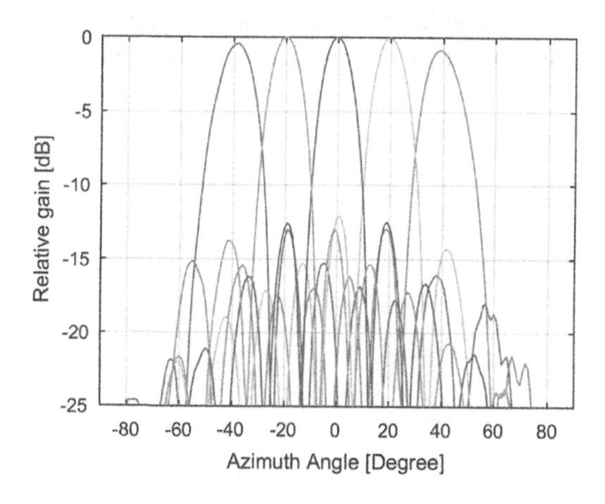

FIGURE 7.8

Beam patterns. Measured azimuth beam patterns for five different beams from one module.

frequencies, however, a hand-held device is large compared to the wavelength. This makes it more difficult to design omni-directional antennas since the device chassis acts as a larger ground plane. On the other hand, since the physical size of an antenna having a certain gain decreases with increasing frequency, it is possible accommodate millimeter-wave antenna arrays to the limited form factor of a hand-held device. One approach to achieve omni-directional coverage at high frequencies is to have several directive antennas that cover different angular sectors. The antennas can be discrete antennas with a fixed beam pattern or antenna arrays performing dynamic beam-forming. If instantaneous omnidirectional coverage is needed each antenna needs its own transceiver, otherwise a single transceiver can be switched to the currently best antenna.

Having antennas at different locations on the device can mitigate the effects of blockage caused by the user putting the hand or finger over an antenna. Furthermore, having different angular coverages of the antennas can reduce the impact of other nearby obstacles such as the user's body, other people, cars, etc. In such cases it is important to be able to quickly switch to another direction to find an alternative propagation path, e.g., a strong reflection [7]. Having different angular coverage of the antennas can also be beneficial for spatial multiplexing and diversity.

If the device is equipped with antenna arrays, dynamic beam-forming can be used to compensate for movements and rotation of the device. Due to stringent requirements on cost and power consumption in hand-held devices, implementation by analog arrays is most likely. Providing support for the UE to dynamically track the best signal direction with analog beam-forming is one of the major new challenges in NR within the multiantenna area. This support is given by the beam management framework, described in more detail in Section 7.3.4.

The feasibility of integrating antenna arrays into mobile phones at millimeter-wave frequencies has been demonstrated with several prototypes. For example, [19] shows a 28 GHz prototype where two 16-element arrays have been integrated on the top and bottom of a cellular phone. Another example of a 28 GHz prototype is shown in [52] where two linear antenna arrays with eight elements each have been integrated on opposite edges of a mobile phone mock-up with metallic back casing. Fig. 7.9 shows a schematic illustration of the antenna array placement on the phone in [19] and [52], respectively.

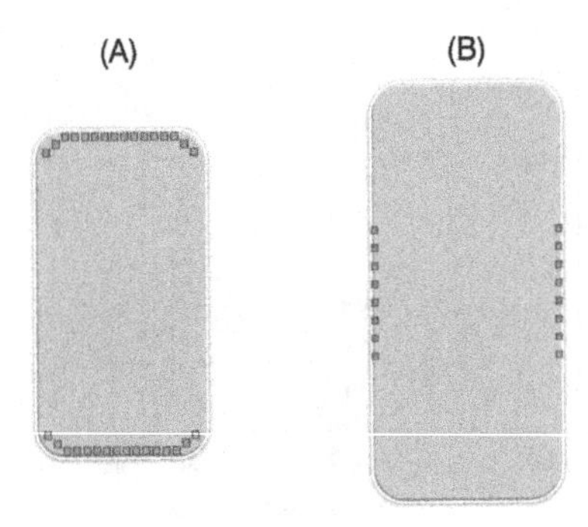

FIGURE 7.9

UE antenna arrays. Prototype millimeter-wave antenna array placement on a phone: (A) in [19], (B) in [52].

A major challenge in UE antenna design is the large number of frequency bands that needs to be supported by the UE. With the exploitation of new bands in millimeter-wave spectrum this number will increase even further. Another challenge is that regulatory requirements on RF electromagnetic field (EMF) exposure of humans are different at millimeter-wave frequencies compared to those in traditional cellular frequency bands. Although the regulations vary between different parts of the world it is common that at frequencies above a transition frequency, typically 6 or 10 GHz, the restriction metric changes from specific absorption rate (SAR) to free-space power density. This change in metric implies that the maximum permissible transmitted power for a UE may be significantly lower at millimeter-wave bands than at frequencies used for current cellular technologies [11].

7.2.5 ANTENNA PORTS AND QCL

Although multiantenna features are treated extensively in 3GPP specifications, the antennas are rarely described as hardware components. Features and procedures are instead referring to so-called *antenna ports*. An antenna port is in the 3GPP specifications an abstract concept that is a logical entity rather than a physical antenna. To ease the understanding of the multiantenna techniques in 3GPP, this section explains the meaning of an antenna port and the related concept of quasicolocation (QCL) and also gives some motivation for their definitions.

In many cases it is important for the receiver to know which assumptions it can make on the channel corresponding to different transmissions. For example, the receiver needs to know which reference signal transmission it can use to estimate the channel in order to decode a transmitted signal. It is also important for the UE to be able to report relevant CSI to the BS which it can use for scheduling and link adaptation purposes. For this purpose, two important concepts were introduced in LTE: antenna port and QCL. An antenna port by definition functions such "that the channel over which a symbol on the antenna port is conveyed can be inferred from the channel over which another symbol on the same antenna port is conveyed" [4]. The receiver can assume that two transmissions correspond to the same radio channel if and only if they use the same antenna port [13].

In practice, the antenna port can be said to be defined by the transmitted reference signal. The reference signal could have been transmitted from a single physical antenna element or using a beam-former applied on a subarray of elements. For example, even if two signals are transmitted using the same physical antennas they will correspond to different antenna ports if they are beam-formed with different weights, since the corresponding effective channels will be different.[21] The receiver can use a reference signal transmitted on an antenna port to estimate the channel for this antenna port and this channel estimate can subsequently be used for decoding data transmitted on the same antenna port. For example, the DM-RS (see Section 2.5) in LTE and NR can be used for channel estimation to decode data transmitted on the same antenna port.

QCL is defined in a similar manner: "Two antenna ports are said to be quasi-co-located if properties of the channel over which a symbol on one antenna port is conveyed can be inferred from the channel over which a symbol on the other antenna port is conveyed". The main difference between the antenna port and QCL definitions is that the former speaks of *the channel* while the latter refers to the *properties of the channel*. Thus QCL is a less stringent requirement than an antenna port since only the properties of the channel and not the channel itself need to be the same for quasi-co-located antenna ports. If two signals have been transmitted on two closely spaced, but different, antennas, they could experience different channels due to fading but the large-scale properties of the two channels will probably be the same. In such a case, the two antennas would be different antenna ports but they would be quasi-collocated. Large-scale properties include second-order statistics of the channel such as delay/Doppler spread, average channel gain, etc. Such information can, for example, be useful to

[21]If the beam-forming weights are known to the receiver, it could be the same antenna ports, since then the beam-forming weights need not be considered as part of the channel.

the UE for performing channel estimation. An example of antenna ports not being QCL is if they use antennas from different locations, for example in multipoint transmission.

The concept of QCL was introduced in LTE Release 11 to support different types of multipoint transmissions. In NR, QCL has a more central role to play since, e.g., the beam management procedures rely heavily on the QCL concept. In particular, spatial QCL assumptions are used to help the receiver to select an analog Rx beam during beam management; see Section 7.3.4.4 for details.

7.2.6 CSI ACQUISITION

Acquiring CSI is one of the most important aspects of multiantenna techniques since the quality of the CSI is often the limiting factor on the performance of multiantenna techniques. CSI in general can be detailed such as the complex channel matrix for every subcarrier in an OFDM system or coarse, such as the direction to a UE in LoS. Advanced multiantenna techniques such as MU-MIMO and interference suppression exploit detailed channel knowledge and therefore put high demands on the CSI acquisition. The performance gain with such techniques can be very high if perfect CSI is available but in practice it must be estimated based on measurements. If a sufficient CSI quality cannot be obtained, the gain with advanced techniques that rely on detailed CSI may vanish.

In 3GPP, the term 'CSI' has a more specific meaning, namely the particular reports from the UE to the BS indicating the channel quality and other channel properties. Most of the CSI parameters in such reports are actually preferred downlink transmission parameters rather than explicit channel parameters. The UE estimates the channel quality and other properties based on measurements on reference signals transmitted in the downlink. The chief downlink reference signal in LTE and NR for computing CSI is the CSI-RS; see Section 2.5. A CSI report can be periodic, semi-persistent, or aperiodic. Periodic reports occur at regular time instants, while aperiodic reports are triggered on a per need basis. Semi-persistent reports are transmitted periodically until further notice. Aperiodic reports are generally more detailed than periodic reports. In NR, CSI consists of the following components:

- The channel quality indicator (CQI). CQI is an index to the highest modulation and coding scheme (MCS) that would result in a block-error probability of at most 10%, conditioned on a certain transmission hypothesis.
- The rank indicator (RI). RI is a recommendation of which transmission rank to use in codebook-based precoding.
- The precoding matrix indicator (PMI). PMI indicates the preferred precoder to use in codebook-based transmission, conditioned on the indicated transmission rank.
- The CSI-RS resource indicator (CRI). CRI is used for indicating the best beam in beam management or beam-formed CSI-RS transmissions. This is described later in this chapter.
- The layer-1 reference signal received power (L1-RSRP). L1-RSRP is the received power as measured by the UE on a configured reference signal, e.g., a CSI-RS. Layer-1 (physical layer) is in contrast to layer-3 (RRC) RSRP for which additional filtering is applied.
- The strongest layer indicator (SLI). SLI indicates which column in the precoding matrix that corresponds to the layer with the highest SINR. This can be used for transmitting PT-RS on the strongest layer to achieve the most accurate phase tracking at millimeter-wave frequencies.

In this section CSI is used in a broader meaning and not necessarily limited to the parameters listed above.

CSI is needed for several purposes such as scheduling, selection of multiantenna scheme, rank and link adaptation (setting MCS), coherent demodulation, and for determining precoding or combining weights in multiantenna transmission and reception. CSI acquisition is facilitated by transmitting predefined reference signals known to the receiver. The receiver can then estimate the channel by correlating the received signal with the corresponding reference signal. Channel state information at the receiver (CSIR) is thereby relatively straightforward to acquire. Acquiring CSIT is more challenging. The two main alternatives for acquiring CSIT are feedback and reciprocity based. In feedback-based CSI acquisition, CSI is obtained by feedback from a receiving node that has performed channel estimation on the transmitted reference signals. Reciprocity-based CSI acquisition relies on the assumption that CSIT can be obtained from measurements on received reference signals. In this section we discuss CSI acquisition aspects with focus on acquiring CSIT at the BS for downlink transmission.

7.2.6.1 Reciprocity Based

Reciprocity may in broad terms be defined as that knowledge about the Tx channel can be inferred from knowledge about the Rx channel. One can think of different degrees of reciprocity, such as:

- The complex channel matrix is the same for the Tx and Rx channels. This is the strongest form of reciprocity and is in general only possible to achieve in TDD operation.
- The channel second-order statistics, e.g., covariance matrix, is similar for Tx and Rx. Wideband and long-term channel properties are similar for the uplink and downlink also for FDD systems which can be utilized for reciprocity-based operation. For example, precoding based on channel covariance can to some extent be based on reciprocity.
- The angles of departures are the same as angle of arrivals. Fast fading is caused by complex superposition of multiple channel rays, and they will sum up differently for the DL and UL carriers in an FDD system. The directions of the rays are, however, typically not dependent on the carrier frequency, so they can be reciprocal in FDD. For example, beam-forming in a LoS channel can be based on reciprocity.

The rest of this subsection is concerned with the strongest form of reciprocity described in the first bullet above. Reciprocity-based CSIT acquisition for downlink transmission can be performed by configuring the UE to transmit SRSs in the uplink. The BS can perform channel estimation on the received SRSs to obtain an estimate of the uplink channel matrix. If reciprocity holds, the uplink channel estimate can then be used as an estimate of the downlink channel. A clear advantage with reciprocity-based CSI acquisition compared to feedback based acquisition is a reduced overhead, since no feedback signaling is needed to obtain CSIT. Also the reference signaling overhead is reduced if the number of BS antenna elements is larger than the number of UE antenna elements, which typically is the case. With feedback-based CSI acquisition, one reference signal per BS Tx antenna element needs to be transmitted, while one reference signal per UE Tx antenna element needs to be transmitted with reciprocity-based CSI acquisition. Another advantage with reciprocity-based CSI acquisition is that it is less reliant on specification support than feedback-based methods in the sense that it can be used for any number of BS antenna elements. Feedback-based methods typically use some kind of quantized channel feedback such as codebooks, and these need to be specified for each supported number of BS antenna ports and need to be known to both the BS and the UE.

A disadvantage with reciprocity-based methods is that they can only be used in TDD systems if the strongest form of reciprocity should be utilized. However, even if the system operates in TDD, sometimes only partial reciprocity can be guaranteed. Although the propagation channel is reciprocal in a TDD system,[22] the BS transceivers may not be. To utilize reciprocity, also the BS transceivers need to be reciprocal, i.e., the Tx and Rx branches need to have the same characteristics. This can be achieved to a certain level of accuracy by reciprocity calibration of the transceivers [50]. Passive components like the antenna radiating elements are typically reciprocal as long as the same elements are used for transmission and reception, which, however, may not always be the case.

Another type of partial reciprocity is when the UE has fewer Tx than Rx branches, which is quite common in hand-held devices in order to save battery time. In this case the UE cannot sound all antennas simultaneously, since there are not enough Tx branches. This implies that not the full channel matrix can be estimated from uplink sounding, only the channels to the antennas which have a connected Tx branch. A possible solution to this problem is to switch the Tx branches between the antennas sequentially in time until all antennas have been sounded. Partial reciprocity may also be due to the number of carriers being different in uplink and downlink, which could be the case in carrier aggregation. Reciprocity can then only be utilized for the carriers that are common for uplink and downlink.

The discussion of reciprocity has so far concerned the channel between a BS and the served UE, i.e., the desired link. Interfering links are also important to take into account in multiantenna transmission. Interference can have a significant impact on link adaptation, scheduling, and precoder design. Even if the propagation is reciprocal, the interference is typically different in uplink and downlink. For example, the downlink interference received by a UE from another, non-serving, BS cannot be measured in uplink by the serving BS. Therefore, even if reciprocity holds for the desired link in a TDD system, it needs to be complemented by feedback containing information regarding the interference. The interference feedback does not have to be the interference in itself; it could be the impact of the interference, e.g., on the quality of the signal after the receiver. Since the BS may not know which interference suppression capability the UE has in its receiver, it can be better to feed back an SINR related quantity instead of the actual interference level. One such quantity that is used in LTE and NR is CQI.

A difference between the downlink and uplink, regardless of TDD or FDD, is that the UE usually has a much lower Tx power than the BS. This may cause problems with reciprocity-based CSI acquisition in some scenarios if the UE Tx power is not high enough to yield sufficient SNR in the channel estimation performed by the BS. Poor channel estimation quality will lead to poor performance of downlink multiantenna transmission schemes that rely on detailed channel knowledge, e.g., MU-MIMO. It may then be better to use feedback-based CSI acquisition, since the UE can perform channel estimation at a higher SNR than the BS due to the higher Tx power in the BS.

7.2.6.2 Feedback Based

In downlink feedback-based CSIT acquisition the BS transmits reference signals that are known to the UE. The UE estimates the downlink channel and feeds back a quantized channel estimate by uplink

[22] Propagation is reciprocal if the medium is linear. Furthermore, the uplink channel estimation and downlink transmission must be performed within the channel coherence time so that the channel does not change due to fading.

signaling. The feedback can be explicit or implicit. With explicit feedback, a quantized and possibly compressed representation of the channel could be reported to the BS, e.g., a quantization of the channel matrix itself or of the principal eigenvectors of the channel covariance matrix. The channel feedback can also be complemented by feedback of the interference experienced by the UE. With implicit feedback, preferred transmission parameters are fed back. One example of implicit feedback is codebook-based CSI acquisition where a number of candidate precoding matrices are collected in a codebook. The UE then evaluates, based on its channel estimates, which precoding matrix in the codebook would give the highest performance if used by the BS [29]. The UE then feeds back an index to the precoder matrix in the codebook, in LTE and NR called a PMI.

Note that the use of codebooks for CSI acquisition does not mean that the BS has to use a precoder from the codebook in the data transmission. This depends on whether the reference signals used for coherent demodulation are precoded in the same way as data or not. If the demodulation reference signals are precoded with the same precoder as the data, the UE does not need to know which precoder the BS has used in order to demodulate the data. The codebook is in this case only used to feed back CSI to the BS. The BS can then use this information to design an arbitrary precoder without informing the UE. An example of a transmission scheme where the demodulation reference signals are precoded in the same way as data is transmission mode (TM) 9 in LTE. If instead the reference signals for coherent demodulation are transmitted per antenna element without any precoding, the UE needs to know which precoder the BS has applied to the data transmission so that the UE can apply this precoder to the estimated channel before demodulating the data. An example of such a transmission scheme is TM 4 in LTE, where non-precoded reference signals, so-called cell-specific reference signals (CRSs) are used for demodulation.

The main advantage of feedback-based CSI acquisition is that it does not rely on reciprocity and can thus be used for both FDD and TDD. A disadvantage is the signaling overhead required to feed back the CSI, especially if high-resolution CSI is needed. This can be prohibitive for multiantenna transmission schemes that need CSI with high resolution, e.g., MU-MIMO, if the number of antenna ports is large. Feedback-based CSI acquisition is also closely tied to the standard, since it must be specified how the channel should be represented. For example, with codebook-based precoding a codebook for each supported number of antenna ports needs to be specified.

A way to reduce the feedback overhead and CSI computation complexity with codebook-based CSI is to design the codebook based on prior knowledge about a certain channel structure. For example, if the channel is expected to be correlated between antenna elements a smaller codebook can be used than if the channel is uncorrelated, since then the channel coefficient for one antenna element is related to the others. A high correlation between antenna elements is a reasonable assumption for closely spaced, co-polarized antenna elements in a channel with limited angular spread, e.g., when using a tower-mounted macro BS antenna. Several codebooks in the LTE and NR specifications have been designed based on this assumption.

To appreciate the construction of such a codebook, consider the array response vector $\mathbf{a}(\theta)$ for a ULA with N elements when a single plane wave from direction θ relative to the array boresight is impinging upon the array. This vector can be written as

$$\mathbf{a}(\theta) = \begin{bmatrix} 1 & e^{-j2\pi \frac{d}{\lambda} \sin\theta} & \cdots & e^{-j2\pi (N-1) \frac{d}{\lambda} \sin\theta} \end{bmatrix}^T \tag{7.36}$$

where λ is the carrier wavelength and d is the distance between two adjacent elements. Making the substitution $\psi = d\sin(\theta)/\lambda$ we obtain

$$\mathbf{a}(\psi) = \begin{bmatrix} 1 & e^{-j2\pi\psi} & \dots & e^{-j2\pi(N-1)\psi} \end{bmatrix}^T. \tag{7.37}$$

Hence, $\mathbf{a}(\psi)$ has the same structure as the vectors in a discrete Fourier transform (DFT) matrix. Beam-forming in a direction θ can be achieved by applying a weight vector $\mathbf{w} = \mathbf{a}^*$, where * denotes complex conjugate, to the array. A codebook designed for a ULA can therefore be constructed by DFT vectors corresponding to beam-forming in a number of hypothesized directions, resulting in a so-called DFT codebook. A set of beams corresponding to beam-formers in different directions is often called a grid-of-beams (GoB).[23]

DFT codebooks are used extensively in LTE and NR for codebook-based CSI acquisition. For a ULA with N elements, the kth precoding vector in such a codebook is given by

$$\mathbf{w}_{\text{ULA}}(k) = \frac{1}{\sqrt{N}}\begin{bmatrix} 1 & e^{j2\pi\frac{k}{QN}} & e^{j2\pi(N-1)\frac{k}{QN}} \end{bmatrix}^T, \; k = 0, 1, \dots, QN-1. \tag{7.38}$$

Here, Q is a so-called oversampling factor which is used to obtain a finer granularity in the angular domain than what a codebook with N orthogonal DFT vectors would provide. For a UPA, a corresponding precoding vector is obtained by the Kronecker product of the 1-D precoding vectors in each dimension, i.e., $\mathbf{w}_{\text{UPA}}(k,l) = \mathbf{w}_{\text{ULA}}(k) \otimes \mathbf{w}_{\text{ULA}}(l)$, where $\otimes$ denotes the Kronecker product. A common BS antenna array architecture is the UPA with dual-polarized elements. Since orthogonal polarizations typically fade independently, a reasonable approach is to have DFT vectors per polarization and a phase difference between the DFT vectors for different polarizations. Such codebooks have been constructed for both LTE and NR and contain vectors of the form

$$\tilde{\mathbf{w}}_{\text{UPA}}(k,l,\phi) = \begin{bmatrix} \mathbf{w}_{\text{UPA}}^T(k,l) & e^{j\phi}\mathbf{w}_{\text{UPA}}^T(k,l) \end{bmatrix}^T. \tag{7.39}$$

DFT-based codebooks are also used when transmitting multiple precoded layers in spatial multiplexing. The codebooks therefore contain different combinations of DFT vectors with one vector for each layer. An entry in the codebook that corresponds to multilayer transmission is therefore a matrix. When the BS has received a PMI from a UE it can use the corresponding precoding vector for subsequent data transmissions. Alternatively, it can interpret the PMI as a quantized representation of the channel and design another precoder based on this and possibly also other information.

The feedback overhead associated with codebook-based CSI acquisition can be reduced further by exploiting that some channel properties are frequency selective, while others may be the same over the system bandwidth. For example, directional properties are typically the same over the bandwidth, while polarization properties are frequency selective. By separating the codebook into a wideband and a frequency-selective part, the feedback overhead can be reduced since the two parts can be reported with different frequency granularity. An example of this is the so-called dual-stage codebook introduced in LTE Release 10, in which a precoding matrix in the codebook is factorized as $\mathbf{W} = \mathbf{W}_1\mathbf{W}_2$, where

[23] The beam-forming vectors in a GoB do not necessarily have to be DFT vectors.

$\mathbf{W}_1$ captures the long-term/wideband properties and $\mathbf{W}_2$ the short-term/frequency-selective part. Such a codebook is useful for dual-polarized arrays with closely spaced antenna elements. In this case, $\mathbf{W}_1$ can perform beam-forming over co-polarized antenna elements while $\mathbf{W}_2$ performs co-phasing of polarizations.

Advances in active array antenna technology leads to the circumstance that the number of digital elements in a BS array antenna is continuously increasing. Obtaining CSI for a large number of antenna ports based on codebooks can be prohibitive if this number becomes too large, since it leads to a large overhead and CSI computational complexity. It will also lead to a lower antenna gain for each antenna port if the port now becomes connected to a single antenna element instead of a subarray of elements. This in turn implies a lower SNR in the channel estimation and thereby a lower CSI quality.

A remedy to these problems is to dynamically beam-form the reference signals in a UE-specific manner. In this way the channel is estimated in beam space instead of element space. This leads to a lower required number of reference signals and also a higher antenna gain for the transmitted reference signals. Reference signals can then be transmitted in a few beams instead of on many antenna elements. The UE then measures the channel quality on each beam-formed reference signal and reports the preferred beam and CSI for the selected beam. However, this approach requires some prior knowledge about the direction to a UE. This can be obtained by, e.g., uplink measurements or from previous downlink reference signal transmissions. Since only directions are of interest, full reciprocity is not necessary, so uplink measurements can be used also for FDD. Previous reference signal transmissions could be transmitted with a low density in time and/or frequency to give coarse estimates about directions without incurring too much overhead. More dense beam-formed reference signals can then be transmitted in these directions to acquire detailed channel knowledge in this angular region. A beam tracking procedure for moving UEs can also be performed where a few candidate beams centered around the active beam are monitored by the UE.

Beam-forming of reference signals can be beneficial when the BS has many antenna elements and when the number of UEs in the cell is low. An advantage for the UE is reduced complexity in the CSI calculations since only a few beams have to be evaluated instead of many precoder candidates. An example of CSI acquisition using beam-formed reference signals is beam-formed CSI-RS introduced in LTE Release 13, also called CSI feedback class B [13].

7.2.7 MASSIVE MIMO

Massive MIMO has received considerable attention in both the academic research and the wireless industry in recent years and is a key technology component for 5G, both for NR and the evolution of LTE. The term has a relatively well-defined meaning in academic circles, while the wireless industry often uses it in broader terms. In academic parlance, massive MIMO is usually associated with a digital array having a large number (hundreds) of antenna elements serving much fewer (tens) UEs using reciprocity-based MU-MIMO. It is often considered to be restricted to TDD since obtaining CSIT in an FDD system is difficult if the number of antenna elements is large. In the wireless industry, the term 'massive MIMO' is often used for large antenna arrays with somewhat less number of elements which are not necessarily used for MU-MIMO. The industry sometimes also uses the term 'massive MIMO' for analog beam-forming with many antenna elements at millimeter-wave frequencies.

Massive MIMO was initially introduced as an asymptotic notion of letting the number of base station antennas grow to infinity [31]. This leads to several interesting theoretical results, such as that simple linear precoding with MRT is optimal and that the impact of fast fading, noise, and some types of interference and hardware impairments vanish thanks to averaging over infinitely many antennas. In order not to clutter the entire radio resource grid with pilots, channel reciprocity is usually assumed, so that the downlink channel state can be inferred from uplink measurements. In theory, this could enable a very high capacity with simple transceivers and scheduling strategies. Of course, in any real-world implementation the number of antenna elements has to be finite. However, massive MIMO does represent a paradigm shift from using a few high-end radio transceivers in traditional base stations to a large number of transceivers with relaxed quality requirements per transceiver. Indeed, several field trials from both academia and the industry have shown that impressive spectral efficiencies can be achieved with massive MIMO; see, e.g., [17,36] and Section 1.3.4.

Regardless of how massive MIMO is defined, antenna arrays with a large number of elements are instrumental to fulfilling the performance requirements for 5G, whether they are used for reciprocity-based MU-MIMO to increase capacity at lower frequencies, or for analog beam-forming providing coverage at millimeter-wave frequencies. We do not pursue the massive MIMO-specific discussion any further here, since many of the issues associated with massive MIMO, such as CSI acquisition and precoder design, are treated in other sections of this chapter.

7.3 MULTIANTENNA TECHNIQUES IN NR

For low frequencies, multiantenna techniques in NR build to a large extent upon those in later releases of LTE. However, NR does contain a number of improvements that can increase the capacity and make it easier to adapt to diverse use cases. A major enhancement is a new flexible, modular, and scalable CSI framework. This framework also includes a high-resolution CSI reporting mode targeting improved MU-MIMO operation. For high frequencies, there are more fundamental changes, since LTE was not designed for millimeter-wave spectrum. The main new feature is the support for analog beam-forming in both the BS and the UE. This is called beam management and is described in Section 7.3.4.

The first part of this section discusses MIMO transmission over digital antenna ports and how to acquire CSI to enable such a transmission. The second part is about establishing and maintaining beam pair links for analog beam-forming, i.e., beam management. While the first part is focused on low frequencies and the second part on high frequencies, the techniques can be combined using, e.g., hybrid beam-forming with multipanel arrays. Beam management can then be used to determine the analog beam-formers in the panels and the CSI acquisition and MIMO techniques to determine the digital precoding across panels.

In LTE, the different downlink multiantenna transmission techniques are specified in ten different transmission modes. There are transmission modes for supporting diversity, beam-forming, spatial multiplexing, and CoMP transmission. Beam-forming and spatial multiplexing can be combined using precoded spatial multiplexing transmission modes. LTE supports open- and closed-loop codebook-based precoding as well as non-codebook-based precoding. There is also a transmission mode for MU-MIMO transmission.

The transmission modes also differ in which reference signals are used for demodulation and how CSI is acquired by the UE and fed back to the network. In the early LTE releases, demodulation and CSI acquisition was based mainly on CRS transmissions. These are transmitted on every antenna, in every subframe and resource block and are common for all UEs in a cell. Initially, LTE was designed for a relatively few number of antennas. As the number of supported antennas was increased in later releases, new reference signals were introduced in order to reduce the large overhead a CRS-based transmission would incur. Separate reference signals for demodulation and CSI acquisition were introduced in Release 10 with DM-RS[24] and CSI-RS.

Reference signals for coherent demodulation typically require higher time-frequency density than what is needed for CSI used for the selection of transmission parameters. By having separate reference signals for demodulation and CSI acquisition the density can be optimized for their respective purpose. Furthermore, reference signals for demodulation need only to be transmitted when there is data to transmit and there is need to occupy only the bandwidth scheduled for the data transmission. Moreover, and perhaps most importantly, the number of DM-RS ports scales with the number of layers rather than the number of antenna elements. By letting DM-RS be a UE-specific signal that is transmitted only when there is data to transmit to the UE, the overhead and intercell interference induced by the reference signals can be reduced compared to the always-on CRS. This is particularly important at low traffic load where CRS interference could limit the data transmission performance. DM-RS can also be precoded with an arbitrary precoder, e.g., with one of the interference suppression MU-MIMO precoders described in Section 7.2.2.2. This is transparent to the UE, so it does not need to know which precoder the BS has applied. Furthermore, CRS is transmitted over the whole bandwidth even if there is no data traffic in the cell, leading to unnecessarily high energy consumption.

For these reasons, CRS has been removed in NR. This provides a more lean, energy-efficient, flexible and scalable design. Furthermore, the different transmission modes have been removed in NR. There is only one downlink data transmission scheme, similar to TM 10 in LTE, in which channel estimation for demodulation and CSI acquisition is based on DM-RS and CSI-RS, respectively. In LTE the different functions of CSI configuration, measurement, reporting, and multiantenna transmission were tightly coupled. This has lead to multiple transmission modes, CSI classes, and configurations and many different options to choose from. In NR, these functions have been decoupled to allow for a more flexible configuration, making it easier to optimize for diverse use cases. This also makes it more scalable and renders the introduction of future enhancements easier. For example, NR provides more flexibility in the reference signal placement and density in the time-frequency grid, dynamic triggering of different types of reports, and fast feedback.

Besides a more flexible CSI acquisition framework, NR also contains enhancements targeting improved MU-MIMO transmission such as increased number of MU-MIMO layers, high-resolution CSI feedback and improved interference measurements. Since TDD is expected to be more common in NR deployments, it also has improved support for reciprocity. This includes increased SRS capacity and coverage by allowing multiple SRS symbols in one slot, SRS switching when a UE has fewer Tx than Rx branches, and so-called non-PMI feedback for improved link adaptation when the BS lacks information regarding the downlink interference (see Section 7.3.1.1).

7.3.1 CSI ACQUISITION

As alluded to previously, the purpose of the new CSI acquisition framework in NR is to decouple different CSI functionalities so that they can be configured independently leading to a modular, flexible,

[24]DM-RS was supported already in the first LTE release but was then only used for single-layer data transmission in TM 7.

and scalable design where it becomes easier to introduce new features and adapt to different use cases. The framework is common for MIMO transmission and beam management.

NR defines one or more report settings and one or more resource settings. A report setting is then linked to one resource setting for channel measurement and one resource setting for interference measurement. A resource setting defines which reference signal resources should be used for measurements, what type of reference signal, and the time-domain behavior of the resource configuration, i.e., if it is periodic, semi-persistent, or aperiodic. Two different types of reference signals are used for CSI acquisition in NR, CSI-RS and the synchronization signal block (SSB). The SSB consists of the PSS, SSS, and physical broadcast channel (PBCH). CSI acquisition using SSB is only used for beam management; see Section 7.3.4. A report setting tells the UE how the reporting shall be performed and what a CSI report shall contain. For example, it can include which CSI parameters to report, how the reporting shall be made in the time and frequency domain, e.g., periodic/semi-persistent/aperiodic and wideband/subband, measurement restrictions, codebook configuration, etc. The network can dynamically select one or multiple report settings to trigger the desired CSI reports.

Different transmission schemes and use cases can have different CSI requirements. SU-MIMO transmission typically has lower requirements on the CSI resolution than MU-MIMO, since there are better abilities to suppress SU-MIMO interlayer interference in the UE receiver. This is due to the fact that the number of transmitted SU-MIMO layers cannot be larger than the number of Rx antennas in the UE. In MU-MIMO transmission the total number of layers can be larger than the number of Rx antennas in the UE, which makes it more difficult to suppress interference in the receiver. Therefore, MU-MIMO transmission needs to some extent to rely on transmitter interference suppression, e.g., ZF or MMSE precoding, and this requires high-resolution CSI in order to be effective.

For downlink MIMO transmission, NR has two different CSI reporting types in order to support different CSI requirements. These are called Type I and II, respectively. Type I is a moderate resolution reporting mode targeting SU-MIMO operation while Type II has higher CSI resolution aimed at supporting MU-MIMO transmission, at the expense of a higher feedback overhead. Type I CSI gives information only about the strongest channel direction, while Type II captures channel multipath by representing the channel as a linear combination of multiple orthogonal DFT vectors. Type II is suitable for FDD since detailed CSI cannot be obtained by reciprocity in this case. It can also be useful for TDD if full reciprocity cannot be achieved, e.g., due to uncalibrated transceivers.

Feedback-based CSI in NR is based on codebooks. A number of codebooks have been defined, supporting up to 32 antenna ports. They all have a similar structure and are based on a dual-stage codebook where a precoder matrix is factorized into two components, $\mathbf{W} = \mathbf{W}_1\mathbf{W}_2$, where $\mathbf{W}_1$ captures the wideband channel properties and $\mathbf{W}_2$ the frequency-selective part. An example of a wideband channel property is the dominant propagation directions in the channel, while frequency-selective fading obviously is a frequency-selective property. The codebooks have been designed for dual-polarized UPAs for which the dual-stage codebook is a suitable design. For such arrays, dominant channel directions are conveniently modeled by 2-D DFT vectors for each polarization. The wideband matrix $\mathbf{W}_1$ is therefore composed of one or multiple such DFT vectors. Applying $\mathbf{W}_1$ as a precoder matrix can then be interpreted as transmitting with beams pointing in the dominant channel directions by using the corresponding DFT vectors as beam-forming weight vectors. The frequency-selective part of the dual-stage codebook, $\mathbf{W}_2$, accounts for co-phasing of polarizations, beams, and/or panels.

Type I CSI reporting consists of a single-panel and a multi-panel codebook. The single-panel codebook is similar to the full-dimension MIMO (FD-MIMO) codebooks in LTE Release 13 [13] and supports transmission ranks up to eight. For rank one, it amounts to selecting a single beam from an oversampled grid of DFT beams and co-phasing of the polarizations. The DFT vector corresponding to the selected beam is then contained in $\mathbf{W}_1$, while the co-phasing factors are in $\mathbf{W}_2$. The multi-panel codebook is an extension of the single-panel codebook by including co-phasing of panels in a multi-panel array. Up to four panels are supported. The co-phasing can be either wideband or frequency selective.

In Type II CSI reporting, $\mathbf{W}_1$ contains DFT beam-forming vectors for multiple selected beams and $\mathbf{W}_2$ is a beam combining matrix. Up to four beam-forming vectors can be selected for $\mathbf{W}_1$ from a set of orthogonal DFT vectors. The set of orthogonal DFT vectors can be rotated so that the corresponding beams are better aligned with the dominant directions in the channel. The $\mathbf{W}_2$ matrix contains frequency-selective co-phasing factors for combining the beams selected in $\mathbf{W}_1$. There is also an amplitude scaling factor for each beam which can be wideband or a combination of wideband and frequency selective. The beam selection for $\mathbf{W}_1$ is common for both polarizations and, in the case of rank two, both layers while the co-phasing and amplitude scaling is selected independently per polarization and layer. The precoding vector for polarization r, layer l, and a particular frequency subband is thus represented as

$$\mathbf{w}_{r,l} = \sum_{k=1}^{K} \mathbf{a}_k b_{r,l,k} e^{j\phi_{r,l,k}} \tag{7.40}$$

where the $\mathbf{a}_k$ are the selected beam-forming vectors, $b_{r,l,k}$ is the amplitude scaling factor, and $\phi_{r,l,k}$ the phase. The $\mathbf{a}_k$ are the same for all subbands, $\phi_{r,l,k}$ is selected for each subband, and the selection of $b_{r,l,k}$ is either wideband or a combination of per subband and wideband. In this way, channel multipath can be conveyed by the CSI report, which makes MU-MIMO precoding with interference suppression more effective. Construction of a precoding beam using Type II CSI reporting is illustrated in Fig. 7.10.

The Type II codebook supports only transmission ranks one and two in order to keep the feedback overhead at a reasonable level. Due to the relatively high feedback overhead of Type II reporting it is suitable for low mobility UEs. Tracking a high-mobility UE with high spatial resolution would also require a high CSI-RS density. Type I reporting may therefore be a better choice for high-mobility UEs. The Type II beam combining codebook is intended to be used with non-beam-formed CSI-RS. Type II also has a port selection codebook for beam-formed CSI-RS. The $\mathbf{W}_1$ matrix is then a port selection matrix and amplitude scaling and co-phasing is performed in the same way as for the beam combining codebook.

Although the CSI report contains indices to parameterized precoding vectors, the BS can use any precoder in the data transmission since the DM-RS is precoded in the same way as data. The CSI report is only a recommendation of transmission parameters which the BS can choose to follow or not. The CSI report can be used by the BS to reconstruct the vector in (7.40), which can be interpreted as an approximation of the channel as estimated by the UE. Based on this, the BS can for example apply one of the interference suppression MU-MIMO precoders described in Section 7.2.2.2.

The Type II codebook is similar to the LTE Release 14 advanced CSI codebook but allows for selection of more beams and has finer granularity in the amplitude and phase quantization. Furthermore, for arrays with more than 16 antenna ports there is a restriction in LTE that beam selection can be made only from a subset of all possible beams, a restriction that has been removed in NR. These enhancements can give a substantial performance gain of the NR codebook compared to the LTE codebook. Simulation results in [38] for MU-MIMO transmission in the 3GPP UMi scenario showed 24%

and 56% increased average and cell edge user throughput, respectively, of the NR Type II codebook over the LTE Release 14 advanced CSI codebook. A Type II codebook for 32 BS Tx antennas using four beams and combined wideband and subband feedback of the amplitude scaling was used in the simulations.

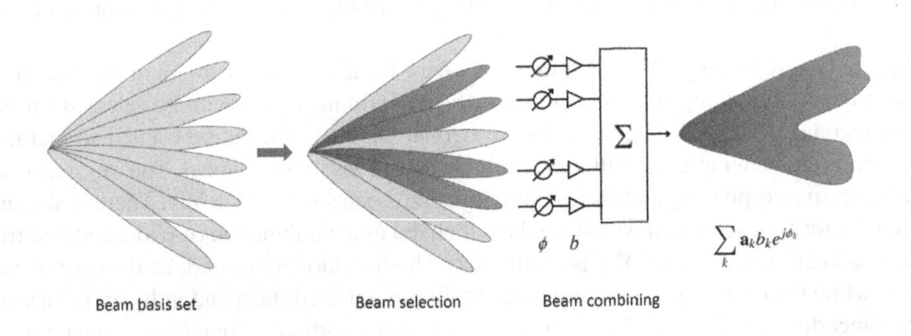

FIGURE 7.10

Type II CSI. Illustration of CSI Type II codebook.

7.3.1.1 Interference Measurements

The discussion of CSI has so far concerned the channel to the served UEs. Another important CSI component is the interference experienced by the UE. In the early LTE releases it was not specified how the UE should estimate the interference; it was up to UE implementation [13]. A typical approach was to estimate the intercell interference based on CRSs. A problem with this approach is that CRS is always transmitted regardless of the traffic load. At low load the intercell interference will therefore be dominated by CRS transmissions from neighboring cells. This means that CRS-based interference estimation will overestimate the interference on the data channel at low traffic load. Another problem is that the network has no control over how the UE averages its interference estimates.

To improve the interference estimation, a so-called CSI-IM configuration was introduced in LTE Release 11 [13]. This allows the network to control in which resource elements the UE should measure interference. A CSI-IM resource is just a CSI-RS resource in which nothing is transmitted from the BS to the UE that has been configured with the resource, a so-called ZP CSI-RS. Since the CSI-IM collides with data transmissions from other cells instead of CRSs, interference estimation at low traffic load can be improved.

In NR, the bandwidth of CSI-IM is configurable so that the UE can be configured to measure interference on only a part of the frequency band. This can be useful when mixing different services in different parts of the frequency band, so that interference is measured only in the part that is occupied by the service that the UE uses.

In LTE, CSI-IM is used for intercell interference estimation in SU-MIMO operation. For MU-MIMO it is important to also consider multiuser interference between the co-scheduled UEs within the cell. NR has therefore introduced support for multiuser interference measurements based on non-zero-power CSI-RS (NZP CSI-RS). Each UE in a multiuser group is then configured with one NZP CSI-RS resource for channel measurement and one NZP CSI-RS resource per co-scheduled UE for multiuser interference measurements. CSI-IM with ZP CSI-RS can still be used for intercell interference estimation in NR.

With reciprocity-based CSI acquisition, the downlink channel can be estimated from uplink sounding. However, in order to choose proper transmission rank and link adaptation parameters, the BS needs to know the SINR in the UE, which also depends on the interference. To convey this information to the BS for reciprocity operation, NR supports so-called non-PMI feedback where the UE reports RI and CQI, but not PMI. The BS can estimate the downlink channel based on SRS transmission from the UE and design a precoding matrix based on the channel estimate. The BS then transmits precoded CSI-RS with the determined precoding matrix and the UE can calculate RI and CQI based on the precoded CSI-RS taking its particular receiver algorithm into account. If the BS performs subsequent data transmission with the same precoding matrix, there is no mismatch between the data precoding and the precoding assumed in the CSI calculation.

7.3.2 DOWNLINK MIMO TRANSMISSION

The downlink MIMO transmission in NR is mainly based on non-codebook-based precoding using CSI-RS for CSI acquisition and DM-RS for coherent demodulation. For reciprocity-based operation, CSI can be obtained by uplink sounding complemented by RI and CQI feedback. For feedback-based CSI, the codebooks described in the previous section can be used. In this case, a typical downlink MIMO transmission consists of the BS first transmitting CSI-RS for CSI acquisition. The CSI-RS could be beam-formed or transmitted per antenna element. The UE estimates the channel and calculates CSI based on the channel estimate. The CSI can consist of RI, PMI, and CQI. For beam-formed CSI-RS, it can also include CRI to indicate the best beam. The UE then sends a CSI report to the BS which uses the report to determine MIMO transmission parameters and performs the data transmission. NR supports MU-MIMO transmission with up to 12 layers with orthogonal DM-RS ports. Up to eight UEs can be spatially multiplexed. In SU-MIMO transmission, a UE can receive up to eight layers. Hybrid beam-forming with multipanel arrays can be performed by using beam management for the analog beam-forming and the multipanel codebook for the digital precoding.

While LTE supports various Tx diversity schemes such as SFBC, FSTD, and large-delay CDD, Tx diversity is currently not explicitly supported in NR and has to be performed in a specification-transparent manner, e.g., using precoder cycling in frequency.

7.3.3 UPLINK MIMO TRANSMISSION

NR Release 15 supports two transmission schemes for the uplink data channel, i.e., physical uplink shared channel (PUSCH): codebook-based transmission and non-codebook-based transmission. Tx diversity for PUSCH is not specified and thus needs to rely on specification-transparent methods. For DFTS-OFDM codebook-based uplink transmission is limited to rank one, since the use case for DFTS-OFDM is coverage limited UEs. For CP-OFDM, codebook-based uplink transmission supports up to rank four.

Codebook-based transmission can be used for both FDD and TDD. Since it is based on CSI feedback from the BS, it can be used when reciprocity does not hold, e.g., for FDD or for TDD when a UE is not reciprocity-calibrated. In codebook-based uplink transmission for NR, the UE transmits non-precoded SRSs. A UE can transmit one or two SRS resources and an SRS resource can have up to

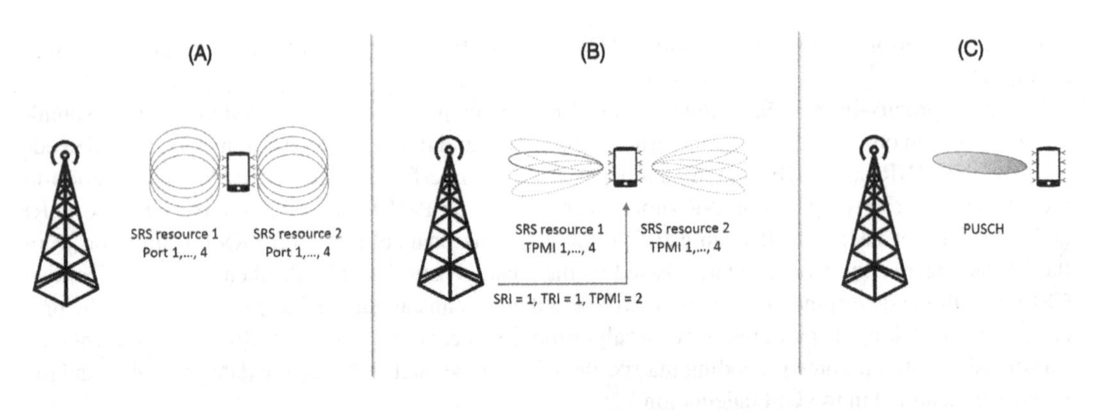

FIGURE 7.11

Codebook-based uplink transmission. An example of codebook-based uplink transmission.

four ports. The BS determines CSI based on the received SRSs and instructs the UE to use the determined CSI parameters. In this case CSI consists of SRS resource indicator (SRI),[25] transmit precoder matrix indicator (TPMI), and transmit rank indicator (TRI). SRI indicates the selected SRS resource, TRI the preferred transmission rank, and TPMI the preferred precoder over the ports in the selected resource, where the precoder is selected from the uplink codebook. The UE then performs the uplink transmission based on the CSI report from the BS.

An example illustrating codebook-based precoding is shown in Fig. 7.11. In this example, the UE has two linear arrays with four antennas each on opposite sides of the device. An SRS resource can in this example correspond to one array and the ports within an SRS resource to the antenna elements within an array. In step (A) in Fig. 7.11, the UE transmits SRS resource 1 on the left array and SRS resource 2 on the right array. The ports within a resource are mapped to the antenna elements in each array. In step (B), the BS evaluates the different precoding matrices in the codebook for the two SRS resources and determines the best SRS resource, transmission rank, and precoding matrix and signals SRI, TRI, and TPMI to the UE. In this example, SRI = 1, TRI = 1, and TPMI = 2. The dashed beams in (B) illustrate all beams in the codebook and the solid beam illustrates the selected beam corresponding to SRI = 1 and TPMI = 2. Finally, in step (C), the UE uses the selected precoding matrix for the PUSCH transmission.

For codebook-based transmission, NR supports three levels of coherence capabilities of a UE: full, partial, and non-coherent. A UE with full coherence can transmit coherently over all antenna ports, i.e., it can control the relative phase between all Tx chains. A UE with partial coherence is able to transmit coherently over pairs of antenna ports, and a non-coherent UE cannot transmit coherently over any antenna ports. The uplink codebook consists of different parts adapted to the different UE coherence capabilities. The part for full coherence is similar to the NR downlink codebook and the part for partial coherence consists of precoders that combine ports within coherent pairs. For non-coherent UEs, a part with one layer per antenna port is used. For example, rank one transmission in this case uses a port

[25] SRI is not reported if only a single SRS resource has been configured.

selection codebook. The BS can configure the UE to use a subset of the entire codebook depending on its coherence capability. The codebook subset can be "full + partial + non-coherent", "partial + non-coherent", or "non-coherent". For example, a fully coherent UE is allowed to use the entire uplink codebook, while a non-coherent UE can only use the non-coherent part.

When reciprocity holds, non-codebook-based transmission is an option. The SRS transmission can then be precoded. The BS can transmit CSI-RS to help the UE to design suitable precoders for the SRS transmission. In this case the UE can transmit up to four SRS resources where each resource has one port. The BS then determines one or multiple SRIs based on the received SRSs and the UE transmits one layer per SRI after it has received the report from the BS. Hence, in this case the transmission rank is equal to the number of SRIs. One way to use non-codebook-based transmission is that the UE designs precoders using channel estimates based on received CSI-RSs and transmits precoded SRS using these precoders; one SRS resource per precoder. The BS then selects one or multiple precoders and indicates these to the UE by reporting the corresponding SRIs. The UE then transmits one layer per indicated SRI using the corresponding precoder. An example that illustrates this is shown in Fig. 7.12. In step (A), the BS transmits CSI-RS from which the UE can estimate the channel. Based on this estimate, the UE designs precoders for the SRS transmission. In step (B), the UE transmits two precoded SRS resources, each with a single port. In step (C), the BS determines the best SRS resource and signals this to the UE with an SRI, in this case SRI = 1. In step (D), the UE performs data transmission using the same precoder as was used for the SRS resource indicated by the BS.

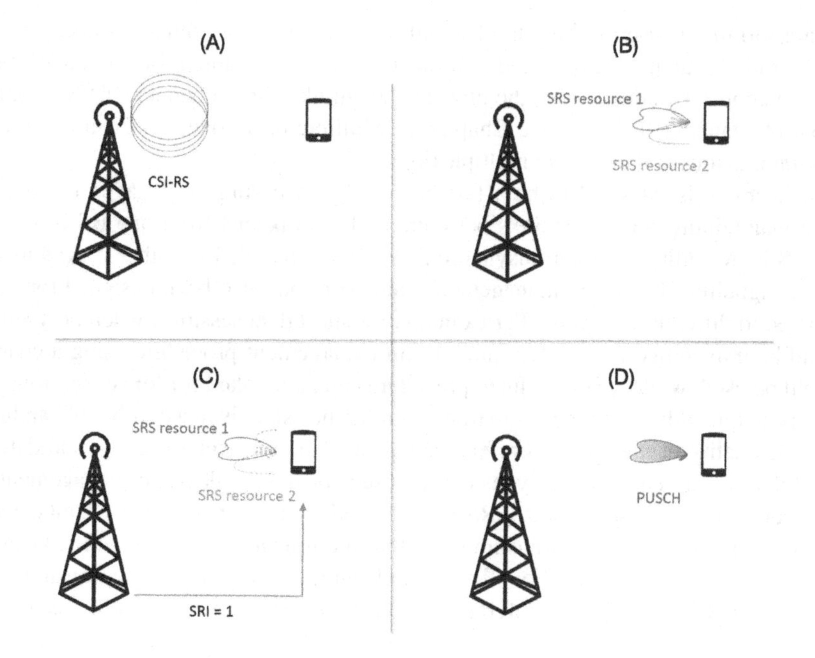

FIGURE 7.12

Non-codebook-based uplink transmission. An example of non-codebook-based uplink transmission.

Uplink Tx diversity is not explicitly specified, but open-loop Tx diversity can be performed transparently by, e.g., frequency hopping. Furthermore, Tx antenna selection diversity can be performed by utilizing SRI or CSI-RS.

7.3.4 BEAM MANAGEMENT

One of the main new features in NR is the support for analog beam-forming, which is foreseen to be prevailing at millimeter-wave frequencies. For this purpose a new framework called beam management has been developed in order to support analog beam-forming at both the BS and the UE side. Beam management has been defined in 3GPP as a set of Layer 1/2 procedures to acquire and maintain a set of BS and/or UE beams[26] that can be used for downlink and uplink transmission/reception [1]. It includes a number of features, such as:

- Sweeping. Covering an angular sector by sweeping analog beams over the sector.
- Measurement. Measuring the quality of different beams.
- Reporting. Reporting beam information such as which beams are best and their measured qualities.
- Determination. Selecting one or a few beams out of a number of candidate beams.
- Indication. Indicating which beam or beams has been or have been selected for data transmission.
- Switching. Switching to another beam if another beam gets higher quality than the current beam.
- Recovery. Finding a new beam if the current beam cannot maintain a communication link due to, e.g., blockage.

If analog beam-forming is used in both the BS and the UE, beams at both sides need to be found so that beam pairs rather than just beams need to be acquired and maintained. Beam management is used for both data and control channels, i.e., the physical downlink shared channel (PDSCH) and physical downlink control channel (PDCCH) (see Chapter 2). Multiple beam pair links can be established for robustness, joint transmission or spatial multiplexing.

Beam management is intended to be a fast process by involving only Layer 1 and 2 signaling. It is aimed at maintaining beam pair links between a UE and beams from one BS or several tightly synchronized BSs. Mobility between unsynchronized BSs is handled by other procedures involving slower Layer 3 signaling. The beam management framework has also been designed for flexibility so it can be adapted to different use cases. For example, for short data sessions, when only small amount of data should be transmitted/received, a quick beam management procedure using a coarse level of accuracy could be used, while a more refined procedure can be invoked for longer sessions.

A general principle of beam finding is to transmit reference signals in a number of candidate beams and estimate the quality of the received reference signal at the receiver for each candidate Tx beam. In NR, two different reference signal types can be used for downlink beam management: SSB and CSI-RS. For uplink beam management, SRS can be used. A reference to the beam(s) with highest quality is then reported back to the transmitter so that the transmitter knows which beam is best to use in subsequent data transmissions. For NR downlink, an index to a previously transmitted reference signal (SSB or CSI-RS) is used as a reference to the best beam, so-called SSB resource indicator

[26] The NR specifications seldom use the term "beam". What we mean by beam herein is what in the NR specifications is referred to as *spatial domain Tx filter* or *spatial domain Rx filter*.

(SSBRI) or CRI. Similarly, the receiver can find its best Rx beam by estimating the quality of the received reference signal for each candidate Rx beam. During the Rx beam finding procedure, the transmitter should repeat the transmission of the reference signal in the same Tx beam so that the receiver can make a fair comparison of the candidate Rx beams. These procedures should be repeated whenever needed to track UE movements and changes in the radio environment.

The main focus of this section is on downlink beam management. In many cases uplink beam management is not needed. If the BS and UE has so-called *beam correspondence*, the beam pair links established by downlink beam management procedures can also be used for uplink without any need for separate uplink procedures. Beam correspondence is a kind of reciprocity that implies that a BS or UE can determine its Tx beam based on measurements on its Rx beams, or vice versa. For example, if a set of beam-forming weights gives the same beam pattern for the Tx and the Rx, beam correspondence holds. If beam correspondence does not hold, separate uplink beam management procedures can be invoked. These are similar to the corresponding downlink procedures. Uplink specific beam management aspects are summarized at the end of this section.

7.3.4.1 Beam Acquisition During Initial Access

Before a UE enters the network, a BS does not have any information about the direction to the UE. The signals in the initial access procedure therefore need to be transmitted and received without any prior knowledge about the direction to the UE. At low frequencies, these signals can be transmitted with a wide beam that covers the entire angular sector of the cell. At millimeter-wave frequencies, however, the antenna gain with such a wide beam may not be sufficient to achieve the desired coverage. Therefore, NR supports beam-forming also of initial access signals. Since the direction to a UE is not known a priori, the set of beams carrying the initial access signals must cover the entire sector, e.g., by beam sweeping.

The first signals that the BS transmits to aid a UE to access the network are contained in the SSB. The purpose with SSB is to provide coarse time and frequency synchronization and basic system information such as how the UE shall access the system, indication of the physical cell identity (ID), and where to find the remaining configurations. The SSB is transmitted periodically and can be beam-formed in order to ensure sufficient cell coverage. The SSB is then transmitted repeatedly in different beams, a so-called SSB burst set. For carrier frequencies above 6 GHz, up to 64 beams can be used within an SSB burst set of 5 ms. This burst set is then repeated with a specified periodicity. For initial cell selection the default value of the SSB burst set periodicity is 20 ms. See Fig. 7.13 for an illustration of beam-forming of SSB.

During initial access, the UE measures the different SSBs in an SSB burst set in order to determine the best BS Tx beam. If the UE has analog beam-forming it can measure multiple SSB burst sets to find a suitable Rx beam, or it can use a wide beam. The UE then transmits the physical random access channel (PRACH) preamble in a resource that is associated to the SSB that corresponds to the determined BS Tx beam. If the UE has analog beam-forming and beam correspondence, the UE can transmit the preamble with a Tx beam that corresponds to the Rx beam it used when receiving the SSB from the best BS Tx beam. The BS can receive the PRACH preamble using a wide beam or the same beams it used for transmitting the SSB burst set. When the BS has received the PRACH preamble it can deduce from the corresponding PRACH resource which BS Tx beam was best for that UE. The BS can then use the identified Tx beam in possible beam refinement procedures or subsequent data and control transmission.

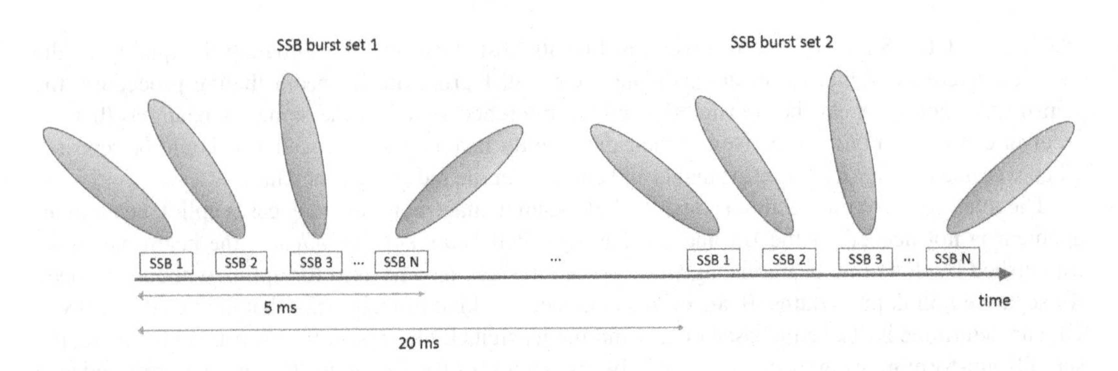

FIGURE 7.13

SSB burst. Two SSB burst sets with N beams in each set.

Since the SSB can be transmitted with a periodic beam sweep, it is also useful for beam management purposes. SSB can be used not only for initial beam acquisition but also for other purposes such as input to beam refinement procedures and discovery of new beams when a UE moves or if changes occur in the radio environment. In some cases, SSB may be all that is needed in beam management. The SSB transmissions can be used both by the BS and the UE to find suitable Tx and Rx beams.

To provide robustness against mobility and blockage and to reduce signaling overhead, the SSB beams can be relatively wide so that the cell can be covered by a few beams. For UEs with low data rate requirements or good channel conditions it may be sufficient to use the wide beams acquired during initial access also for the data transmission. In other cases, narrower beams with higher gain can be acquired by subsequent beam refinement procedures. The beams found during initial access can then be useful as input to the beam refinement procedures, e.g., to trigger a beam sweep with narrow beams centered around the direction estimated during initial access.

7.3.4.2 Beam Management Procedures

Although not explicitly stated in the specifications, downlink beam management has been divided into three procedures [1]:

- P-1. The purpose of P-1 is to find initial BS Tx beam(s) and possibly also UE Rx beam(s) by performing a beam sweep over a relatively wide angular sector.
- P-2. This is used for beam refinement of the BS Tx beam(s) by performing a beam sweep in a more narrow angular sector than in P1.
- P-3. This is used for performing an Rx beam sweep at the UE. In P-3, the BS Tx beam is fixed during the UE Rx beam sweep.

There are similarities between the procedures and not all procedures are needed. Furthermore, P-2 can be a special case of P-1. An example of how the P-1, P-2, and P-3 procedures can be performed is schematically illustrated in Fig. 7.14. In P-1, the BS performs a beam sweep over an angular sector that covers the entire cell by transmitting a unique reference signal in each beam. To limit the number of beams in such a wide beam sweep the beams could be relatively wide to give an initial, coarse estimate of the best beam direction. The reference signal could be, e.g., the SSBs during initial access or a periodic CSI-RS transmission that has been configured for beam management. The UE measures the power of the received reference signals from all BS Tx beams using a wide Rx beam and reports to the BS which beam has the highest received power. In P-2, the BS performs beam refinement by an aperiodic CSI-RS transmission using narrower beams in an angular sector around the best beam reported by the UE in P-1. The UE measures the power of the received CSI-RSs from these BS TX beams, still using a wide Rx beam, and it reports to the BS which of the narrow beams has the highest received power. In P-3, the BS transmits CSI-RS repeatedly in the best narrow beam reported by the UE in P-2 so that the UE can perform an Rx beam sweep to find its best Rx beam by measuring the power of the received CSI-RS in each Rx beam. In the data transmission, the BS uses the best BS Tx beam found during P-2 and the UE uses the best UE Rx beam found during P-3.

Note that this is just one example of how to perform beam management and other ways are possible. For example, P-1 could be a joint BS Tx/UE Rx beam sweep in which the UE sweeps its Rx beams for each BS Tx beam. The BS then has to repeat the reference signal transmissions in each BS Tx beam so that the UE can evaluate different Rx beams for every BS Tx beam. Therefore, this approach is more costly in terms of reference signaling overhead and beam acquisition time.

To provide robustness against blocking, a UE can be configured to monitor PDCCH on multiple beam pair links. For example, while data transmission is being performed on an active beam pair link, the UE can monitor PDCCH on another beam pair as a backup link for swift fallback if there should be a sudden blockage of the active link.

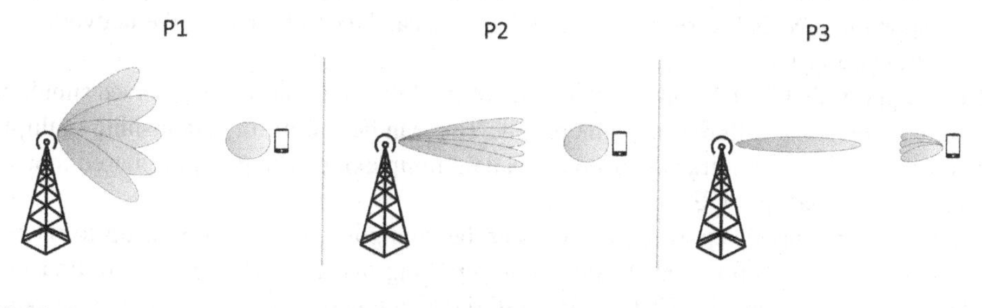

FIGURE 7.14

Beam management procedures. Schematic illustration of the beam management procedures P-1, P-2, and P-3.

Which beam management process to use is not explicitly configured. Instead, the CSI acquisition framework described previously is used to configure a UE with the pertinent report and resource settings. For beam management, a report setting can for example contain information regarding the number of beams to report on, which CSI parameters to report (e.g., L1-RSRP), time-domain behavior, and frequency granularity of the reporting. A resource setting can contain information regarding which reference signal type (CSI-RS, SSB) the UE shall measure on, time-domain behavior, and one or multiple resource sets each containing multiple CSI-RS resources.

7.3.4.3 Beam Measurement and Reporting

A central component in beam management is measuring and reporting of the quality of different candidate beams. In downlink beam management, measurements are performed by the UE on SSB or CSI-RS transmitted by the BS. Beam sweeping using CSI-RS is more flexible than SSB since there is more freedom in configuring the CSI-RS transmission. For example, the number of beams and their coverage area can be configured more flexibly. A CSI-RS beam sweep is UE-specifically configured, so it is possible to tailor a beam sweep for a particular UE. For example, the beams in a CSI-RS beam sweep can be narrow beams centered around a wide beam acquired from SSB during initial access. A CSI-RS beam sweep does not need to be transmitted periodically but can be triggered whenever needed. Other advantages with CSI-RS over SSB for beam management are that CSI-RS is more wideband and can use two antenna ports, which potentially could lead to more reliable beam measurements. For example, two antenna ports can be used for transmitting with two orthogonal polarizations in order to reduce polarization mismatch losses. An advantage with using SSB for beam management is that SSB is transmitted for initial access purpose anyway, so using it also for beam management incurs no additional reference signal overhead.

CSI-RS for beam management can be periodic, semi-persistent, or aperiodic. Periodic and semi-persistent CSI-RS can be efficient in wide beam sweeps for acquiring coarse beams when there is little a priori information as regards the directions to UEs, or for beam sweeps at high load for covering a service area containing multiple UEs. To save signaling overhead, aperiodic CSI-RS can be more efficient at low load to perform a refined, UE-specific beam sweep when there is some prior knowledge about the direction to the UE.

The quality metric used for beam management is L1-RSRP for both SSB and CSI-RS. In the case that CSI-RS resources with two ports have been configured, the UE shall report the linear average over both ports. Information as regards which beam the UE has selected is conveyed by reporting an index to the corresponding CSI-RS resource, i.e. CRI. The UE can be configured by the network to report a number of strongest beams.

NR also supports that the UE can report a group of BS Tx beams that can be received simultaneously by the UE, e.g., by using different antenna panels. This can be useful for establishing multiple beam pair links for robustness, joint transmission or spatial multiplexing. The grouping-based reporting can be turned on and off on a per UE basis.

For non-grouping-based reporting, the UE can be configured to measure on up to 64 beams in a beam sweep and report up to four beams in an reporting instance. The reports are transmitted on physical uplink control channel (PUCCH) or PUSCH depending on the reporting being periodic, semi-persistent, or aperiodic.

Measurement and reporting for beam management and CSI acquisition have a common framework so it can be used for both purposes by different configurations. At low frequencies with digital beam-forming, beam management is typically not needed. In this case, the UE can be configured to measure on a number of CSI-RS resources, select the best resources, and report CSI for these resources. At high frequencies with hybrid beam-forming the optimal approach would be to jointly optimize the analog beam-formers and the digital precoder. However, this may not be feasible due to high computational complexity and signaling overhead. A more practical approach is to determine the analog beam-formers and digital precoder separately. The analog beam-formers can then be determined using a beam management procedure followed by design of the digital precoder using the selected analog beams and the CSI acquisition framework. For example, the BS can first transmit SSB or CSI-RS configured for beam management in order to select beam pair links. The UE measures and reports indices and qualities of the preferred beam pair links. The BS can then transmit CSI-RS for CSI acquisition using the selected beam pair links. The UE measures on the CSI-RSs and reports CQI, PMI, and RI. The BS then transmits data using the selected analog beams and digital precoder.

7.3.4.4 Beam Indication

For downlink, NR supports beam management with and without beam indication. Beam indication is an indication to the UE which Tx beam the BS will use in coming data or control transmission so that the UE can update its Rx beam accordingly. This may be needed when analog beam-forming is used in the UE. If the BS changes its Tx beam the UE may need to change its Rx beam at the same time so that it matches the new BS Tx beam, or the link may be lost. It can also be needed for the UE to know which beam to receive a CSI-RS beam sweep in a limited sector (a P-2 procedure). To be able to do this with analog beam-forming, the UE needs to know beforehand which Rx beam to use so that it can switch to that beam when the data is transmitted with the new Tx beam. The downlink beam indication provides information to the UE so that it can use an Rx beam that is suitable for the new Tx beam.

Beam indication is needed when several beam pair links are maintained, e.g., when monitoring multiple PDCCHs, or when a joint BS Tx and UE Rx beam sweep is performed. In that case all combinations of BS Tx beams and UE Rx beams are evaluated, the UE reports the best beam pairs and the BS selects which Tx beam to use. A beam indication is then needed since the UE needs to know which Tx beam the BS uses in order to set its Rx beam.

In other cases, explicit beam indication may not be needed. For example, if a single beam pair link is used for PDSCH and PDCCH transmission and the UE reports only the best BS Tx beam in a beam sweep, the UE does not need any beam indication provided that the BS uses the reported beam in the next transmission. When operating without beam indication, different BS Tx beams are evaluated under the assumption that the UE holds its Rx beam fixed and different UE Rx beams are evaluated under the assumption that the BS holds its Tx beam fixed. The UE can be informed that the BS keeps its Tx beam fixed by a CSI-RS resource set configuration that contains an information element which indicates that the BS repeats its CSI-RS transmission in the same Tx beam. For example, in a sequential P-2/P-3 procedure the BS first performs a P-2 Tx beam sweep and the UE keeps its Rx beam fixed (or uses a wide beam). After the P-2 procedure, the BS updates its Tx beam without informing the UE and triggers a P-3 UE Rx beam sweep. After the P-3 procedure, the UE updates its Rx beam without informing the BS. The UE only needs to remember which Rx beam was best and use that information in the next reception without any beam indication from the BS. An advantage with operation without beam indication is the reduced signaling overhead and delay associated with beam indication.

A beam indication is a reference to a previously transmitted reference signal in the form of a spatial QCL relation. More specifically, it is an indication that the UE can assume that the DM-RS of PDSCH/PDCCH is spatially quasi-co-located with a previously transmitted downlink reference signal, e.g., a particular SSB or CSI-RS resource. This means that the UE can use the same Rx beam for the coming data/control transmission as it used when it received the indicated reference signal. Beam indication can also be used to provide a spatial QCL relation between different reference signals, e.g., that an aperiodic CSI-RS is spatially quasi-co-located with an SSB. This could be used for, e.g., beam refinement when an aperiodic CSI-RS beam sweep should be performed with narrow beams within a wide SSB beam.

Downlink beam indication is performed by signaling a so-called transmission configuration indicator (TCI) to the UE which provides a spatial QCL reference that the UE can use to set its Rx beam. This is similar to the PDSCH rate matching and quasicolocation indicator (PQI) used for CoMP operation in LTE. The UE is configured with a list of TCI states by higher-layer signaling, where each TCI state is configured with a set of CSI-RS or SSB IDs. In each state, one CSI-RS or SSB ID that should be used as spatial QCL reference is selected. The beam indication is then performed by signaling a selected

TCI state to the UE that it should use for obtaining a QCL reference for the coming PDSCH/PDCCH transmission. By maintaining multiple TCI states, the BS can dynamically switch between different Tx beams. Before the TCI states have been configured and activated, the UE can assume that the DM-RS of PDSCH is spatially quasi-co-located with the SSB determined in the initial access procedure. This means that the UE can use the same Rx beam it used when it received the SSB during initial access. A TCI state can also contain a QCL reference for time/frequency parameters such as delay and Doppler spread. Reference signal for time/frequency QCL can be, e.g., a TRS (see Section 2.5). For frequencies below 6 GHz, a TCI state may contain only a time/frequency QCL reference and no spatial QCL information.

7.3.4.5 Beam Recovery

Narrow-beam transmission and reception is useful for improving the link budget at millimeter-wave frequencies but this may become susceptible to so-called beam failure. A beam failure means that the quality of the beam pair links for all control channels becomes too low for maintaining communication. This can be caused by, e.g., sudden blockage or failure in a beam management process. This may be lead to radio link failure (RLF) wherein a costly higher-layer reconnection procedure is needed. To avoid this, NR supports a faster procedure using lower layer signaling to recover from beam failure, referred to as beam recovery. For example, instead of initiating a cell reselection when a beam pair link quality becomes too low, a beam pair reselection within the cell can be performed.

Beam failure is detected by monitoring a beam failure detection reference signal and assessing if a beam failure trigger condition has been met. The beam failure detection reference signal can be SSB or a periodic CSI-RS configured for beam management and is associated with the Tx beam with which a control channel is transmitted. Beam failure detection is triggered if a number of consecutive detected beam failure instances exceeds a maximum value, where a beam failure instance occurs if the block error rate (BLER) of a hypothetical PDCCH transmission is above a threshold.

To find candidate new beams, the UE monitors a beam identification reference signal, which can be SSB or a periodic CSI-RS configured for beam management. These reference signals can be transmitted with wider beams than the ones used for data. They can be used both for finding a new BS Tx beam and a new UE Rx beam. When a UE has declared beam failure and found a new beam it transmits a beam recovery request message to the serving BS. The BS responds to the request by transmitting a recovery response over PDCCH to the UE and the UE monitors the control channel for the response. If the response is received successfully, the beam recovery is completed and a new beam pair link has thus been established. If the UE cannot detect any response within a specified time, the UE may perform a retransmission of the request. If the UE cannot detect any response after a specified number of retransmissions, then it notifies higher layers, potentially leading to RLF and cell reselection.

7.3.4.6 Uplink Beam Management

As mentioned previously, if the BS and UE have beam correspondence, separate procedures for uplink beam management are not needed, since the beam determination can be based on downlink procedures. If beam correspondence does not hold, separate uplink procedures similar to P-1, P-2, and P-3 described previously can be used. These are referred to as U-1, U-2, and U-3, and are based on uplink reference signals.

In uplink beam management, SRS is used in a similar way as CSI-RS is used in downlink beam management. An uplink beam management procedure is initiated by the BS by triggering one or several

SRS resource sets, where each resource set contains a number of SRS resources. One SRS resource is transmitted per UE Tx beam, the BS measures the received SRSs, determines a preferred UE Tx beam, and reports an SRI to the UE. The UE may also repeat an SRS transmission in a Tx beam so that the BS can find a suitable Rx beam. Multiple SRS resource sets can be used to train multiple beam-formers in parallel by triggering one set per beam-former.

Uplink beam indication may be needed so that the UE can adjust its Tx beam based on the Rx beam used by the BS. This may be needed also when beam correspondence holds and no separate uplink beam management is performed. Beam indication for uplink is supported for PUCCH and SRS. PUCCH can be spatially related to SRS or downlink signals such as SSB or CSI-RS. Such spatial relations can be used if the UE has beam correspondence. For example, if a UE receives a downlink RS in an Rx beam pointing in a certain direction it can perform an uplink transmission in a Tx beam pointing in the same direction. If the UE does not have beam correspondence, it can use a previously transmitted SRS for determining its Tx beam.

7.4 EXPERIMENTAL RESULTS

In order to lend some empirical support to the theoretical descriptions in previous sections, this section shows results from two different measurement campaigns that demonstrate the effectiveness of beam-forming at millimeter-wave frequencies. In the first campaign, an assessment of the achievable beam-forming gain using analog beam-forming in different environments was made and in the second campaign, successful beam tracking of a high speed UE was demonstrated. Furthermore, we present some results from system simulations using 3GPP models that illustrate what SINR levels and beam-forming gains can be expected in an urban macro scenario at 30 GHz for different antenna sizes at the BS and the UE.

7.4.1 BEAM-FORMING GAIN

In [42], measurements of beam-forming gain using analog beam-forming at 15 GHz in different environments were reported. The measurements were performed with a 5G test-bed using a planar antenna array with 8×8 antenna elements divided into 32 subarrays of 2×1 antenna elements each. Analog beam-forming using a 2-D GoB with 48 beams was applied over the subarrays to generate beams with an azimuth and elevation half-power beam width (HPBW) of 14°. Reference signals were transmitted sequentially in each beam and a UE prototype measured and logged average received power of the reference signals for all beams every 100 ms along a drive route. The analog beam grid is depicted in Fig. 7.15.

Fig. 7.16 shows estimated beam-forming gain along a drive route in an outdoor environment consisting of a square with surrounding buildings and streets. Buildings and the BS position are indicated in the sketch of the environment. The BS is mounted 7 m above the ground and faces the 110 m × 60 m square, which is surrounded by buildings of two to eight storeys. The measurement positions on the square are in LoS and the positions on the four streets out from the square are mainly in NLoS from the BS. The beam-forming gain was estimated by comparing the received power in the strongest beam with the received power averaged over all beams. The brightness of each dot on the route shows

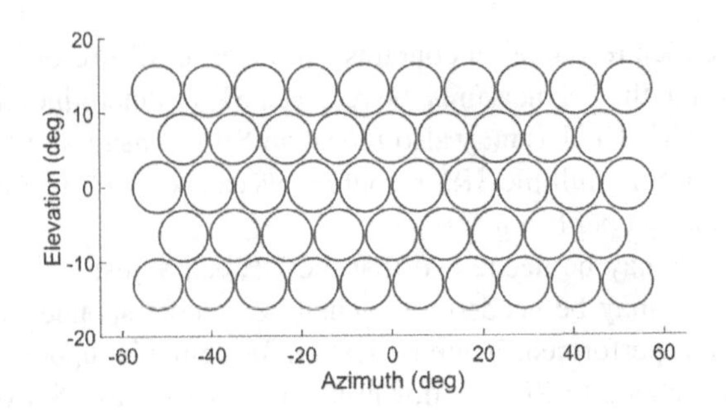

FIGURE 7.15

GoB. Beam grid used in the measurements.

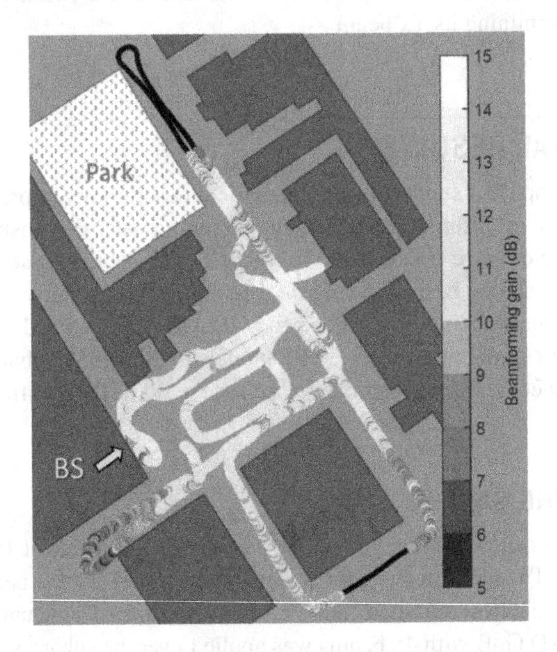

FIGURE 7.16

Beam-forming gain. Estimated beam-forming gain along an outdoor drive route.

the estimated beam-forming gain for that position.[27] Since the analog beam-forming is performed over 32 subarrays, the maximum beam-forming gain is 15 dB. It is clear from this figure that the LoS positions on the open square in front of the BS have higher beam-forming gain than the NLoS positions on the streets behind buildings. It can also be seen that the beam-forming gain is somewhat higher in

[27]The thinner black parts of the route correspond to lost connections.

the center of the square than in the positions close to the surrounding buildings where reflections have a negative impact on the beam-forming gain.

Fig. 7.17 shows cumulative distribution functions (CDFs) of the estimated beam-forming gain from the drive route shown in Fig. 7.16. The measurement results have been separated into LoS and NLoS positions along the route. The results show that the beam-forming gain is high in the LoS positions, mainly in the range 10–13 dB. As expected, the beam-forming gain in the NLoS positions is lower, around 7–12 dB. Results of measurements in an indoor environment with rich multipath were also reported in [42]. In this environment, the beam-forming gain was around 6–11 dB. Although the gain with analog beam-forming is substantial also in the reflective environments, there is potential for higher gain with digital precoding that can utilize several different propagation paths. The potential gain with hybrid beam-forming was also assessed in [42] by assuming a perfect coherent combining of the best analog beams. A substantial improvement was found by combining only a few beams. To achieve a certain total beam-forming gain, more beams were required in the reflective environments.

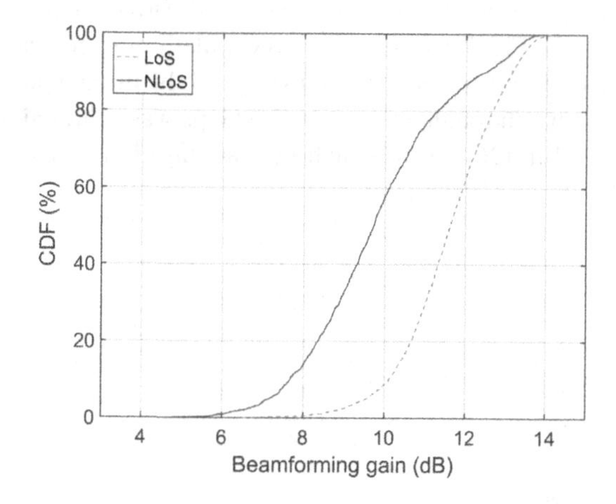

FIGURE 7.17

CDF of beam-forming gain. CDF of estimated beam-forming gain.

7.4.2 BEAM TRACKING

NR should support use cases with high mobility such as users traveling with high speed in, e.g., trains and cars. At millimeter-wave frequencies it may be essential to track moving users with narrow beams using analog beam-forming. Experimental results of beam tracking at 28 GHz of a car moving at high speed have been reported in [26]. The measurements were performed with a test system consisting of four transmission points deployed along a race track in Republic of Korea and one UE mounted on top of a car. Each transmission point had a rectangular antenna array with $16 \times 4 = 64$ antenna elements per polarization. Analog beam-forming was performed by using a GoB shown in Fig. 7.18. The HPBW of the array was 6° in azimuth and 24° in elevation. The UE had four dual-polarized, directive antennas covering different angular sectors. The bandwidth of the system was 800 MHz and supported MIMO transmission of two layers providing a peak rate of about 7.8 Gbps.

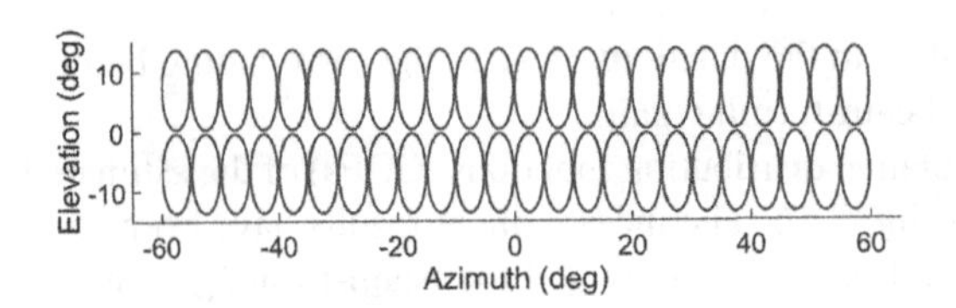

FIGURE 7.18

GoB. Beam grid used in the measurements.

Beam tracking was performed by transmitting beam-formed reference signals which were measured upon and reported by the UE. In the beam selection, beams from all four transmission points were evaluated and the best beam was selected regardless from which antenna the beam was transmitted. The transmission points were connected to the same baseband. Drive tests were performed on the race track at speeds up to 170 km/h. Fig. 7.19 shows the downlink throughput and UE speed as a function of time along the drive route. It can be seen that at low speed the throughput exceeded 6 Gbps and at the highest speed of 170 km/h, a throughput of up to 3.6 Gbps was achieved. Furthermore, successful beam switching was reported in [26], even though the car stayed in a beam for only 25 ms at high speeds.

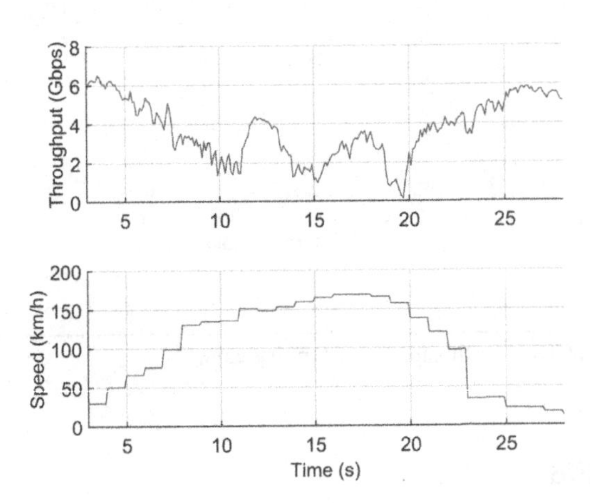

FIGURE 7.19

Throughput. Downlink throughput and UE speed vs. time.

7.4.3 SYSTEM SIMULATIONS

Finally, this section provides some results from system simulations using a 3GPP NR scenario and channel model that show how analog and hybrid beam-forming can increase the SINR at millimeter-wave frequencies.

Ideally, the analog beam-forming gain is $10\log_{10}(N_T N_R)$ (dB), where N_T and N_R is the number of Tx and Rx antenna elements, respectively. However, as discussed in Section 7.2.1, this can only be achieved in highly correlated channels when a LoS or specular reflection direction is known both at the transmitter and the receiver so that a beam can be steered in this direction. In practice angular spread will limit the analog beam-forming gain. Furthermore, with analog beam-forming a limited number of hypothesized directions must be tested sequentially in time to find the best beam direction, typically by using a fixed GoB. Since the true direction will not coincide exactly with an angle grid point, a so-called straddling loss will incur. To evaluate the beam-forming gains that are more realistic to achieve, we show results from system simulations based on models that take these effects into account.

We evaluate the downlink SINR as a function of the number of antenna elements in a BS panel for an urban macro scenario at 30 GHz carrier frequency using the 5G channel model[28] in [2]. The inter-site distance is 200 m and other simulation assumptions are as in [37]. We assume that the BS has one, two, or four square panels with $N_T \times N_T$ dual-polarized antenna elements each. Furthermore, we assume that analog beam-forming is performed per polarization within each panel and that the panels and polarizations are combined to a single transmit layer using SVD precoding assuming perfect CSIT. Two different UE antenna configurations are considered: one with and one without beam-forming. For the case without beam-forming the UE is assumed to have one dual-polarized isotropic antenna and for the case with beam-forming the UE is assumed to have two panels mounted back-to-back, where each panel has 2×4 dual-polarized antenna elements. During data reception only the UE panel with highest received power during the beam management procedure is used, i.e., panel combining is not used in the UE. It assumed that analog beam-forming is performed per polarization within this panel and that the polarizations are combined with interference rejection combining.

An ideal beam management procedure is simulated by finding the best pair of analog Tx beam at the BS and analog Rx beam at the UE. The beams are selected from a GoB constructed from DFT beams without angular oversampling. The zenith angles in the BS GoB are given by the angles $180° \cdot (n - 1/2)/N, n = 1, \dots, N$, where N is the number of antenna rows in a panel, that lie in the interval [90° 160°]. The azimuth angles in the BS GoB are given by the angles $180° \cdot (n - 1/2)/N - 90°$, $n = 1, \dots, N$, where N is the number of antenna columns in a panel that lie in the interval [−60° 60°]. The same analog beam is used in both polarizations and in all BS panels in the case that it has multiple panels.

Fig. 7.20 shows the 5- and 50-percentile, respectively, in the SINR CDF over all UEs in the network as a function of the number of BS antenna elements, N_T^2, per panel assuming square panels each having $N_T \times N_T$ elements. The different curves are for different numbers of BS and UE panels. The solid curves are for the cases with one, two, or four BS panels and two UE panels. The dashed curves are the corresponding results for a single dual-polarized antenna at the UE.

The results show that without any BS or UE beam-forming the performance is not acceptable, since the 5-percentile SINR is below −16 dB. With a single 8×8 BS panel and two 2×4 panels in the UE, the 5-percentile SINR is increased to −4 dB, which is a significant improvement. The results also show that doubling the number of BS panels gives roughly a 3 dB increase in SINR. Doubling the number of antenna elements within a panel gives slightly less than 3 dB 5-percentile SINR gain for

[28]However, recall from Chapter 3 that the highly resolved properties of the channel model have not been experimentally validated.

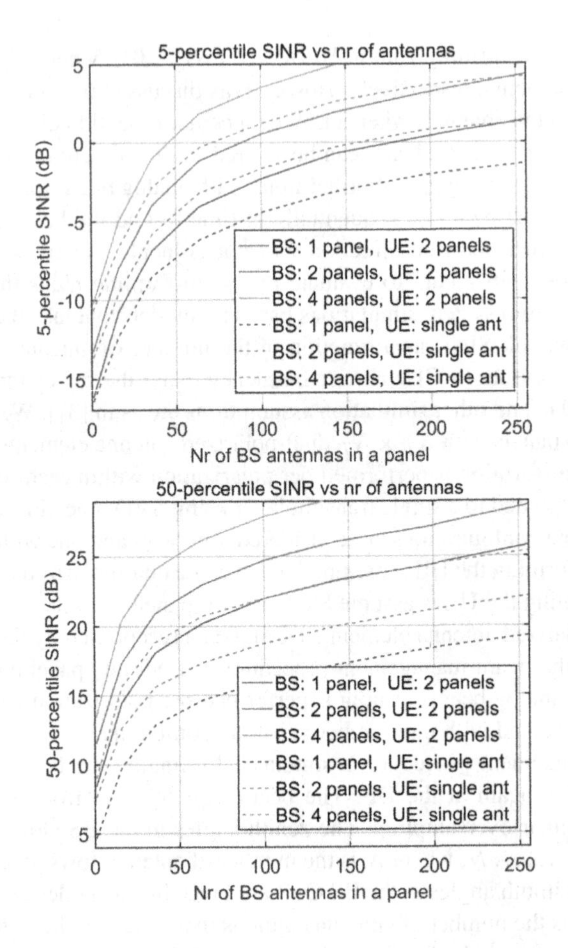

FIGURE 7.20

Simulation results. SINR vs. the number of antenna elements in a BS antenna panel obtained from system simulations using the 3GPP urban macro 5G channel model.

large panels. This may be due to the angular spread in the channel combined with narrow analog beams. Note that the multiple panels in the BS have in these simulations been used for coherent combining to increase coverage. Alternatively, the panels could be used to increase capacity by, e.g., MU-MIMO transmission.

The SINR gain of beam-forming in the UE, i.e., going from a single dual-polarized antenna in the UE to switching between two 2 × 4 panels, is around 2–3 dB at the 5-percentile and around 5–6 dB at the median. A possible explanation of the lower SINR gain at the 5-percentile is that UEs with low SINR are probably located indoors and/or in NLoS and therefore experience a larger angular spread which will limit the analog beam-forming gain.

REFERENCES

[1] 3GPP TR 38.802, Study on New Radio Access Technology, Physical Layer Aspects, 2017.
[2] 3GPP TR 38.900, Study on channel model for frequency spectrum above 6 GHz, 2017.
[3] 3GPP TR 38.901, Study on channel model for frequencies from 0.5 to 100 GHz, 2017.
[4] 3GPP TS 36.211, Physical channels and modulation, 2018.
[5] 3GPP TS 38.101-1, NR; User Equipment (UE) radio transmission and reception; Part 1: Range 1 Standalone, 2017.
[6] S.M. Alamouti, A simple transmit diversity technique for wireless communications, IEEE Journal on Selected Areas in Communications 16 (8) (1998, Oct.) 1451–1458.
[7] X. An, C.S. Sum, R.V. Prasad, J. Wang, Z. Lan, J. Wang, R. Hekmat, H. Harada, I. Niemegeers, Beam switching support to resolve link-blockage problem in 60 GHz WPANs, in: 2009 IEEE 20th International Symposium on Personal, Indoor and Mobile Radio Communications, 2009, Sept., pp. 390–394.
[8] J.B. Andersen, Array gain and capacity for known random channels with multiple element arrays at both ends, IEEE Journal on Selected Areas in Communications 18 (11) (2000, Nov.) 2172–2178.
[9] E. Björnson, M. Bengtsson, B. Ottersten, Optimal multiuser transmit beamforming: a difficult problem with a simple solution structure, IEEE Signal Processing Magazine 31 (4) (2014, July) 142–148.
[10] G. Caire, S. Shamai, On the achievable throughput of a multiantenna Gaussian broadcast channel, IEEE Transactions on Information Theory 49 (7) (2003, July) 1691–1706.
[11] D. Colombi, B. Thors, C. Törnevik, Implications of EMF exposure limits on output power levels for 5G devices above 6 GHz, IEEE Antennas and Wireless Propagation Letters 14 (2015) 1247–1249.
[12] M. Costa, Writing on dirty paper, IEEE Transactions on Information Theory 29 (3) (1983, May) 439–441.
[13] E. Dahlman, S. Parkvall, J. Sköld, 4G LTE-Advanced Pro and The Road to 5G, third ed., Academic Press, 2016.
[14] R.F.H. Fischer, C. Windpassinger, A. Lampe, J.B. Huber, Space–time transmission using Tomlinson–Harashima precoding, in: Proc. 4th Int. ITG Conf. Source and Channel Coding, 2002, pp. 139–147.
[15] X. Gu, D. Liu, C. Baks, O. Tageman, B. Sadhu, J. Hallin, L. Rexberg, A. Valdes-Garcia, A multilayer organic package with 64 dual-polarized antennas for 28 GHz 5G communication, in: 2017 IEEE MTT-S International Microwave Symposium (IMS), 2017, June, pp. 1899–1901.
[16] H2020-ICT-671650-mmMAGIC/D5.2, mmMAGIC Deliverable D5.2, Final multinode and multiantenna transmitter and receiver architectures and schemes, 2017.
[17] P. Harris, W.B. Hasan, H. Brice, B. Chitambira, M. Beach, E. Mellios, A. Nix, S. Armour, A. Doufexi, An overview of massive MIMO research at the University of Bristol, in: Radio Propagation and Technologies for 5G (2016), 2016, Oct., pp. 1–5.
[18] B.M. Hochwald, C.B. Peel, A.L. Swindlehurst, A vector-perturbation technique for near-capacity multiantenna multiuser communication-part II: perturbation, IEEE Transactions on Communications 53 (3) (2005, March) 537–544.
[19] W. Hong, K.H. Baek, Y. Lee, Y. Kim, S.T. Ko, Study and prototyping of practically large-scale mmWave antenna systems for 5G cellular devices, IEEE Communications Magazine 52 (9) (2014, September) 63–69.
[20] A. Huebner, F. Schuehlein, M. Bossert, E. Costa, H. Haas, A simple space-frequency coding scheme with cyclic delay diversity for OFDM, in: 2003 5th European Personal Mobile Communications Conference (Conf. Publ. No. 492), 2003, April, pp. 106–110.
[21] IEEE Std 145-1993, IEEE Standard Definitions of Terms for Antennas, 1993.
[22] ITU-R M.2135-1, Guidelines for evaluation of radio interface technologies for IMT-Advanced, 2009.
[23] S.A. Jafar, A. Goldsmith, Transmitter optimization and optimality of beamforming for multiple antenna systems, IEEE Transactions on Wireless Communications 3 (4) (2004, July) 1165–1175.
[24] M. Joham, W. Utschick, J.A. Nossek, Linear transmit processing in MIMO communications systems, IEEE Transactions on Signal Processing 53 (8) (2005, Aug.) 2700–2712.
[25] J. Jose, A. Ashikhmin, T.L. Marzetta, S. Vishwanath, Pilot contamination and precoding in multicell TDD systems, IEEE Transactions on Wireless Communications 10 (8) (2011, August) 2640–2651.
[26] K. Larsson, B. Halvarsson, D. Singh, R. Chana, J. Manssour, M. Na, C. Choi, S. Jo, High-speed beam tracking demonstrated using a 28 GHz 5G trial system, in: VTC 2017 Fall, 2017.
[27] X. Li, E. Björnson, E.G. Larsson, S. Zhou, J. Wang, A multicell MMSE precoder for massive MIMO systems and new large system analysis, in: 2015 IEEE Global Communications Conference (GLOBECOM), 2015, Dec., pp. 1–6.
[28] X. Li, E. Björnson, E.G. Larsson, S. Zhou, J. Wang, Massive MIMO with multicell MMSE processing: exploiting all pilots for interference suppression, EURASIP Journal on Wireless Communications and Networking (2017, June), https://doi.org/10.1186/s13638-017-0879-2.

[29] D.J. Love, R.W. Heath, V.K.N. Lau, D. Gesbert, B.D. Rao, M. Andrews, An overview of limited feedback in wireless communication systems, IEEE Journal on Selected Areas in Communications 26 (8) (2008, October) 1341–1365.
[30] R.J. Mailloux, Phased Array Antenna Handbook, second ed., Artech House, 2005.
[31] T. Marzetta, Noncooperative cellular wireless with unlimited numbers of base station antennas, IEEE Transactions on Wireless Communications 9 (11) (2010, November) 3590–3600.
[32] A. Paulraj, R. Nabar, D. Gore, Introduction to Space–Time Wireless Communications, Cambridge University Press, ISBN 9780521826150, 2003.
[33] C.B. Peel, B.M. Hochwald, A.L. Swindlehurst, A vector-perturbation technique for near-capacity multiantenna multiuser communication-part I: channel inversion and regularization, IEEE Transactions on Communications 53 (1) (2005, Jan.) 195–202.
[34] 5G collaboration in lead up to 2018 winter sporting events, press release, available at https://www.ericsson.com/en/news/2017/5/5g-collaboration-in-lead-up-to-2018-winter-games, 2017, May.
[35] IBM & Ericsson announce research advance for 5G communications networks, press release, available at https://www.ericsson.com/en/press-releases/2017/2/ibm--ericsson-announce-research-advance-for-5g-communications-networks, 2017, Feb.
[36] Going Massive with MIMO, available at https://www.ericsson.com/en/news/2018/1/massive-mimo-highlights, 2018, Jan.
[37] R1-1703536, Evaluation assumptions for Phase 2 NR MIMO system level calibration, 3GPP TSG RAN WG1 Meeting #88, 2017, February.
[38] R1-1708688, Codebook design for Type II CSI feedback, 3GPP TSG RAN WG1 Meeting #89, 2017, May.
[39] M. Sadek, A. Tarighat, A.H. Sayed, A leakage-based precoding scheme for downlink multiuser MIMO channels, IEEE Transactions on Wireless Communications 6 (5) (2007, May) 1711–1721.
[40] B. Sadhu, Y. Tousi, J. Hallin, S. Sahl, S. Reynolds, Ö. Renström, K. Sjögren, O. Haapalahti, N. Mazor, B. Bokinge, G. Weibull, H. Bengtsson, A. Carlinger, E. Westesson, J.E. Thillberg, L. Rexberg, M. Yeck, X. Gu, D. Friedman, A. Valdes-Garcia, A 28 GHz 32-element phased-array transceiver IC with concurrent dual polarized beams and 1.4 degree beam-steering resolution for 5G communication, in: 2017 IEEE International Solid-State Circuits Conference (ISSCC), 2017, Feb., pp. 128–129.
[41] H. Sampath, Linear precoding and decoding for multiple input multiple output (MIMO) wireless channels, Ph.D. thesis, Stanford University, 2001.
[42] A. Simonsson, M. Thurfjell, B. Halvarsson, J. Furuskog, S. Wallin, S. Itoh, H. Murai, D. Kurita, K. Tateishi, A. Harada, Y. Kishiyama, Beamforming gain measured on a 5G test-bed, in: 2017 IEEE 85th Vehicular Technology Conference (VTC Spring), 2017, June, pp. 1–5.
[43] V. Tarokh, H. Jafarkhani, A.R. Calderbank, Space–time block codes from orthogonal designs, IEEE Transactions on Information Theory 45 (5) (1999, Jul.) 1456–1467.
[44] S.H. Tsai, Transmit equal gain precoding in Rayleigh fading channels, IEEE Transactions on Signal Processing 57 (9) (2009, Sept.) 3717–3721.
[45] S.H. Tsai, An equal gain transmission in MIMO wireless communications, in: 2010 IEEE Global Telecommunications Conference GLOBECOM 2010, 2010, Dec., pp. 1–5.
[46] D. Tse, P. Viswanath, Fundamentals of Wireless Communications, 2005.
[47] A.M. Tulino, A. Lozano, S. Verdu, Capacity-achieving input covariance for single-user multiantenna channels, IEEE Transactions on Wireless Communications 5 (3) (2006, March) 662–671.
[48] D. Tuninetti, On the capacity of the AWGN MIMO channel under per-antenna power constraints, in: 2014 IEEE International Conference on Communications (ICC), 2014, June, pp. 2153–2157.
[49] M.K. Varanasi, T. Guess, Optimum decision feedback multiuser equalization with successive decoding achieves the total capacity of the Gaussian multiple-access channel, in: Conference Record of the Thirty-First Asilomar Conference on Signals, Systems and Computers (Cat. No.97CB36136), vol. 2, 1997, Nov., pp. 1405–1409.
[50] J. Vieira, F. Rusek, O. Edfors, S. Malkowsky, L. Liu, F. Tufvesson, Reciprocity calibration for massive MIMO: proposal, modeling, and validation, IEEE Transactions on Wireless Communications (ISSN 1536-1276) 16 (5) (2017, May) 3042–3056.
[51] H. Weingarten, Y. Steinberg, S.S. Shamai, The capacity region of the Gaussian multiple-input multiple-output broadcast channel, IEEE Transactions on Information Theory 52 (9) (2006, Sept.) 3936–3964.
[52] B. Yu, K. Yang, C.Y.D. Sim, G. Yang, A novel 28 GHz beam steering array for 5G mobile device with metallic casing application, IEEE Transactions on Antennas and Propagation 66 (1) (2018, Jan.) 462–466.
[53] X. Zheng, Y. Xie, J. Li, P. Stoica, MIMO transmit beamforming under uniform elemental power constraint, in: 2007 IEEE 8th Workshop on Signal Processing Advances in Wireless Communications, 2007, June, pp. 1–5.

CHAPTER

CHANNEL CODING

8

Forward error correction (FEC) schemes play a fundamental role in every digital communication system, because they provide robustness against noise and other channel uncertainties (e.g., imperfect channel-state information). The large range of use cases envisaged for 5G NR (see Chapter 1) makes the problem of designing good FEC schemes particularly challenging. Indeed, together with the traditional *enhanced mobile broadband* (eMBB) use case, which involves providing a high data rate to mobile users, 5G NR will also support two new use cases: *massive machine-type communication* (mMTC), which aims at guaranteeing connectivity to a very large number of low-cost and low-energy devices, and *ultrareliable low-latency communications* (URLLC), which deals with providing connectivity with latency and reliability levels that are orders of magnitude more stringent than in eMBB and mMTC [28]. As we shall see, the FEC schemes that are optimal for data transmission in eMBB may be suboptimal in URLLC applications, where the latency constraint imposes severe restrictions on the code blocklength. Furthermore, whereas the problem of designing good FEC schemes for large blocklength values is well understood both from a theoretical and a practical implementation viewpoint, designing good FEC schemes for short blocklength is—as we shall see—still an area of research.

In this chapter, we shall provide an introduction to FEC schemes for 5G NR. Focusing first on a simple binary-input additive white Gaussian noise (AWGN) channel, we shall first characterize the fundamental trade-off between rate, packet error probability, and blocklength using recently developed nonasymptotic tools in information theory [29].

We shall then review the family of low-density parity-check (LDPC) and polar codes that have recently been selected for data and control channel, respectively, in 5G new-radio (NR) eMBB transmission [1], and benchmark their performance against the theoretical bounds given by nonasymptotic information theory. We will also review other FEC schemes that are competitive for short blocklengths and may be relevant for URLLC applications.

Finally, looking beyond what is currently standardized in NR, we shall consider the problem of communicating over multiantenna fading channels, and we shall use nonasymptotic information theory to shed light on the role of frequency and spatial diversity (see Chapter 1), with specific focus on the URLLC use case. We shall see that the use of space-frequency codes is necessary to achieve the required ultrahigh reliability. All the performance results reported in this chapter can be reproduced using the numerical routines described in [10,20], which are available online.

5G Physical Layer. https://doi.org/10.1016/B978-0-12-814578-4.00013-8

8.1 FUNDAMENTAL LIMITS OF FORWARD ERROR CORRECTION

8.1.1 THE BINARY AWGN CHANNEL

We shall focus for simplicity of exposition on the binary-input AWGN (bi-AWGN) channel, i.e., a memoryless discrete-time AWGN channel

$$y_k = \sqrt{\rho} x_k + w_k, \quad k = 1, \ldots, n, \tag{8.1}$$

whose input symbols $\{x_k\}$ belong to the binary alphabet $\{-1, 1\}$. We assume that the additive noise samples $\{w_k\}$ are independent and identically distributed zero mean, unit variance Gaussian random variables. Hence, ρ can be thought of as the signal-to-noise ratio (SNR). Finally, n in (8.1) denotes the number of discrete-time channel uses that can be employed to transmit a packet of information bits.

8.1.2 CODING SCHEMES FOR THE BINARY-AWGN CHANNELS

We shall next review the fundamental limits on the rate at which we can communicate over this channel for a given latency (expressed in terms of number of channel uses) and reliability constraint. To do so, we first introduce the concept of a channel coding scheme for the channel (8.1). We shall focus on coding schemes with codewords whose entries belong to the binary field $\mathbb{F}_2$ and assume that each binary coded symbol c_k is mapped into the binary phase-shift-keying (BPSK) symbol $x_k = 2c_k - 1$ in the Euclidean space.

Definition 1. A (n, M, ϵ) (binary) coding scheme for the channel (8.1) consists of:

- An encoder $f : \{1, 2, \ldots, M\} \mapsto \mathbb{F}_2^n$ that maps the information message $j \in \{1, 2, \ldots, M\}$, where $M = 2^k$ and k is the number of information bits, to a codeword in the set $\{\mathbf{c}_1, \ldots \mathbf{c}_M\}$, with $\mathbf{c}_m \in \mathbb{F}_2^n$, $m = 1, \ldots, M$. The set of M codewords is commonly referred to as the channel code or the codebook.
- A decoder $g : \mathbb{R}^n \mapsto \{1, 2, \ldots, M\}$ that maps the received sequence $\mathbf{y} \in \mathbb{R}^n$ into a message $\hat{\jmath} \in \{1, 2, \ldots, M\}$, or, possibly, it declares an error. This decoder satisfies the average packet error probability constraint

$$\mathbb{P}\{\hat{\jmath} \neq j\} \leq \epsilon. \tag{8.2}$$

□

The rate R of a (n, M, ϵ) coding scheme is $R = \log_2(M)/n = k/n$. A remark on terminology is at this point appropriate. Following [26], we differentiate between a *code* (i.e., the list of codewords) and a *coding scheme* (i.e., the code together with the encoder and the decoder), because a given code can be decoded using different decoding algorithms (often of drastically different complexity), yielding different coding schemes. We warn the reader that this distinction is frequently omitted in the coding-theory literature.

8.1.3 PERFORMANCE METRICS

We define the maximum coding rate $R^*(n, \epsilon)$ as the largest rate achievable with (n, M, ϵ) coding schemes. Mathematically,

$$R^*(n, \epsilon) = \sup\left\{\frac{\log_2(M)}{n} : \exists\,(n, M, \epsilon) \text{ coding scheme}\right\}. \tag{8.3}$$

Determining $R^*(n, \epsilon)$, which describes the fundamental trade-off between rate, blocklength, and packet error probability in the transmission of information, is a fundamental problem in information theory, with a long history. In a seminal and groundbreaking contribution, Shannon characterized the asymptotic behavior of $R^*(n, \epsilon)$ in the limit $n \to \infty$ for general memoryless channels [35]. Specifically, he showed that, for every $0 < \epsilon < 1$, the maximum coding rate $R^*(n, \epsilon)$ converges to a constant C—the so-called channel capacity—that depends on the characteristics of the channel. The consequence of this result is that, for every transmission rate R less than C, there exists a sequence of coding schemes with rate R and vanishing packet error probability as $n \to \infty$. Conversely, one can show that if $R > C$, then the packet error probability over most memoryless channels of practical relevance (including the bi-AWGN channel (8.1)) goes to 1. This means that reliable communication is possible only at rates less than C.

For the bi-AWGN channel in (8.1), the channel capacity is given by

$$C = \frac{1}{\sqrt{2\pi}} \int e^{-z^2/2}\left(1 - \log_2\left(1 + e^{-2\rho + 2z\sqrt{\rho}}\right)\right) dz. \tag{8.4}$$

The proof of Shannon's coding theorem is based on a random-coding argument, and it does not provide a constructive way to approach the channel capacity. Indeed, it took about 50 years from the publication of Shannon's paper for the coding community to demonstrate near-capacity performance with practical coding schemes [11]. We shall review some of these coding schemes in Section 8.2.

It is worth stressing at this point that the channel capacity C is an asymptotic performance metric describing the behavior of the maximum coding rate $R^*(n, \epsilon)$ in the limit $n \to \infty$. This means that capacity cannot be used to benchmark the performance of coding schemes in which the blocklength n is short, as it is expected in some 5G use cases, due, for example, to a latency constraint.

This observation has renewed the interest in nonasymptotic characterizations of the maximum coding rate $R^*(n, \epsilon)$. The exact computation of such a quantity is a formidable task unless the number M of codewords is very small (e.g., $M = 4$; see [39]). However, tight upper (converse) and lower (achievability) bounds on $R^*(n, \epsilon)$ that can be computed efficiently can be obtained for a variety of channels for practical interest for 5G, including the bi-AWGN channel (8.1), using the finite-blocklength information-theoretic tools recently introduced in [29].

Such upper and lower bounds are depicted in Fig. 8.1 for the bi-AWGN channel. Specifically, Fig. 8.1A illustrates the tightest known bounds as a function of the blocklength n, for a target error probability of 10^{-4} and $\rho = 0$ dB. Equivalently, we can use the bounds to study the minimum packet error probability $\epsilon^*(n, R)$ achievable for a fixed blocklength n and rate R:

$$\epsilon^*(n, R) = \min\left\{\epsilon : \exists\,(n, \lceil 2^{nR}\rceil, \epsilon) \text{ coding scheme}\right\}. \tag{8.5}$$

This is illustrated in Fig. 8.1B where the bounds on $\epsilon^*(n, R)$ are plotted as a function of the minimum energy per bit E_b normalized with respect to the noise power spectral density N_0, which for the bi-AWGN channel, is given by

$$\frac{E_b}{N_0} = \frac{\rho}{2R}. \tag{8.6}$$

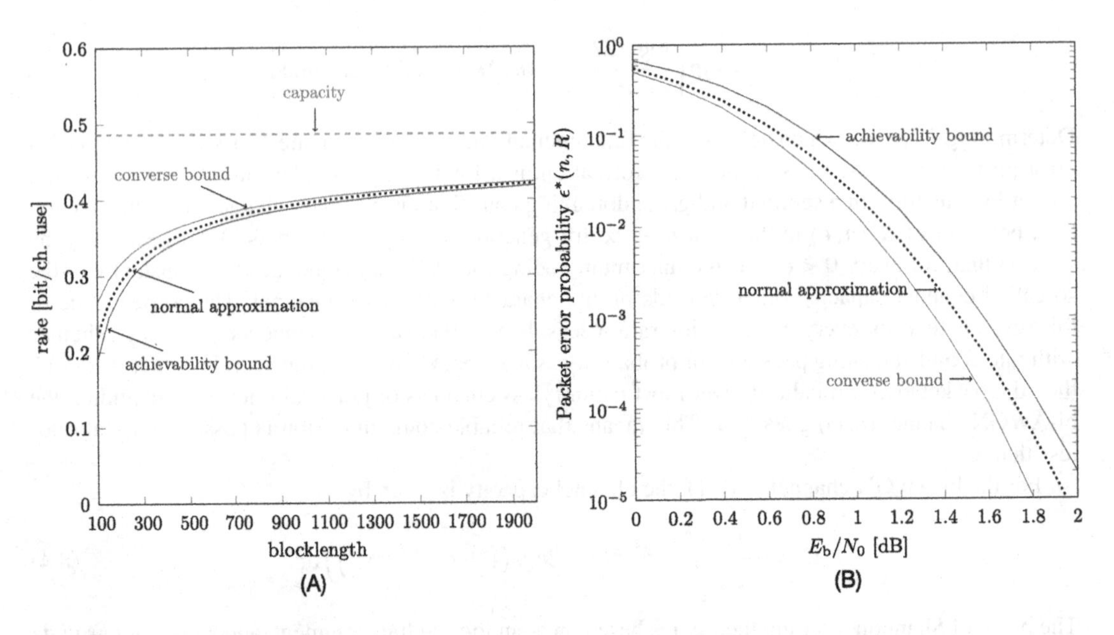

FIGURE 8.1

Bounds on the maximum coding rate $R^*(n, \epsilon)$ and on the minimum probability of error $\epsilon^*(n, R)$ achievable on the bi-AWGN channel (8.1). (A) $R^*(n, \epsilon)$ as a function of n for $\epsilon = 10^{-4}$, $\rho = 0$ dB. (B) $\epsilon^*(n, R)$ as a function of E_b/N_0 for $R = 0.5$, $n = 512$.

In both figures, the converse bound (which is an upper bound on $R^*(n, \epsilon)$ and a lower bound on $\epsilon^*(n, R)$) is based on the minimax converse [29, Thm. 27] (see [14] for details). The achievability bound (which is a lower bound on $R^*(n, \epsilon)$ and an upper bound on $\epsilon^*(n, R)$) is the random-coding union bound with parameter s (RCUs) [23, Thm. 1], a relaxation of the RCU bound [29, Thm. 16] that lends itself to efficient numerical evaluation. The dotted curve in Fig. 8.1A is the so-called normal approximation [29, Eq. (223)] to the maximum coding rate $R^*(n, \epsilon)$, which is given by

$$R^*(n, \epsilon) \approx C - \sqrt{\frac{V}{n}} Q^{-1}(\epsilon) + \frac{1}{2n} \log_2 n. \tag{8.7}$$

Here, V is the channel dispersion, which, for the bi-AWGN channel, is

$$V = \frac{1}{\sqrt{2\pi}} \int e^{-z^2/2} \left(1 - \log_2\left(1 + e^{-2\rho + 2z\sqrt{\rho}}\right) - C\right)^2 \mathrm{d}z, \tag{8.8}$$

and $Q^{-1}(\cdot)$ is the inverse of the Gaussian Q-function. In Fig. 8.1B, the normal approximation is

$$\epsilon^*(n, R) \approx Q\left(\frac{C - R + (2n)^{-1} \log_2 n}{\sqrt{V/n}}\right). \tag{8.9}$$

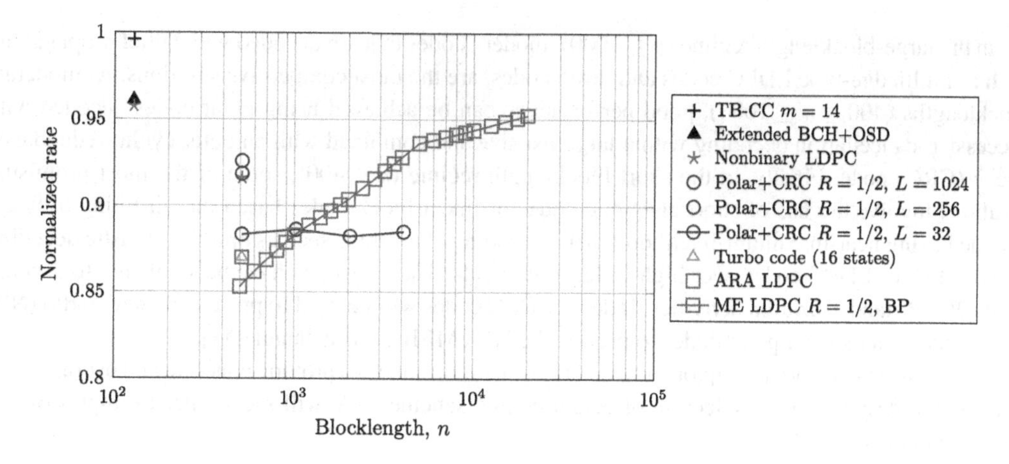

FIGURE 8.2

Normalized rate R_{norm} of some of the coding schemes reviewed in this chapter when used over the bi-AWGN channel (8.1). Here, $\epsilon = 10^{-4}$.

As can be seen from Fig. 8.1, the bounds characterize tightly the nonasymptotic performance metrics of interest, i.e., $R^*(n, \epsilon)$ and $\epsilon^*(n, R)$. Furthermore, the normal approximation, which is simple to evaluate numerically, turns out to be accurate in the parameter range considered in the figure. It is worth highlighting that this approximation is typically inaccurate in scenarios where both the required packet error probability and the SNR are low. In such scenarios, approximations based on saddle-point methods [23] should be used instead.

As we shall see in Section 8.3, the nonasymptotic bounds depicted in Fig. 8.1 can be generalized to include the presence of fading, and the use of pilot-aided transmission, and multiple antennas. Such generalizations provide valuable insights on how to optimally communicate over multiantenna fading channels.

The nonasymptotic bounds in Fig. 8.1 provide a natural way to benchmark the performance of practical coding schemes, which is more informative than the classic error probability vs. E_b/N_0 curves. Specifically, one fixes a code of a given blocklength n and rate R, and determines the minimum SNR $\rho_{\min}(\epsilon)$ needed to achieve a target error probability ϵ. Then one defines the normalized rate

$$R_{\text{norm}} = \frac{R}{R^*(n, \epsilon)} \tag{8.10}$$

where the maximum coding rate $R^*(n, \epsilon)$ can be evaluated, for example, using the normal approximation (8.7) with ρ in (8.4) and (8.8) replaced by $\rho_{\min}(\epsilon)$. Note that R_{norm} is a normalized performance metric that allows one to compare coding schemes of different blocklengths and different rates. The larger R_{norm}, the better the coding scheme.

Fig. 8.2 summarizes the performance over the bi-AWGN channel of some of the coding schemes we will review in the next sections of this chapter. As we shall see, there exist three different regimes in which different coding schemes are preferable: the short-blocklength regime, the moderate-blocklength regime, and the large-blocklength regime.

In the large-blocklength regime ($n \geq 1000$), modern codes that are decoded with belief propagation, such as multiedge-type LDPC codes and turbo codes, are the most competitive solutions. At moderate blocklengths ($400 \leq n \leq 1000$), good performance can be achieved using polar codes decoded with successive-cancellation decoding with a large list size and combined with an outer cyclic-redundancy check (CRC) code. Finally, in the short-blocklength regime ($n \leq 400$), some of the most promising solutions involve the use of short algebraic codes or linear block codes based on tail-biting trellises, decoded using near-maximum-likelihood (ML) decoding algorithms such as ordered-statistic decoding (OSD) [17], or LDPC codes over high-order finite fields. These insights have been taken into account in the 3GPP standardization activities. Indeed, LDPC codes will be used to protect the new-radio (NR) eMBB data channel and polar codes to protect the NR eMBB control channel [1].[1]

This confirms that nonasymptotic information-theoretic analyses provide concrete and useful guidelines on the design and the selection of actual coding schemes. We will see further examples of this principle in Section 8.3.

8.2 FEC SCHEMES FOR THE BI-AWGN CHANNEL

8.2.1 INTRODUCTION

Designing codes for large blocklengths is a well-investigated problem and effective solutions are available. Indeed, modern codes (e.g., turbo and LDPC codes) offer excellent performance under suboptimal but low-complexity iterative decoding algorithms such as belief propagation (BP). The design problem is more open for short blocklengths. On the one hand, the BP decoding performance becomes increasingly unsatisfactory when the blocklength decreases; on the other hand, a reduction in the blocklength makes it feasible to use near-maximum-likelihood decoding algorithms, which, when applied to, e.g., classical algebraic codes, yield performances that are sometimes superior to what can be achieved by modern codes with BP decoding.

In this section, we shall review some of the code constructions that are of interest for NR. Our emphasis will be mainly on the short- and moderate-blocklength regimes. For simplicity, we shall focus exclusively on the bi-AWGN channel. Also, in the spirit of the book, we shall highlight the general principles of each coding scheme, without delving too much into the many additional features (e.g., rate flexibility and suitability to hybrid automatic-repetition-request protocols) that are required for a coding scheme to be compatible with the requirements set in NR. Since our focus is on the bi-AWGN channel, we will not discuss coded-modulation techniques. We just highlight that the approach used in LTE to map coded bits into modulation symbols, which relies on bit-interleaved coded modulation (BICM), is suboptimal because it yields a well-known shaping loss for high-order constellations. This is particularly relevant for NR, which will support constellations belonging to sets of cardinality as large as 256. A different approach, which is becoming increasingly popular in fiber-optic applications, is to use the probabilistic shaping method proposed in [9]—an ingenious technique that provides rate adaptation and reduces significantly the shaping loss.

[1] The standardization of the coding schemes to be used in the other use cases is still ongoing.

8.2.2 SOME DEFINITIONS

We start our review by collecting here some standard definitions concerning linear block codes that will turn out to be useful in the remainder of this chapter (see, e.g., [31] for more details).

We say that the list of codewords $\mathcal{C}$ of an $(n, 2^k, \epsilon)$ binary coding scheme (see Definition 1) is a (n, k) *linear block code* if the 2^k codewords are a k-dimensional subspace of the vector space of all binary n-tuples. Here, addition and multiplication are the ones of the binary field $\mathbb{F}_2$. It follows by this definition that the codewords of a (n, k) linear block code can be expressed as a linear combination of k linearly independent codewords $\mathbf{g}_1, \mathbf{g}_2, \ldots, \mathbf{g}_k$. In other words, the set $\{\mathbf{g}_1, \mathbf{g}_2, \ldots, \mathbf{g}_k\}$ forms a basis for $\mathcal{C}$. We can use this basis to perform encoding as follows. Let the so-called $k \times n$ *generator matrix* of the code be defined as[2]

$$\mathbf{G} = \begin{bmatrix} \mathbf{g}_1 \\ \mathbf{g}_2 \\ \vdots \\ \mathbf{g}_k \end{bmatrix}. \tag{8.11}$$

Then the encoder output $\mathbf{c} = f(j)$ corresponding to the input message $j \in \{1, \ldots, 2^k\}$ is

$$\mathbf{c} = \mathbf{bG} \tag{8.12}$$

where $\mathbf{b}$ is the k-dimensional binary representation of j.

The $(n-k)$-dimensional dual space $\widetilde{\mathcal{C}}$ of $\mathcal{C}$ is the set of all binary n-tuples $\widetilde{\mathbf{c}}$ satisfying

$$\widetilde{\mathbf{c}}\mathbf{c}^T = 0, \quad \forall \mathbf{c} \in \mathcal{C}. \tag{8.13}$$

Let $\{\mathbf{h}_1, \mathbf{h}_2, \ldots, \mathbf{h}_{n-k}\}$ be a basis of $\widetilde{\mathcal{C}}$. The parity check matrix (PCM) $\mathbf{H}$ of $\mathcal{C}$ is the $(n-k) \times n$ binary matrix

$$\mathbf{H} = \begin{bmatrix} \mathbf{h}_1 \\ \mathbf{h}_2 \\ \vdots \\ \mathbf{h}_{n-k} \end{bmatrix}. \tag{8.14}$$

Note that if $\mathbf{c} \in \mathcal{C}$ then

$$\mathbf{c}\mathbf{H}^T = \mathbf{0}. \tag{8.15}$$

This highlights the important role of the PCM for error detection at the decoder. A linear block code is defined uniquely by the matrices $\mathbf{G}$ and $\mathbf{H}$.

For a given (n, k) linear block code, let $\mathcal{X} = \{\mathbf{x}(1), \mathbf{x}(2), \ldots, \mathbf{x}(2^k)\}$ be the set of BPSK coded sequences in the n-dimensional Euclidean space corresponding to the 2^k codewords. Under the assumption that the information message j is drawn uniformly from $\{1, 2, \ldots, 2^k\}$, the decoding rule that minimizes the packet error probability ϵ is the maximum-likelihood (ML) decoding rule

[2]Following the standard convention in coding theory, all vectors in the remainder of the chapter are row vectors.

$$\hat{\jmath} = \underset{m \in \{1,\ldots,2^k\}}{\arg\max}\; p(\mathbf{y} \,|\, \mathbf{x}(m)) \tag{8.16}$$

where $p(\mathbf{y} \,|\, \mathbf{x})$ denotes the channel law (8.1),

$$p(\mathbf{y} \,|\, \mathbf{x}) = \frac{1}{\sqrt{2\pi}} \exp\left(-\frac{\|\mathbf{y} - \sqrt{\rho}\mathbf{x}\|^2}{2} \right). \tag{8.17}$$

It follows from (8.17) that the ML decoding rule (8.16) can be equivalently expressed as

$$\hat{\jmath} = \underset{m \in \{1,\ldots,2^k\}}{\arg\min}\; \|\mathbf{y} - \sqrt{\rho}\mathbf{x}(m)\|^2. \tag{8.18}$$

Stated explicitly, the ML decoder selects the message whose corresponding BPSK coded sequence is closest to the received signal $\mathbf{y}$ in the Euclidean space.

Assume that the coded sequence $\mathbf{x}(1)$, corresponding to $j = 1$, is transmitted. The probability that a different coded sequence $\mathbf{x}(\ell)$ with $\ell \neq 1$ is closer to the received signal $\mathbf{y}$ than $\mathbf{x}(1)$ depends on the Euclidean distance between $\mathbf{x}(\ell)$ and $\mathbf{x}(1)$. Specifically,

$$\mathbb{P}\{\hat{\jmath} = \ell \,|\, j = 1\} = Q\left(\frac{\sqrt{\rho}\|\mathbf{x}(\ell) - \mathbf{x}(1)\|}{2} \right). \tag{8.19}$$

Let now K_d be the average number of coded sequences that have a neighbor at distance d. It follows from (8.19) and from an application of the union bound that [16, Eq. (2.32)]

$$\epsilon \leq \sum_d K_d Q\left(\frac{\sqrt{\rho} d}{2} \right). \tag{8.20}$$

The upper bound in (8.20) is typically dominated by the first term

$$K_{d_{\min}} Q\left(\frac{\sqrt{\rho} d_{\min}}{2} \right) \tag{8.21}$$

where $d_{\min}$ is the minimum Euclidean distance between any two coded sequences. This quantity is equal to twice the minimum Hamming weight of the nonzero codewords in $\mathcal{C}$. The bound (8.20) highlights the dependence of the packet error probability ϵ on the distance spectrum of the code. More sophisticated and tighter bounds on the packet error probability of linear block codes under ML decoding are described in [33]. It is worth highlighting that the evaluation of the ML rule (8.18) is in practice unfeasible already for values of k larger than a few tens of bits because of the complexity, unless the code possesses structures that facilitate it.

8.2.3 LDPC CODES

8.2.3.1 Fundamentals of LDPC Codes

LDPC codes are a class of linear block codes characterized by a PCM that is sparse, i.e., it contains only few nonzero entries. Originally proposed by Gallager [18] in the 1960s, and later rediscovered and generalized in the 1990s [22,5], LDPC codes provide a performance close to capacity for a large set of communication channels. These codes are currently deployed in several standards, including

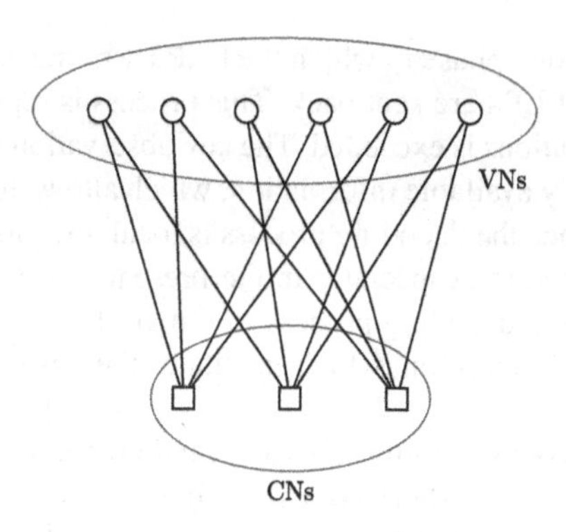

FIGURE 8.3

Tanner graph of the linear code with parity-check matrix given in (8.22).

IEEE802.11n, IEEE802.16e (WiMAX), IEEE 802.11ad (WiGig) and DVB-S2. As we shall briefly review, the sparseness of the PCM enables the use of low-complexity iterative decoding algorithms, which provide often near-ML performance. Throughout this section, we shall focus on binary LDPC codes. Extensions to higher fields are discussed in Section 8.2.5.3.

An (n, k) binary LDPC code is defined in terms of a $m \times n$ sparse PCM. Here, $m \geq n - k$, which implies that the PCM may be rank deficient. A convenient way to represent the PCM is through its Tanner graph representation [37]. A Tanner graph of an LDPC code is a bipartite graph, i.e., a graph in which the nodes are of two different types, and the edges connect only nodes of different types. These two types of nodes are commonly referred to by variable nodes (VNs), which are as many as the codeword length n, and by check node (CN), which are as many as the number of rows m in the PCM, i.e., as many as the parity-check equations. The Tanner graph is constructed from the PCM by drawing an edge between the ith check node and the jth variable node whenever the entry $[\mathbf{H}]_{ij}$ of the PCM contains a 1. As an example, the linear block code with PCM given by

$$\mathbf{H} = \begin{bmatrix} 1 & 1 & 0 & 1 & 1 & 0 \\ 1 & 0 & 1 & 0 & 1 & 1 \\ 0 & 1 & 1 & 1 & 0 & 1 \end{bmatrix} \tag{8.22}$$

is equivalently described by the Tanner graph depicted in Fig. 8.3.

We say that a LDPC code is regular if all VNs have the same degree, i.e., they are connected to the same number of edges, and the same occurs for the CNs. In the example of Fig. 8.3, all VNs have degree 2 and all CNs have degree 4. Allowing for VNs and CNs of different degrees, which results in irregular LDPC codes, turns out to be beneficial from a performance viewpoint [30]. Irregular LDPC codes are typically described in terms of their degree distributions, which provide the fraction of all edges connected to VNs/CNs of each degree.

Coarsely speaking, the decoding of LDPC codes is an iterative process, often referred to as BP, in which log-likelihood ratios (LLRs) about the coded bits are exchanged along the edges of the Tanner graph. Each decoding iteration consists of two phases: a first phase in which, at each VN, the LLRs from the channel and from the upcoming edges are processed and transformed in updated LLRs sent to the neighboring CNs; and a second phase in which the LLRs arriving at each CN from the neighboring VNs is processed and updated LLRs are sent back. This process is repeated until a codeword is found or the maximum number of iterations is exceeded. The key observation is that the processing at the CNs and VNs depends only on locally available information, which allows for an efficient and parallelizable decoding process. However, since the decoding process is local, the globally optimal ML solution may not be found through this procedure, especially in the presence of short cycles in the Tanner graph. Indeed, such cycles constrain the decoding process to remain local. This observation also shows the importance of the low-density assumption, which facilitates the design of Tanner graphs free of short cycles.

One way to design LDPC codes is to use a pseudorandom algorithm that constructs a PCM with given degree distributions and avoids short cycles. Such an approach, despite yielding LDPC codes with extremely good performance [30], is impractical from an implementation viewpoint, because the absence of further structures in the PCM makes both the encoding and the decoding complexity too high for practically relevant blocklengths and rates.

A more practically appealing approach is to design structured LDPC codes constructed from a smaller protograph. The LDPC codes that one obtains through this construction form a subset of the more general class of MET-LDPC codes, whose performances are illustrated in Fig. 8.2. The PCM of protograph-based LDPC codes can be specified in terms of a small base matrix. The actual PCS is constructed from the base matrix by replacing each entry in the base matrix by a $Q \times Q$ binary matrix whose rows and columns have a weight equal to the corresponding entry in the base matrix. Here, Q is the so-called *lifting factor*. It is particularly convenient to pick as binary matrix a $Q \times Q$ cyclic permutation matrix, whose row and column weights are one. The resulting code is quasi-cyclic, a property that allows for simplified encoding and decoding, with negligible performance loss [21,31]. As an example, consider the following base matrix $\mathbf{H}_\text{b}$:

$$\mathbf{H}_\text{b} = \begin{bmatrix} 1 & 0 & 0 & 1 \\ 0 & 1 & 0 & 1 \end{bmatrix} \tag{8.23}$$

and assume that 3×3 cyclic permutation matrices are used to extend the base matrix to a PCM. Such a PCM may have the following structure:

$$\mathbf{H} = \left[\begin{array}{ccc|ccc|ccc|ccc} 1 & 0 & 0 & 0 & 0 & 0 & 0 & 0 & 0 & 0 & 0 & 1 \\ 0 & 1 & 0 & 0 & 0 & 0 & 0 & 0 & 0 & 1 & 0 & 0 \\ 0 & 0 & 1 & 0 & 0 & 0 & 0 & 0 & 0 & 0 & 1 & 0 \\ \hline 0 & 0 & 0 & 0 & 0 & 1 & 0 & 0 & 0 & 1 & 0 & 0 \\ 0 & 0 & 0 & 0 & 1 & 0 & 0 & 0 & 0 & 0 & 0 & 1 \\ 0 & 0 & 0 & 1 & 0 & 0 & 0 & 0 & 0 & 0 & 1 & 0 \end{array}\right]. \tag{8.24}$$

8.2.3.2 The LDPC-Code Solution Chosen for 5G NR

The LDPC codes chosen for the data channel in 5G NR are quasi-cyclic and have a rate-compatible structure that facilitates their use in hybrid automatic-repetition-request (HARQ) protocols.[3]

[3] This section is largely based on [32], where further information about the LDPC-code family chosen for 5G NR, including the rate-matching procedure and their use in combination with HARQ, can be found.

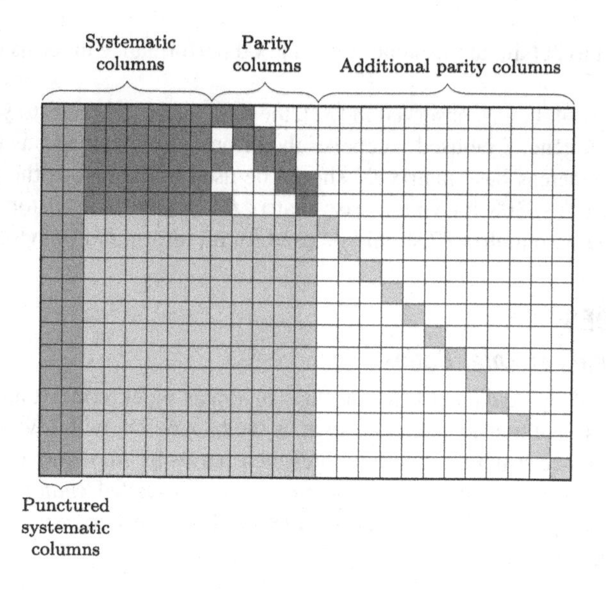

FIGURE 8.4

The general structure of the base matrix used in the quasi-cyclic LDPC codes selected for the data channel in NR.

To cover the large range of information payloads and rates that need to be supported in 5G NR, two different base matrices are specified. The general structure of these base matrices is provided in Fig. 8.4. In the figure, each white square represents a zero in the base matrix and each nonwhite square represents a one. The first two columns in gray correspond to punctured systematic bits that are actually not transmitted. Their addition is known to improve the threshold of the resulting code, i.e., its minimum SNR operating point [13]. The blue (dark gray in print version) part constitutes the kernel of the base matrix, and it defines a high-rate code. The dual-diagonal structure of the parity subsection of the kernel enables efficient encoding. Transmission at lower code rates is achieved by adding additional parity bits, i.e., by including an appropriately chosen subset of rows and columns in the base matrix containing entries marked in pink (light gray in print version). To enable maximum parallelism, the rows of the base matrix outside the kernel are designed so as to be orthogonal or quasi-orthogonal. The maximum lift factor $Q_{\max}$ is 384. This number is chosen to trade optimally between the parallel processing opportunities enabled by a large Q and the performance loss in terms of threshold due to the resulting higher amount of structure.

The base matrix #1, which is optimized for high rates and long blocklengths, supports LDPC codes of a nominal rate between $1/3$ and $8/9$. This matrix is of dimension 46×68 and has 22 systematic columns. Together with a lift factor of 384, this yields a maximum information payload of $k = 8448$ bits (including CRC).

The base matrix #2 is optimized for shorter blocklengths and smaller rates. It enables transmissions at a nominal rate between $1/5$ and $2/3$, it is of dimension 42×52, and it has 10 systematic columns. This implies that the maximum information payload is $k = 3840$.

The choice of each of the $Q \times Q$ circulant matrices to be substituted into the entries of the base matrix to form the full PCM is specified in [1]. These circulant matrices are selected so as to ensure efficient encoding and to obtain, at the same time, a good performance in terms of both error floor and threshold.

It is worth pointing out that, as observed in [32], the base matrix #2 tends to yield lower-complexity decoding and should in general be used whenever the information payload k is less than 3840 and the rate is less than 2/3, whereas base matrix #1 should be used in the rest of the parameter range. Two exceptions are the case $k \leq 308$ for which base matrix #2 should be used for all rates, and the case $R \leq 1/4$, for which the base matrix #2 should be used for all information-payload sizes k.

8.2.4 POLAR CODES

8.2.4.1 Fundamentals of Polar Codes

Polar codes, introduced by Arıkan [6], are a class of linear block codes that provably achieve the capacity of memoryless symmetric channels, such as the bi-AWGN, with low encoding and decoding complexity, and a recursive structure that facilitates their hardware implementation.

To introduce the core idea behind polar codes, i.e., the so-called *channel polarization*, we start by observing that, among all bi-AWGN channels (8.1), there are two extreme types, for which the communication problem is trivial[4]:

- The perfect (noiseless) channel $y_k = \sqrt{\rho} x_k$.
- The useless channel $y_k = w_k$.

Uncoded transmission is sufficient to achieve the capacity of the first channel, whereas no information can be transmitted on the second channel. Arıkan's polarization technique is a lossless and low-complexity method to convert any binary-input symmetric channel into a mixture of extremal binary-input channels.

Polarization is achieved through the *polar transform*, which operates as follows: given two copies of a binary-input channel $W : \mathbb{F}_2 \to \mathcal{Y}$, where $\mathcal{Y}$ denotes the output set ($\mathcal{Y} = \mathbb{R}$ for the bi-AWGN channel), the polar transform creates, under the assumption of uniformly distributed inputs, two new *synthetic channels* $W^- : \mathbb{F}_2 \to \mathcal{Y} \times \mathcal{Y}$ and $W^+ : \mathbb{F}_2 \to \mathcal{Y} \times \mathcal{Y} \times \mathbb{F}_2$, defined as follows:

$$W^-(y_1, y_2 \mid u_1) = \frac{1}{2} \sum_{u_2 \in \mathbb{F}_2} W(y_1 \mid u_1 \oplus u_2) W(y_2 \mid u_2), \tag{8.25}$$

$$W^+(y_1, y_2, u_1 \mid u_2) = \frac{1}{2} W(y_1 \mid u_1 \oplus u_2) W(y_2 \mid u_2). \tag{8.26}$$

Here, $\oplus$ is the addition in $\mathbb{F}_2$. Such a transform is illustrated in Fig. 8.5. Note that the channel W^+, which has input u_2 and output (y_1, y_2, u_1), contains the channel W. Hence, its capacity under uniform inputs (which we shall assume throughout this section) is no smaller than the capacity of the channel W. Since the transformation

$$x_1 = u_1 \oplus u_2, \tag{8.27}$$

$$x_2 = u_2, \tag{8.28}$$

[4] See, e.g., [26, Ch. 14] and Telatar's plenary talk at the 2017 International Symposium on Information Theory https://goo.gl/zQz6nB for a more comprehensive introduction to polar codes.

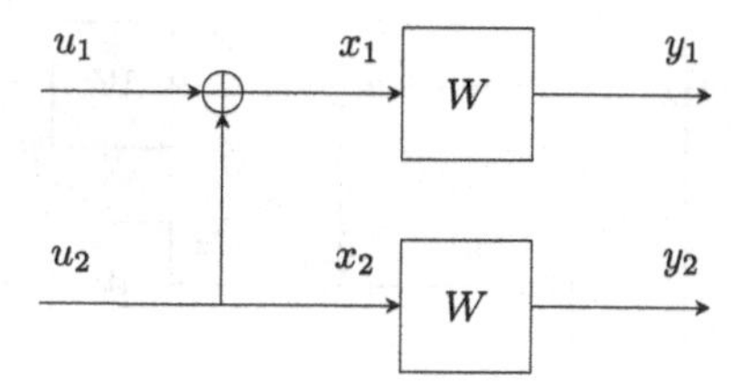

FIGURE 8.5

The polar transform.

is invertible, the sum of the capacities of W^+ and W^- must be equal to twice the capacity of W. This implies that the capacity of W^- must be smaller than that of the original channel W. To summarize, we started from two identical copies of W. Through the application of the polar transform we obtained two new synthetic channels: W^-, whose capacity is smaller than W, and W^+, whose capacity is larger than W.

Example. Consider a binary erasure channel (BEC), i.e., a discrete memoryless channel with input–output relation

$$y = \begin{cases} x & \text{with prob. } 1-p, \\ ? & \text{with prob. } p, \end{cases} \tag{8.29}$$

where x denotes the binary input and the symbol "?" denotes an erasure. In words, with probability $(1-p)$ the input symbol x is received correctly, and with probability p it is erased. One can verify that the synthetic channels W^- and W^+ induced by the polar transform are also BECs, with erasure probability $p^- = p(2-p) \geq p$ and $p^+ = p^2 \leq p$, respectively.

Some comments on the synthetic channels W^- and W^+ are in order. The channel W^-, which has input u_1 and output (y_1, y_2), is indeed a genuine channel, because the receiver has access to both y_1 and y_2. The channel W^+, however, has u_1 as output, which is not available at the receiver. However, W^+ can be synthesized by imposing a decoding order. Namely, we first decode u_1 using y_1 and y_2. Then we use y_1, y_2, and the estimate $\hat{u}_1$ of u_1 to decode u_2. This corresponds to successive-cancellation decoding. One can show that the block-error probability $\Pr\{(\hat{u}_1, \hat{u}_2) \neq (u_1, u_2)\}$ achievable with successive-cancellation decoding coincides with the one attainable by a genie-aided successive-cancellation decoder that uses u_1 instead of $\hat{u}_1$ in the second step of the decoding procedure. In other words, error propagation is not an issue if we measure performance in terms of block-error probability.

The polarization transform can now be applied again to the inputs of W^- and W^+. This results in the four channels W^{--}, W^{-+}, W^{+-}, and W^{++} illustrated in Fig. 8.6. This process can be applied recursively N times to synthesize 2^N channels out of 2^N copies of W. Applying this process to a BEC with erasure probability $p = 0.3$ for the case $N = 10$ yields 1024 synthetic channels, whose capacities, which are equal to one minus the corresponding erasure probability, are illustrated in Fig. 8.7. This figure exemplifies the polarization phenomenon: roughly 30% of the synthetic channels have capacity close to zero, i.e., they are useless, whereas 70% have capacity close to 1, i.e., they are almost perfect. Observe now that the capacity of the underlying BEC is exactly $1 - p = 0.7$. Consequently, Fig. 8.7

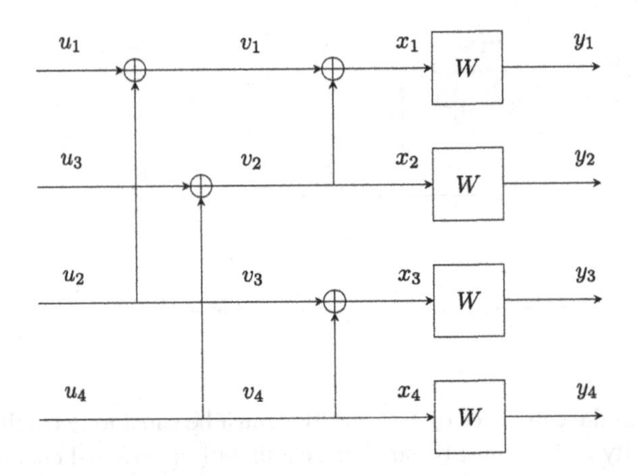

FIGURE 8.6

The four synthetic channels obtained by applying twice the polar transform.

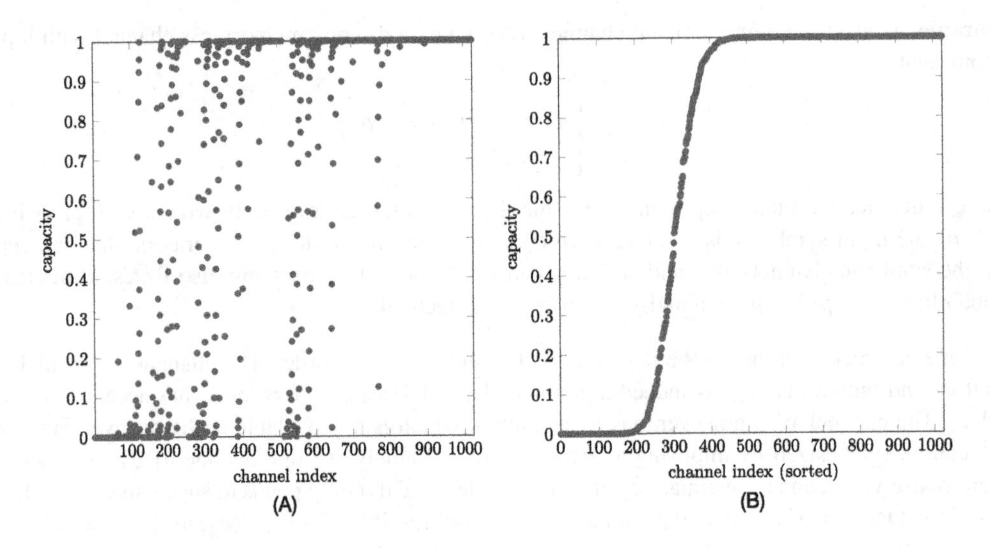

FIGURE 8.7

Capacity of the 1024 synthetic channels obtained by applying recursively the polar transform $N = 10$ times to a BEC with erasure probability p. (A) Unsorted. (B) Sorted.

suggests that, to achieve capacity, one can use a simple rate 0.7 binary code in which the k information bits are mapped to the almost perfect channels; the codeword entries corresponding to the remaining almost useless channels are frozen, i.e., they contain symbols known to the decoder (i.e., zeros).

More formally, we define a polar code as follows. Fix an integer N and let the blocklength be $n = 2^N$. Denote by $\mathbf{G} = \mathbf{F}^{\otimes n}$ the $n \times n$ polar transform matrix, where $\otimes$ stands for the Kronecker product, and

$$\mathbf{F} = \begin{bmatrix} 1 & 0 \\ 1 & 1 \end{bmatrix} \tag{8.30}$$

is the polar transform. Let also $\mathcal{U} \subset \{0, 1, \dots 2^N - 1\}$ be a set of indices of cardinality k, where k is the size of the payload, and let $\mathcal{F}$ be the complementary set of size $n - k$. The k information bits are mapped to the entries belonging to $\mathcal{U}$ of a vector $\mathbf{u}$ of size n. The remaining entries, i.e., the ones with indices belonging to the set $\mathcal{F}$, are frozen to zero. In the remainder of the section, we shall refer to $\mathcal{F}$ as the frozen set. This set is constructed so as to contain the $n - k$ synthetic channels that have the smallest capacity. Finally, the resulting binary codeword $\mathbf{c}$ is obtained by performing the polarization mapping $\mathbf{c} = \mathbf{uG}$.

The receiver uses successive-cancellation decoding; the decoding order is obtained by applying a bit-reversal permutation to the channel index set: specifically, we number the channels from 0 to $n - 1$, we obtain a new index for each channel by reversing the binary representation of its index, and we decode following the ordering induced by the new indices.

One can formally show [7] that the block-error probability ϵ achievable with the polar-coding scheme just described decays roughly as $2^{-\sqrt{n}}$, where $n = 2^N$ is the blocklength, for every rate below capacity. This proves that polar codes are indeed capacity achieving. Furthermore, they have the additional practical benefit over modern codes decoded with BP of not suffering from an error floor [24].

As far as complexity is concerned, the recursive nature of the polar transform allows one to perform encoding and decoding with a complexity that scales as $n \log_2 n$. Furthermore, these operations are naturally parallelizable.

One disadvantage of polar codes is that, when combined with successive-cancellation decoding, they offer mediocre performance in the short- and moderate-blocklength regimes that are of interest for control-channel applications. The performance of polar codes in this regime can be improved significantly by using instead a successive-cancellation list decoder [36]. This decoder considers simultaneously, at each decoding stage, L alternative paths, where L is the list size. The complexity of such a decoder is of order $Ln \log_2 n$. Performance can be further improved by concatenating the polar code with a high-rate CRC code. This code is used at the end of the list-decoding process to select the final codeword among the ones contained in the list. Concatenating a polar code with a CRC code turns out to be beneficial, because it improves the minimum distance of the polar code. As shown in Fig. 8.2, such an approach yields a state-of-the-art performance at moderate-blocklength values.

8.2.4.2 The Polar-Code Solution Chosen for 5G NR

Polar codes in 5G NR will be used to protect control signaling, with the exception of the transmission of payloads of up to $k = 11$ bits, for which Reed–Muller codes will be used, similar to 4G. Polar codes of different lengths and rates are supported. Such codes are obtained from an underlying parent code of length $2^{10} = 1024$. As recently summarized in [40], the specific polar code adopted in 5G NR relies on three innovations compared to what was described in the previous section, which allow the resulting coding scheme to satisfy the flexibility, processing latency, and complexity desiderata in 5G NR. Such innovations are:

- the offline computation of a deterministic reliability ordering of the synthetic channels;

- the use of parity-check bits to assist successive-cancellation list decoding also during its intermediate stages;
- a low-complexity rate-matching algorithm providing the needed rate and blocklength flexibility.

Deterministic Reliability Ordering

As explained in the previous section, one crucial step in the design of polar codes is the construction of the frozen-bit set $\mathcal{F}$. As shown in Fig. 8.7, the useless channels do not seem to follow a regular pattern. It turns out that, apart from the BEC, no efficient algorithm is known to rank the synthetic channels according to their reliability.[5] Since the reliability of the synthetic channels and their relative ordering depends on time-varying parameters such as the SNR, adaptive offline and online algorithms, which are able to order the synthetic channels as a function of the current values of channel parameters, appear to be unfeasible because of latency and/or memory requirements. The solution adopted in 5G NR is to assign to each synthetic channel of the parent code a deterministic, i.e., channel-parameter-independent, polarization weight that expresses its reliability. The polarization weights induce an ordering on the subchannels that is prestored so as to avoid online computations. Using techniques such as the β-expansion in number theory [19,40], these polarization weights can be chosen so as to satisfy the natural universal partial ordering existing among the synthetic channels [34].

To save complexity, the polarization weights of the parent code are used also for transmission involving shorter codes of length $n = 2^N$, $N < 10$. Specifically, let $\mathbf{q}^{n_{\max}}$ be the vector containing the ordered sequence of the 1024 synthetic-channel indices of the parent code, ordered according to increasing polarization weights. Then a reliability index list $\mathbf{q}^n$ for the shorter code is obtained by removing from $\mathbf{q}^{n_{\max}}$ all indices larger or equal to n. Although this choice is suboptimal, the significant complexity reduction justifies the resulting moderate performance loss. The polarization weights in 5G NR have been chosen through extensive simulations to guarantee good performance for all blocklengths and code rates.

Parity-Check Coding

The polar-code structure selected in NR relies on the addition of a more general, yet hardware friendly, outer code than just a CRC. Specifically, n_{pc} parity-check bits are appended to the information payload. Some of them are assigned to the $k + n_{\text{pc}}$ unfrozen synthetic channels with lowest polarization weight. This improves the error performance of the code. Some of them are assigned to the unfrozen-bit positions in the vector $\mathbf{u}$ that correspond to the rows in the polar transform matrix $\mathbf{G}$ with smallest Hamming weights. This improves the distance spectrum of the resulting concatenated code.

To aid the intermediate steps of the successive-cancellation list-decoding algorithm, each parity-check bit is designed to depend only on preceding information bits. Specifically, the parity bits are computed through a length 5 shift register that evaluates the $\mathbb{F}_2$-sum of information bits that are five positions apart; see [40, Algorithm 3] for details. Empirical evidence suggests that enforcing such spacing makes the resulting scheme robust against error propagation. The same shift-register architecture can be used by the successive-cancellation list decoder to prune all paths that result in an erroneous parity-check bit.

---

[5] See, however, [34,25,19] for recent progress on this problem, based on universal partial ordering, and the β-expansion in number theory.

Rate Adaptation

The polar-code construction reviewed so far allows one to generate codes of blocklength $n = 2^N$ for some integer $N \leq 10$. Additional blocklength values can be obtained by *puncturing* or *shortening* [8].

In puncturing, one transforms the original (n, k) polar code into a $(n - p, k)$ polar code by removing $p < n - k$ bits from each codeword $\mathbf{c}$. At the receiver, the channel LLRs corresponding to the punctured coded bits are set to 0 (the bits are assumed erased), and then the decoder of the original (n, k) polar code is applied. The presence of p zero-valued channel LLRs induces, through the decoding process, zero-valued LLRs for p of the entries of the n-dimensional input vector $\mathbf{u}$. To avoid poor performance, these entries must be set to frozen bits. This can be achieved by puncturing the bits of $\mathbf{c}$ with indices corresponding to the reverse-bit permutations of $\{1, 2, \ldots, p\}$ and by freezing the corresponding bits of $\mathbf{u}$. The remaining frozen bits are chosen among the ones with lowest polarization weight, as usual.

We next discuss shortening. Let us assume that the original (n, k) code is systematic. Shortening allows one to obtain a $(n - p, k - p)$ code, $p < k$, by setting p systematic bits to zero and by not transmitting them. At the decoder side, the channel LLRs corresponding to the punctured systematic bits are set to infinity, and the decoder of the original code is applied. For nonsystematic polar codes, a natural way to select the coded bits to shorten is by requiring them to be linear combinations of frozen bits. This can be achieved by shortening the bits of $\mathbf{c}$ with indices corresponding to the reverse-bit permutations of $\{n - p + 1, \ldots, n\}$ and by freezing the corresponding bits of $\mathbf{u}$. As before, the remaining frozen bits are chosen among the ones with lowest polarization weight.

8.2.5 OTHER CODING SCHEMES FOR THE SHORT-BLOCKLENGTH REGIME

To conclude our overview, we shall next present a selection of additional coding schemes that exhibit a favorable performance/complexity trade-off in the short-blocklength regime ($n < 400$). Even though these schemes are not standardized in 5G, some of them may enter future releases due to their suitability for URLLC.

8.2.5.1 Short Algebraic Linear Block Codes With Ordered-Statistics Decoding

As already mentioned, when the blocklength is short, classic algebraic codes such as BCH and extended BCH codes can be decoded using near ML decoding algorithms. OSD, which we shall review next, is one such algorithm.

Recall that the evaluation of the ML decoding rule (8.16), which reduces to (8.18) in the bi-AWGN case, has in general a prohibitive complexity, already for small values of the information-payload size k. The idea behind OSD is to replace (8.18) with

$$\widehat{\jmath} = \arg\min_{m \in \mathcal{L}} \|\mathbf{y} - \sqrt{\rho}\mathbf{x}(m)\|^2, \tag{8.31}$$

where the optimization is performed over a list $\mathcal{L}$ of much smaller cardinality than 2^k. OSD constructs such a list through the following steps.

Let us consider the problem of decoding a (n, k) linear block code $\mathcal{C}$ that is used over the bi-AWGN channel (8.1). Let $\mathbf{G}$ be the generator matrix of the code, $\mathbf{u}$ the k-dimensional information vector, $\mathbf{c}$ the corresponding codeword, and $\mathbf{x}$ its vector representation in the Euclidean space after BPSK modulation.

For a given received vector $\mathbf{y}$, we let $\mathbf{r} = [|y_1|, \ldots, |y_n|]$; furthermore, we construct the additional vector $\mathbf{r}'$ by ordering the entries of $\mathbf{r}$ in decreasing order. Note that the scalars $\{y_\ell\}_{\ell=1}^n$ are proportional to the channel LLRs. Hence, the vector $\mathbf{r}'$ contains a scaled version of the channel LLRs ordered according to their reliability. Let π_1 be the permutation that maps $\mathbf{r}$ into $\mathbf{r}'$. If we apply this permutation to the columns of the generator matrix $\mathbf{G}$, we obtain a new generator matrix:

$$\mathbf{G}' = \pi_1(\mathbf{G}). \tag{8.32}$$

It turns out to be convenient to associate to each column of $\mathbf{G}'$ a reliability value, which is given by the corresponding entry of $\mathbf{r}'$. Next we rearrange the columns of $\mathbf{G}'$ so that the first k columns of this matrix are the k linear independent columns with the highest reliability, ordered in decreasing order of reliability, and the remaining $n-k$ columns are also ordered in decreasing order of reliability. We denote the resulting matrix by $\mathbf{G}''$ and the corresponding column permutation by π_2. Finally, we put $\mathbf{G}''$ in the systematic form $\mathbf{G}''' = [\mathbf{I}_k \; \mathbf{P}]$ by performing standard row operations. Here, $\mathbf{P}$ is of size $k \times n-k$. Note that the codes generated by $\mathbf{G}$ and by $\mathbf{G}'''$ are equivalent.

To construct the list, we now apply the second permutation π_2 to $\mathbf{r}'$ and obtain $\mathbf{r}'' = \pi_2(\mathbf{r}')$. Next, we perform a hard decision on the first k entries of $\mathbf{r}''$, to obtain the k-dimensional vector $\hat{\mathbf{u}}$. Specifically, we set $\hat{u}_i = 0$ if $z_i > 0$ and $\hat{u}_i = 1$ otherwise. It is worth remarking that the permutations ensure that the hard decision is performed on the most reliable linear-independent set of channel outputs.

For a given integer t, the list $\mathcal{L}$ is finally constructed by considering all codewords,

$$\hat{\mathbf{c}} = \pi_1^{-1}(\pi_2^{-1}(\tilde{\mathbf{u}}\mathbf{G}''')) \tag{8.33}$$

where $\tilde{\mathbf{u}}$ spans all k-dimensional vectors whose Hamming distance from $\hat{\mathbf{u}}$ is smaller or equal to t. The final decision is taken by computing the Euclidean distance between the BPSK vectors corresponding to each codeword in $\mathcal{L}$ and $\mathbf{y}$, and by selecting the codeword closest to $\mathbf{y}$.

The complexity of this procedure grows with the size of the list

$$|\mathcal{L}| = \sum_{m=0}^{t} \binom{k}{m}. \tag{8.34}$$

Clearly, the larger t, the larger the list and the better the performance of OSD; but also the greater the decoding complexity. Indeed, in the extreme case $t = k$, OSD coincides with ML decoding. In general, the value of t needed to approach ML decoding performance grows with the blocklength n. The decoding complexity can be reduced if the minimum distance $d_{\min}$ of the code is known. In such a case, one can stop the list construction procedure as soon as one finds a codeword whose corresponding BPSK vector has a Euclidean distance from $\mathbf{y}$ less than $\sqrt{\rho d_{\min}}$, since there cannot be a codeword that is closer to $\mathbf{y}$ than this.

8.2.5.2 Linear Block Codes With Tail-Biting Trellises

Short linear block codes can sometimes be represented efficiently by finite-length trellises with multiple initial and final states, in which each codeword corresponds to a tail-biting (TB) path with the same initial and final state. One example of codes that have this property and have also good distance spectra are linear block codes obtained through a tail-biting termination of suitably chosen convolutional codes.

The Viterbi algorithm can be used to decode such codes. However, since the initial state is not known, ML decoding requires running the Viterbi algorithm as many times as the number of initial states.

A suboptimal but lower-complexity decoding method is to use the so-called wrap-around Viterbi algorithm (WAVA). One assumes that all initial states are equiprobable and then runs the Viterbi algorithm one time. If the most likely path returned by the algorithm is a tail-biting path, decoding stops and this path is given as output. If the returned path is not tail-biting, the final state of the trellis is copied to the initial state and the Viterbi algorithm is run again.

The process is repeated until a tail-biting path is returned, or a maximum number of iterations is exceeded, in which case the decoder declares an error.

8.2.5.3 Nonbinary LDPC Codes

The performance under iterative decoding of LDPC codes in the short-blocklength regime can be significantly improved by constructing such codes on higher-order fields than $\mathbb{F}_2$ [12]. Low-complexity implementations of the required decoding algorithm are, however, still an active area of research.

8.2.5.4 Performance

We illustrate next the performance of the coding schemes described in Sections 8.2.5.1–8.2.5.3. We consider a short code of length 128 and rate $R = 1/2$. Hence, $k = 64$.

Fig. 8.8A, illustrates the performance of a $(128, 64)$ extended BCH code with minimum distance 22 [17]. The t parameter in the OSD decoder is set to 4, which results in a list of size 679121. In Fig. 8.8B, we consider three tail-biting convolutional codes with different memory m. Specifically, we analyze a memory 8 convolutional code with generator polynomial (given in octal form) $[515, 677]$, a memory 11 convolutional code with generator polynomial $[5537, 6131]$ and a memory 14 convolutional code with generator polynomial $[75063, 56711]$. We see from the figure that both the eBCH and the tail-biting convolutional codes with memory 11 and 14 operate remarkably close to the finite-blocklength normal approximation limit. The memory-8 convolutional code exhibits a loss of about 0.5 dB at 10^{-4} packet error probability. In Fig. 8.8C, we consider the performance of a nonbinary LDPC code constructed over a finite field of order 256. The PCM of the code has a constant row weight of 4 and a constant column weight of 2. We consider both iterative decoding with 200 iterations, and OSD with $t = 4$. One sees that the performances with OSD are close to the normal approximation, whereas the performance gap with iterative decoding is about 0.6 dB at 10^{-4}. Finally, in Fig. 8.8D we present the performance of a $(128, 71)$ polar code, combined with a $(71, 64)$ shorted cyclic code that serves as CRC. The list size is limited to 32. The performance gap to the normal approximation is about 0.4 dB at 10^{-4}.

8.3 CODING SCHEMES FOR FADING CHANNELS

So far, we have focused on the problem of transmitting information over the bi-AWGN channel (8.1). In this final section, we shall instead consider the more practically relevant scenario of communications over a multiantenna fading channel. The purpose is to illustrate the additional design challenges brought about by fading. Our focus will be on the short-packet regime and on the URLLC use case. We shall first discuss the single-input single-output (SISO) case and then move to multiple-input multiple-output (MIMO) transmissions.

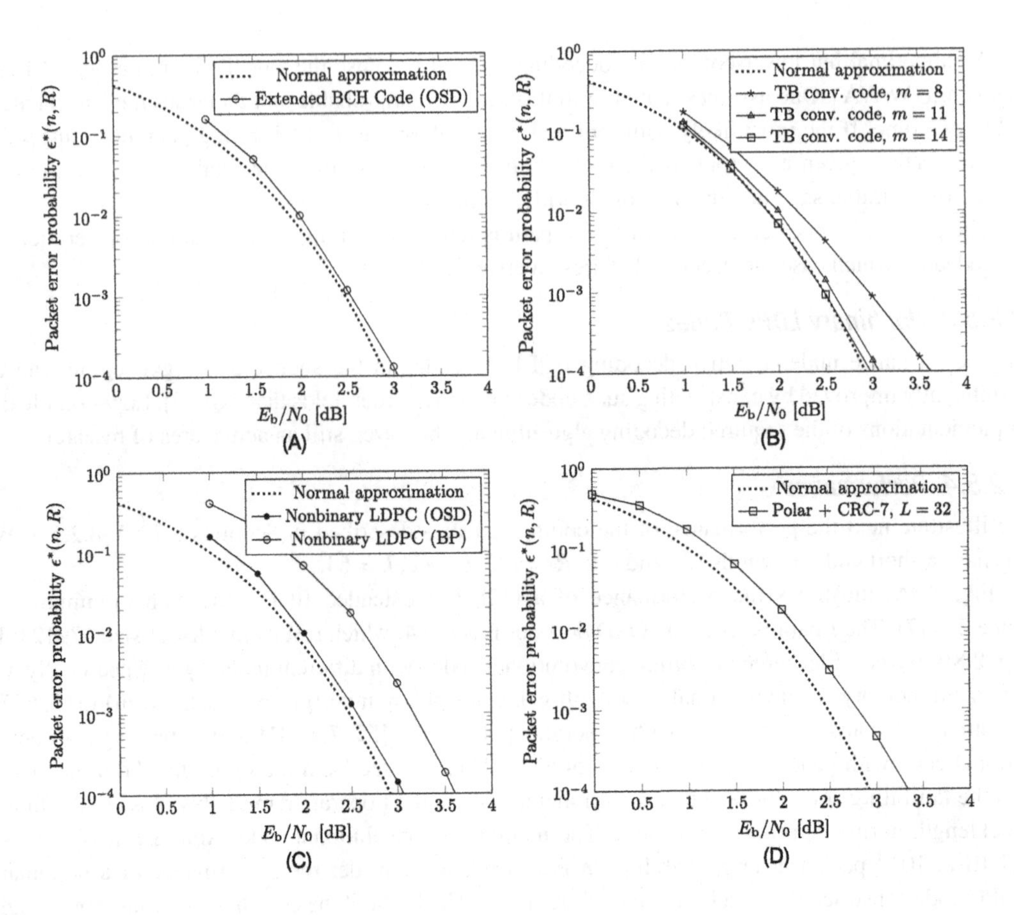

FIGURE 8.8

Performance of the coding schemes described in this section. Here, $R = 1/2$ and $n = 128$.

8.3.1 THE SISO CASE

We assume orthogonal frequency-division multiplexing (OFDM) operations and assume that each codeword spans multiple resource blocks (RBs), which are transmitted in the same time slot but at different frequencies. We assume that each RB contains d OFDM symbols consisting of u subcarriers. Hence, a RB conveys $n_c = d \times u$ complex-valued symbols. Low latency is achieved by selecting a sufficiently small value for d. For example, in downlink control channels, d may be chosen from the set $\{1, 2, 3\}$.

We assume that the coherence time T_c of the channel is larger than the transmission duration, and we let $L_{max} = \lfloor B/B_c \rfloor$ be the ratio between the transmission bandwidth B and the coherence bandwidth of the channel B_c. Hence, L_{max} corresponds to the maximum number of frequency diversity branches offered by the channel. As we shall see, exploiting diversity is fundamental to achieve the reliability constraints set in URLLC.

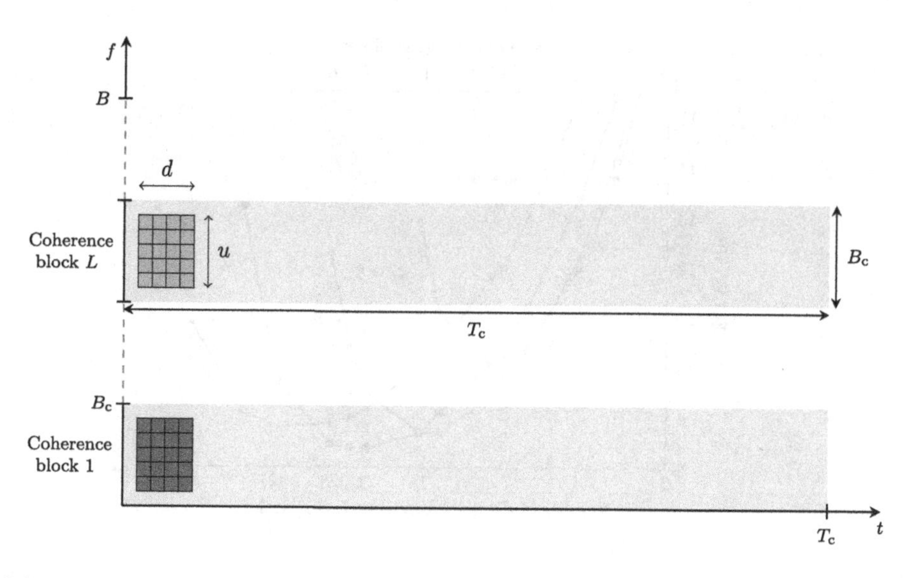

FIGURE 8.9

Signaling strategy in the time-frequency plane.

For simplicity, we model channel variations in frequency using the block-memoryless fading assumption. According to this assumption, the channel stays constant over each coherence interval and changes independently across diversity branches. This means that the fading channel is fully characterized by $L_{\max}$ complex channel coefficients.

As illustrated in Fig. 8.9, we assume that each RB fits within a coherence interval, i.e., all the complex symbols contained in the RB experience the same fading gain. We also assume that different RBs are allocated at different frequency branches. Finally, we assume that each codeword consists of $L \leq L_{\max}$ RBs.

An example is in order. Assuming $B = 20$ MHz, as in LTE, we obtain $L_{\max} = 4$ for the extended pedestrian type-A (EPA) 5 Hz channel model [2], whose coherence bandwidth is 4.4 MHz, whereas $L_{\max} = 30$ for the tapped-delay-line type-C (TDL-C) 300 ms–3 km/h channel model [3], whose bandwidth is 0.66 MHz. In both cases, the coherence interval is about 85 ms, which exceeds by far the duration of a RB in all practically relevant scenarios. Indeed, recall that, with 15 kHz subspacing, the duration of an OFDM symbol is just 66.7 µs. We shall denote by $[\mathbf{x}_1, \mathbf{x}_2, \dots \mathbf{x}_L] \in \mathbb{C}^{Ln_c}$ the vector of the transmitted complex symbols, where $n = L_c$ is the blocklength. The corresponding received vector is $[\mathbf{y}_1, \mathbf{y}_2, \dots, \mathbf{y}_L]$ where $\mathbf{y}_\ell$ is the received signal corresponding to the ℓth RB; we have

$$\mathbf{y}_\ell = \sqrt{\rho} h_\ell \mathbf{x}_\ell + \mathbf{w}_\ell, \quad \ell = 1, \dots, L. \tag{8.35}$$

Here, $\{h_\ell\}$ are i.i.d. fading coefficients, in agreement with the memoryless block-fading assumption. We shall assume that the variance of the $\{h_\ell\}$ is normalized to one. Furthermore, $\mathbf{w}_\ell$ is the additive white complex-Gaussian noise whose entries are i.i.d. and have zero mean and unit variance.

We shall focus on pilot-assisted transmission [38] according to which n_p out of the n_c entries of a RB are allocated to pilot symbols known to the receiver, and the remaining entries are reserved for

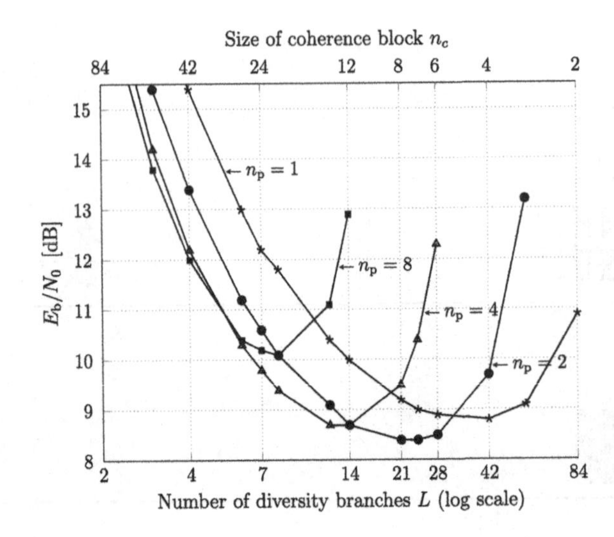

FIGURE 8.10

Minimum energy per bit E_b/N_0 to achieve $\epsilon = 10^{-3}$ for the Rayleigh-fading case, as a function of the number of diversity branches L. Here, $k = 81$ and $n = 168$.

coded data symbols. Specifically, we assume that $\mathbf{x}_\ell = [\mathbf{x}_\ell^{(p)} \mathbf{x}_\ell^{(d)}]$ where $\mathbf{x}_\ell^{(p)}$ is the n_p-dimensional pilot-symbol vector and $\mathbf{x}_\ell^{(d)}$ is the $n_c - n_p$ data vector. Throughout, we will focus on the scenario in which both pilot and data symbols are transmitted at the same power. More precisely, we assume that the entries of $\mathbf{x}_\ell^{(p)}$ and $\mathbf{x}_\ell^{(d)}$ are QPSK symbols with unit energy. QPSK modulation is indeed suitable for the low-rate low-power scenarios that are relevant for URLLC.

At the receiver side, we assume a practically relevant mismatch-decoding structure, in which the n_p-dimensional received vector $\mathbf{y}_\ell^{(p)}$ that corresponds to the pilot symbols is used to obtain a ML estimated $\hat{h}_\ell$ of the fading channel h_ℓ according to

$$\hat{h}_\ell = \mathbf{y}_\ell^{(p)} \frac{(\mathbf{x}_\ell^{(p)})^H}{n_p \sqrt{\rho}}. \tag{8.36}$$

Then the channel estimate is fed to a scaled minimum-distance decoder that treats it as perfect and produces a message estimate as follows:

$$\hat{j} = \underset{m \in \{1,2,\ldots,2^k\}}{\arg\min} \sum_{\ell=1}^{L} \|\mathbf{y}_\ell^{(d)} - \hat{h}_\ell \mathbf{x}_\ell^{(d)}(m)\|^2. \tag{8.37}$$

The performance of this transceiver architecture has been recently studied in [27,15] using finite-blocklength information-theoretic methods similar to the ones described in Section 8.1. This kind of theoretical analyses provides useful insights on the optimal number of diversity branches L one should code over for a given fixed blocklength n, and on the optimal number of pilot symbols n_p that should be allocated in each resource block. This is illustrated in Fig. 8.10 where we have depicted

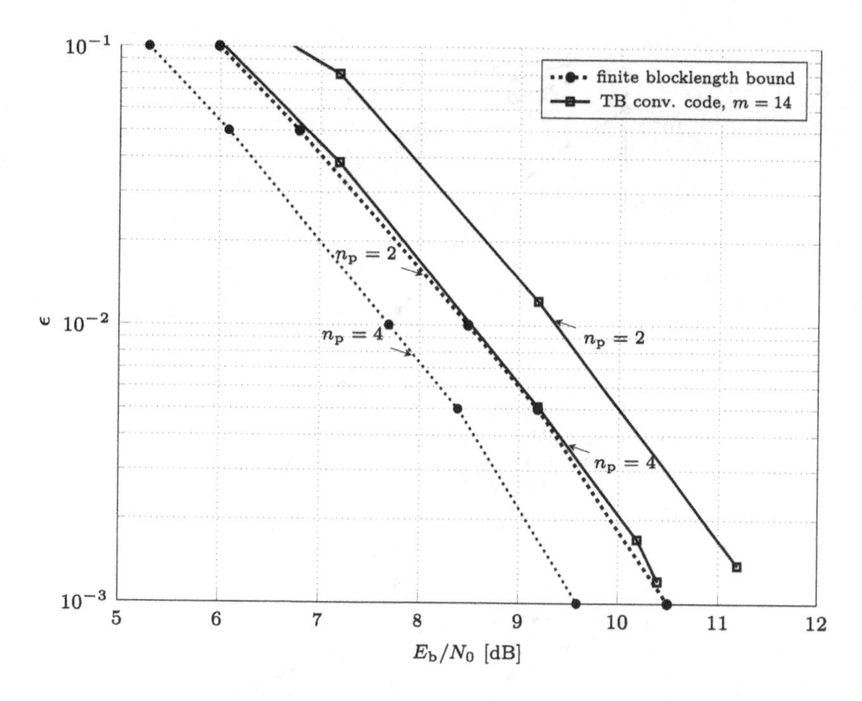

FIGURE 8.11

Packet error probability versus energy per bit for the case of Rayleigh fading, $k = 81$, $n = 186$, $L = 7$. Information-theoretic bounds [27, Th. 3] and performance of an actual coding scheme based on tail-biting convolutional codes and OSD.

the information-theoretic upper bound [27, Th. 3] on the minimum energy per bit E_b/N_0 required to achieve $\epsilon = 10^{-3}$ when transmitting $k = 81$ information bits. The blocklength $n = Ln_c$ is 168 and a different number of diversity branches L is considered (the larger L, the smaller n_c).

We see that there exists an optimum number of diversity branches $L = 21$ that minimizes the energy per bit. When $L < 21$, the system is penalized by the insufficient amount of frequency diversity available, which makes reliable transmission costly from an energy viewpoint. When $L > 21$, and, hence $n_c = n/\ell < 8$, the system suffers from the large pilot-symbol overhead, which is required to track the fast channel variations. We also observe that the number n_p of pilot symbols per resource block needs to be chosen carefully as a function of the number L of frequency diversity branches. Indeed, setting n_p to a suboptimal value yields a significant performance loss.

We next consider the performance of an actual coding scheme and benchmark it against the information-theoretic bounds. Specifically, we choose a (324, 81) binary quasi-cyclic code obtained by tail-biting termination of a rate 1/4 convolutional code with memory $m = 14$. The output of the encoder is passed through a pseudorandom interleaver and then some of the coded symbols are punctured to accommodate the desired number of pilot symbols per resource block (RB) after QPSK modulation. Decoding is performed via OSD with $t = 3$. The performance of this coding scheme when $L = 7$ is illustrated in Fig. 8.11 for the case $n_p = 2$ and $n_p = 4$. In both cases, the gap to the information-theoretic bound is about 1 dB.

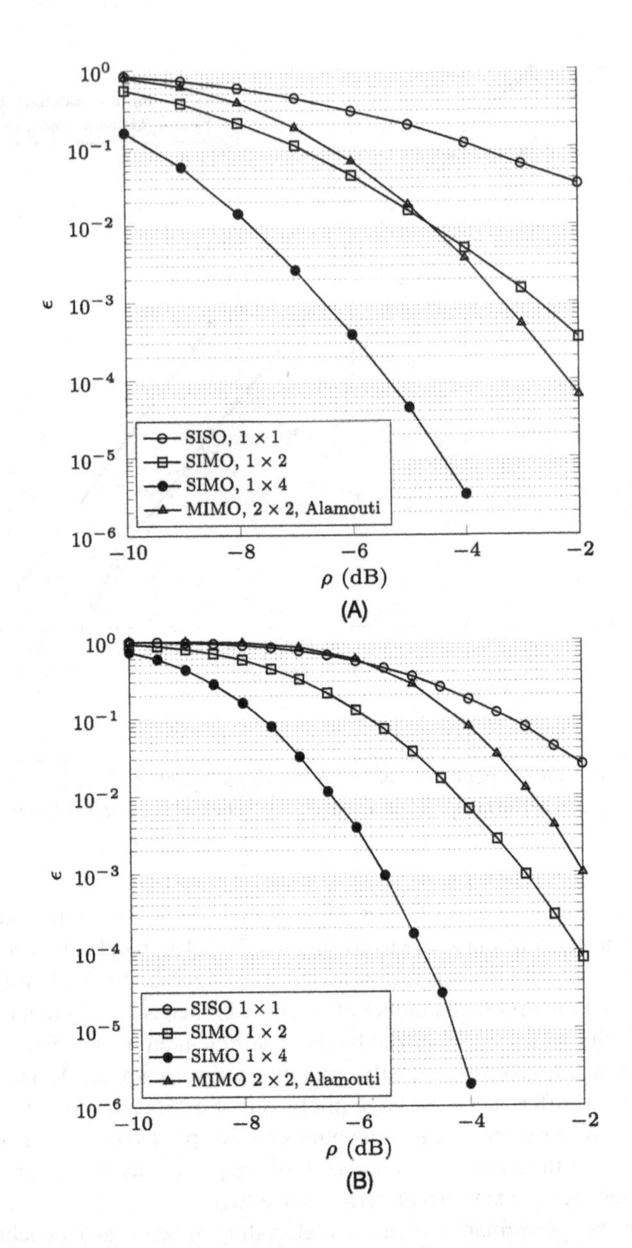

FIGURE 8.12

Packet error probability versus E_b/N_0; $k = 30$, $n = 288$, spatially white Rayleigh fading. (A) $L = 4$. (B) $L = 12$.

8.3.2 THE MIMO CASE

Exploiting the additional spatial diversity provided by MIMO transmission and reception is crucial to achieve the reliability level targeted by URLLC. The information-theoretic bounds depicted in Fig. 8.10

can be extended to MIMO communications, which allows one to explore the benefit of multiple antennas. As discussed in [15], the information-theoretic bounds can be extended to cover the case in which a space-frequency code is used at the transmitter to let the available antennas provide spatial diversity, in the absence of channel-state information at the transmitter.

In Fig. 8.12, we report the performance of different MIMO configurations for the case $k = 30$, which is relevant for downlink control-information transmission, and $n = 288$. We compare the performance of single-input single-output (SISO), 1×2 and 1×4 SIMO, and 2×2 MIMO with Alamouti encoding [4]. In Fig. 8.12A, we consider the case $n_c = 72$ and $L = 4$ (EPA 5 Hz channel model), whereas in Fig. 8.12B we set $n_c = 24$ and $L = 12$ (TDL-C channel model). In both cases, the number of pilot symbols is optimized to minimize the error probability. Furthermore, QPSK transmission is assumed.

We observe that, within the range of SNR values considered in the figure, only the 1×4 SIMO configuration is able to achieve an error probability below 10^{-5}, a common requirement in URLLC. Although the 2×2 MIMO Alamouti configuration offers the same amount of spatial diversity as 1×4 single-input multiple-output (SIMO), it is more sensitive to channel-estimation errors. This is particularly evident in Fig. 8.12B, where the small value of n_c results in a noisy channel estimate. As a consequence, the 2×2 MIMO Alamouti configuration performs worse than 1×2 SIMO over the range of SNR values considered in the figure.

REFERENCES

[1] 3GPP, TS 38.212 V15.0.0: multiplexing, and channel coding, 2017, Dec.

[2] 3GPP, TS 36.104: Technical specification group radio access network, 3GPP, 2012.

[3] 3GPP, TR 38.901: Study on channel model for frequencies from 0.5 to 100 GHz, 3GPP, 2017.

[4] S. Alamouti, A simple transmit diversity technique for wireless communications, IEEE Journal on Selected Areas in Communications (ISSN 0733-8716) 16 (8) (1998, Oct.) 1451–1458, https://doi.org/10.1109/49.730453.

[5] N. Alon, M. Luby, A linear time erasure-resilient code with nearly optimal recovery, IEEE Transactions on Information Theory 42 (11) (1996, Nov.) 1732–1736.

[6] E. Arıkan, Channel polarization: a method for constructing capacity-achieving codes for symmetric binary-input memoryless channels, IEEE Transactions on Information Theory 55 (7) (2009, Jul.) 3051–3073.

[7] E.E. Arıkan, I. Telatar, On the rate of channel polarization, in: Proc. IEEE Int. Symp. Inf. Theory (ISIT), Seoul, Korea, 2009, Jul., pp. 1493–1495, https://arxiv.org/abs/0807.3806.

[8] V. Bioglio, F. Gabry, I. Land, Low-complexity puncturing and shortening of polar codes, in: Proc. IEEE Wireless Commun. Netw. Conf., San Francisco, CA, U.S.A., 2017, Mar.

[9] G. Böcherer, F. Steiner, P. Schulte, Bandwidth efficient and rate-matched low-density parity-check coded modulation, IEEE Transactions on Communications 63 (12) (2015, Dec.) 4651–4665.

[10] A. Collins, G. Durisi, T. Erseghe, V. Kostina, J. Östman, Y. Polyanskiy, I. Tal, W. Yang, SPECTRE: short-packet communication toolbox, v2.0, https://github.com/yp-mit/spectre, 2016, Sep.

[11] D.J. Costello Jr., G.D. Forney Jr., Channel coding: the road to channel capacity, Proceedings of the IEEE 95 (6) (2007, Jun.) 1150–1177.

[12] M.C. Davey, D. MacKay, Low density parity-check codes over GF(q), IEEE Communications Letters 2 (6) (1998, Jun.) 165–167.

[13] D. Divsalar, S. Dolinar, C.R. Jones, K. Andrews, Capacity approaching protograph codes, IEEE Journal on Selected Areas in Communications 27 (6) (2009, Aug.) 876–888.

[14] T. Erseghe, Coding in the finite-blocklength regime: bounds based on Laplace integrals and their asymptotic approximations, IEEE Transactions on Information Theory 62 (12) (2016, Dec.) 6854–6883.

[15] G.C. Ferrante, J. Östman, G. Durisi, K. Kittichokechai, Pilot-assisted short-packet transmission over multiantenna fading channels: a 5G case study, in: Conf. Inf. Sci. Sys. (CISS), Princeton, NJ, 2018, Mar.

[16] G.D. Forney Jr., G. Ungerboeck, Modulation and coding for the linear Gaussian channels, IEEE Transactions on Information Theory 44 (6) (1998, Oct.) 2384–2415.
[17] M. Fossorier, S. Lin, Soft-decision decoding of linear block codes based on ordered statistics, IEEE Transactions on Information Theory (ISSN 0018-9448) 41 (5) (1995, Sep.) 1379–1396, https://doi.org/10.1109/18.412683.
[18] R. Gallager, Low-density parity-check codes, IRE Transactions on Information Theory 8 (1) (1962, Jan.) 21–28.
[19] G. He, J.C. Belfiore, X. Liu, Y. Ge, R. Zhang, I. Land, Y. Chen, R. Li, J. Wang, G. Yang, W. Tong, β-expansion: a theoretical framework for fast and recursive construction of polar codes, in: Proc. IEEE Global Telecommun. Conf. (GLOBECOM), Singapore, 2017, Dec., https://arxiv.org/abs/1704.05709.
[20] G. Liva, F. Steiner, pretty-good-codes.org: online library of good channel codes, http://pretty-good-codes.org/, 2017.
[21] G. Liva, W.E. Ryan, M. Chiani, Quasi-cyclic generalized LDPC codes with low error floors, IEEE Transactions on Communications 56 (1) (2008, Jan.) 49–57.
[22] D. MacKay, R. Neal, Good codes based on very sparse matrices, in: C. Boyd (Ed.), IMA Conf. Cryptography and Coding, Springer-Verlag, 1995, Oct.
[23] A. Martinez, A. Guillén i Fàbregas, Saddlepoint approximation of random-coding bounds, in: Proc. Inf. Theory Applicat. Workshop (ITA), San Diego, CA, U.S.A., 2011, Feb.
[24] M. Mondelli, S.H. Hassani, R. Urbanke, Unified scaling of polar codes: error exponent, scaling exponent, moderate deviations, and error floors, IEEE Transactions on Information Theory 62 (12) (2016, Dec.) 6698–6712.
[25] M. Mondelli, S.H. Hassani, R. Urbanke, Construction of polar codes with sublinear complexity, https://arxiv.org/abs/1612.05295, 2017, Jul.
[26] S.M. Moser, Information Theory (Lecture Notes), fifth ed., ETH Zurich/National Chiao Tung University, Switzerland/Taiwan, 2017, Mar.
[27] J. Östman, G. Durisi, E.G. Ström, M.C. Coşkun, G. Liva, Short packets over block-memoryless fading channels: pilot-assisted or noncoherent transmission?, https://arxiv.org/abs/1712.06387, 2017, Dec.
[28] S. Parkvall, E. Dahlman, A. Furuskär, M. Frenne, NR: the new 5G radio access technology, IEEE Communications Standards Magazine (2017, Dec.) 24–30.
[29] Y. Polyanskiy, H.V. Poor, S. Verdú, Channel coding rate in the finite blocklength regime, IEEE Transactions on Information Theory 56 (5) (2010, May) 2307–2359.
[30] T.J. Richardson, R. Urbanke, Modern Coding Theory, Cambridge Univ. Press, Cambridge, U.K., 2008.
[31] W.E. Ryan, S. Lin, Channel Codes: Classical and Modern, Cambridge Univ. Press, 2009.
[32] S. Sandberg, M. Andersson, A. Shirazinia, Y. Blankenship, LDPC Codes for 5G New Radio, 2018.
[33] I. Sason, S. Shamai (Shitz), Performance analysis of linear codes under maximum-likelihood decoding: a tutorial, Foundations and Trends in Communications and Information Theory (2006), https://doi.org/10.1561/0100000009.
[34] C. Schürch, A partial order for the synthesized channels of a polar code, in: Proc. IEEE Int. Symp. Inf. Theory (ISIT), Barcelona, Spain, 2016, Jul., pp. 220–224.
[35] C.E. Shannon, A mathematical theory of communication, The Bell System Technical Journal 27 (1948, July and October) 379–423, 623–656.
[36] I. Tal, A. Vardy, List decoding of polar codes, IEEE Transactions on Information Theory (ISSN 0018-9448) 61 (5) (2015, May) 2213–2226, https://doi.org/10.1109/TIT.2015.2410251.
[37] R.M. Tanner, A recursive approach to low complexity codes, IEEE Transactions on Information Theory 27 (9) (1981, Sep.) 533–547.
[38] L. Tong, B.M. Sadler, M. Dong, Pilot-assisted wireless transmissions, IEEE Signal Processing Magazine 21 (6) (2004, Nov.) 12–25, https://doi.org/10.1109/MSP.2004.1359139.
[39] G. Vazquez-Vilar, A.T. Campo, A.G. i Fàbregas, A. Martinez, Bayesian M-ary hypothesis testing: the meta-converse and Verdú-Han bounds are tight, IEEE Transactions on Information Theory 62 (5) (2016, May) 2324–2333.
[40] H. Zhang, R. Li, J. Wang, S. Dai, G. Zhang, Y. Chen, H. Luo, J. Wang, Parity-check polar coding for 5G and beyond, in: IEEE Int. Conf. Commun. (ICC), 2018, 01, https://arxiv.org/abs/1801.03616.

CHAPTER

SIMULATOR

9

For further research and development of the NR physical layer, we complement this book with a MATLAB-based link level simulator. We shall present an open-source simulator, including several waveforms, i.e., cyclic prefix based orthogonal frequency division multiplexing (CP-OFDM), windowed-OFDM (W-OFDM), universal filtered OFDM (UF-OFDM), discrete Fourier transform spread OFDM (DFTS-OFDM), offset QAM filter-bank multicarrier (FBMC-OQAM), and QAM filter-bank multicarrier (FBMC-QAM), various hardware impairments, i.e., power amplifier (PA) nonlinearity, oscillator phase noise (PN), and carrier frequency offset (CFO), and a geometry-based stochastic channel model supporting millimeter-wave channel emulation up to 80 GHz.

The simulator is developed by Qamcom Research & Technology AB. We release it with this book as an open-source simulator (with permission from Qamcom Research & Technology AB). The simulator can be accessed at **www.qamcom.se/research/5gsim**.

The simulator was originally developed in the EU funded mmMAGIC project [15] to assess performance of various waveforms in a common environment. As discussed in Chapter 5, several waveforms were proposed for 5G NR. Due to lack of a common waveform simulator, sometimes contradictory results were observed. Therefore, the mmMAGIC project realized importance of a transparent simulator for fair comparisons. Furthermore, the simulator was used to develop receiver algorithms for NR. At present (August 2018), the simulator has the following key components:

- A channel model up to 80 GHz (discussed in Chapter 3);
- A phase-noise model for a free-running oscillator and a phase-noise model for a phased-lock loop based oscillator (discussed in Chapter 4);
- A power-amplifier model based on a generalized memory polynomial (discussed in Chapter 4);
- Modulation/demodulation modules for various waveforms (discussed in Chapter 5);
- Algorithms for channel estimation & equalization, synchronization, and phase-noise compensation (discussed in Chapter 6).

It is likely that the simulator will evolve in the future with contributions from the authors and the open-source community.

This chapter is organized as follows. Section 9.1 provides an overview of the simulator, followed by details of different functional modules and waveforms in Section 9.2 and Section 9.3, respectively. Our ambition is not to discuss fundamentals (which are already covered in previous chapters); rather the focus is on implementation specific details. Finally, Section 9.4 shows a number of simulation exercises with various waveforms subject to the above-mentioned impairments and receiver algorithms. With these exercises, important observations are made.

5G Physical Layer. https://doi.org/10.1016/B978-0-12-814578-4.00014-X

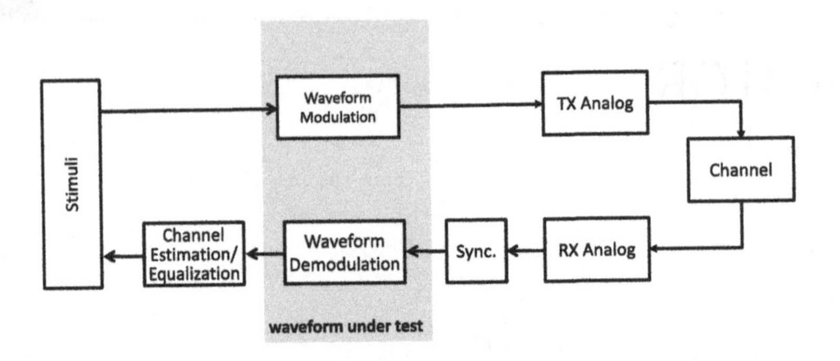

FIGURE 9.1

Block diagram of the simulator.

9.1 SIMULATOR OVERVIEW

A functional diagram of the simulator is shown in Fig. 9.1. The stimuli module sends (receives) QAM symbols to be modulated (demodulated) by the waveform modulator (demodulator). The modulated signal goes through analog modules at the transmitter (Tx) and receiver (Rx), and the radio channel. The Tx analog module adds hardware impairments, such as power amplifier (PA) nonlinearity, oscillator phase noise (PN), and carrier frequency offset (CFO) to the transmitted signal. The transmitted signal then experiences delay spread, Doppler spread, and interference, incurred by the channel model. After the radio channel, the received signal is further impaired by additive white Gaussian noise (AWGN) and PN that occur in the Rx analog module (before demodulation). At the receiver, timing and frequency synchronization are performed first. The synchronized signal is transformed to the frequency (subcarrier) domain by the waveform demodulator. Channel estimation and equalization are performed afterwards. (Note that it is possible to estimate the channel in the time domain before waveform demodulation. Nevertheless, the conventional frequency-domain channel estimation is chosen in this simulator.) At the moment, included waveforms are cyclic prefix (CP) based orthogonal frequency division multiplexing (OFDM) [5], windowed-OFDM (W-OFDM) [20], universal filtered OFDM (UF-OFDM) [22], discrete Fourier transform spread OFDM (DFT-S-OFDM) [4], offset QAM filter-bank multicarrier (FBMC-OQAM) [21], and QAM filter-bank multicarrier (FBMC-QAM) [17]. Other waveforms can be integrated into the open-source simulator as well.

We describe each module of the simulator separately in the following sections. Exemplary simulations are conducted to compare different waveforms and to illustrate the functionality of the simulator.

9.2 FUNCTIONAL MODULES

We present all the functional modules (except for waveforms) of the simulator separately in this section. The presentation of different waveforms is deferred to the next section.

9.2.1 CHANNEL MODEL

Both geometry-based channel models and 5G stochastic channel models [11–16] have been discussed in Chapter 2. Among the 5G stochastic channel model, the mmMAGIC (also known as QuaDRiGa) channel model [16] is the only one that has been made open-source. Therefore, we integrate the mmMAGIC channel model into the simulator. For the sake of completeness, we briefly present the mmMAGIC channel model here. The mmMAGIC channel model is an extension of the well-known WINNER channel model [1]. While the WINNER model is valid up to 6 GHz, the current version of the mmMAGIC channel model supports millimeter-wave channel up to 80 GHz. In addition, the mm-MAGIC channel model can well predict the time variance of the channel (as compared to the WINNER model). The data flow of the mmMAGIC channel model is given as follows:

- define network layout, terminal trajectories, propagation scenarios along the trajectories, and antenna patterns of the terminals and base stations;
- calculate the maps of large-scale parameters (i.e., RMS delay spread, Rician K-factor, shadow fading, azimuth spread of departure/arrival, elevation spread of departure/arrival);
- calculate the initial delays, the path powers, and the angles of departure/arrival;
- calculate the polarized channel coefficients including the transmission loss, the shadow fading, and the K-factor.

A detailed description of the channel model can be found in [14]. The channel model can be switched off, in which case the channel boils down to a simple AWGN channel. This can be useful if one is only interested in making manifest the effect of a certain hardware impairment.

9.2.2 POWER AMPLIFIER MODEL

As discussed in Chapter 4, the PA effects can be accurately modeled by the Volterra series. Nevertheless, the Volterra series model can be quite complicated. By ignoring all the memory and crossterms, the Volterra series model boils down to the polynomial model. The polynomial model is very simple and allows analytical study. However, it cannot capture the dynamic or memory effect of the PA. As a trade-off, the memory polynomial and the generalized memory polynomial are introduced in modeling the PA effects. The former omits all the crossterms and, therefore, captures only the first degree of dynamic effects, whereas the latter only ignores part of the crossterms and is very close to the Volterra series model. The generalized memory polynomial (GMP) model is a good compromise between accuracy and complexity. As the nonlinear order increases, the variances of the estimated parameters also increases. To overcome this problem, local basis functions are used. Specifically, a vector-switched GMP (VS-GMP)-based PA model [2] is included in the simulator. The interested reader may refer to [2] for a detailed description of the PA model.

Note that the PA nonlinearity causes spectral regrowth, i.e., power leakage to adjacent channels or increased out-of-band (OOB) emissions. In order to correctly examine the PA effects, interpolation and decimation ought to be performed before and after the PA model. The implemented PA model [2] requires an interpolation with an oversampling factor of 5 to 7. Otherwise, the OOB emissions caused by the PA will incur a (nonphysical) inband distortion due to the spectral periodicity of time-discrete signals. The Parks–McClellan FIR filter [13] is used as anti-aliasing filter in the interpolation and decimation. Even though the Parks–McClellan FIR filter is close to an ideal low-pass filter, there will inevitably be some distortion in the transition around the cut-off frequency. As a result, guard bands (i.e., zero subcarriers in the beginning and end of each multicarrier symbol) [10] are needed. The Parks–McClellan FIR filter has another effect: delaying the signal by half of the filter length. Since it is used twice (i.e., before and after the PA model), the filter delay equals its filter length. The filter delay and the delay caused by the random multipath channel will be estimated in the synchronization module.

The PA model can be switched off, in which case the Parks–McClellan FIR filter will be switched off automatically.

9.2.3 PHASE-NOISE MODEL

The PN of an oscillator is a multiplicative noise. It causes a random phase rotation of the received signal. For a multicarrier waveform, the PN causes common phase error (CPE) and inter-carrier interference (ICI). While the former represents a common phase rotation to all the subcarrier, the latter refers to the interference for each subcarrier caused by all the other ones. The CPE can be readily corrected using scattered pilots, whereas the ICI may be difficult to eliminate. There are ICI correction algorithms, e.g., [19], [7], which can effectively mitigate the ICI effect. However, these algorithms are computationally heavy.

As discussed in Chapter 3, the PN of a free-running oscillator can be modeled by a random walk (or Wiener) process, whose discrete-time expression is given by $\varphi(n+1) = \varphi[n] + w[n]$, where φ is the PN and w is a zero-mean Gaussian random variable whose variance is $4\pi\beta/f_s$ with f_s denoting the sampling frequency and β representing the 3-dB bandwidth of the phase noise. The Wiener PN model is perhaps the most popular PN model in the literature due to its simplicity and well-known statistical properties. However, its variance increases linearly with time. Hence, in practice, a phase-locked loop (PLL) is usually added to regularize the PN. A PN model of the PLL-based oscillator has been proposed in mmMAGIC [12]. This model was developed specifically for millimeter-wave communications. It is a more realistic (yet complicated) model as compared with the simple Wiener PN model. The PLL PN model is briefly mentioned in Chapter 3.

Both PN models are included in the simulator. The parameter β is used to specify the quality of the free-running oscillator. The PLL PN model involves many parameters. For simplicity, "low" and "high" PN modes were implemented for good and bad PLL-based oscillators. The detailed modeling parameters of the PN of the PLL-based oscillator are listed in Table 9.1.

Either PN model can be chosen at the transmitter and the receiver. The PN model at either side can be switched on and off independently.

9.2.4 SYNCHRONIZATION

As mentioned before, the multipath channel and the filters will cause delay (timing offset) of the received signal. The timing offset has to be estimated before further processing of the received signal. The CFO represents the frequency difference between the carrier frequencies at the transmitter and at the receiver. If left uncompensated, it will cause a phase rotation of the received signal and the phase increases linearly with time. The synchronization module performs timing and frequency synchronization. By assuming common preamble, the timing offset, and CFO are estimated using the methods in [24] and [23], respectively. One can also choose perfect timing and frequency synchronization. In the former case, the true value of the timing offset is used for timing synchronization. In the latter case, the CFO is set to zero and the frequency synchronization module is disabled (since perfect frequency synchronization results in a CFO-free signal).

Table 9.1 Parameter of "low" and "high" phase-noise settings

Component	Parameter	Value (f_c is the carrier frequency in GHz)	
		Low	High
Reference	S_0	$-60 + 20log_{10}(f_c)$ (dBc/Hz)	$-60 + 20log_{10}(f_c)$ (dBc/Hz)
	f_p	1 Hz	1 Hz
	f_x	10 kHz	3980 Hz
PLL	S_0	$-140.35 + 20log_{10}(f_c)$ (dBc/Hz)	$-122.35 + 20log_{10}(f_c)$ (dBc/Hz)
	f_p	∞	∞
	f_p	∞	∞
	f_{fc} (flicker corner)	2 MHz	0.8 MHz
	k (flicker exponent)	1	1
	f_{fp} (flicker pole)	200 Hz	80 Hz
VCO	S_0	$-24.34 + 20log_{10}(f_c)$ (dBc/Hz)	$-3.35 + 20log_{10}(f_c)$ (dBc/Hz)
	f_p	1 Hz	1 Hz
	f_x	20 MHz	80 MHz
Transfer function from reference and loop noise to output	$\frac{K_D K_{VCO} Z(s)}{sN_D + K_D K_{VCO} Z(s)}$	$6.366 \times 10^{-7}s + 1$ (Numerator)	$1.592 \times 10^{-6}s + 1$ (Numerator)
		$7.076 \times 10^{-29}s^4 + 5.780 \times 10^{-21}s^3 + 1.013 \times 10^{-13}s^2 + 6.366 \times 10^{-7}s + 1$ (Denominator)	$6.214 \times 10^{-20}s^3 + 6.333 \times 10^{-13}s^2 + 1.592 \times 10^{-6}s + 1$ (Denominator)
Transfer function from reference and loop noise to output	$\frac{sN_D}{sN_D + K_D K_{VCO} Z(s)}$	$7.076 \times 10^{-29}s^4 + 5.780 \times 10^{-21}s^3 + 1.013 \times 10^{-13}s^2$ (Numerator)	$6.214 \times 10^{-20}s^3 + 6.333 \times 10^{-21}s^3 + 1.013 \times 10^{-13}s^2$ (Numerator)
		$7.076 \times 10^{-29}s^4 + 5.780 \times 10^{-21}s^3 + 1.013 \times 10^{-13}s^2 + 6.366 \times 10^{-7}s + 1$ (Numerator)	$6.214 \times 10^{-20}s^3 + 6.333 \times 10^{-13}s^2 + 1.592 \times 10^{-6}s + 1$ (Denominator)

9.2.5 CHANNEL ESTIMATION AND EQUALIZATION

The frequency-domain channel estimation (equalization) is used in the simulator. The exact channel estimation methods may differ slightly for different waveforms. Yet, in essence, the channel transfer functions (CTFs) are estimated at the pilot subcarriers using the least-square (LS) estimator and the CTFs at other subcarriers are obtained by filtering [6].

Like the synchronization, the simulator allows for perfect channel estimation, in which case the true value of the channel transfer function is used for channel equalization.

9.3 WAVEFORMS

Several waveforms have been presented in Chapter 5. In the following subsections, we will briefly introduce these waveforms from an implementation point of view in the simulator. For all waveforms, the received (equivalent baseband) signal subject to phase noise, CFO, and multipath channel with additive noise is modeled as

$$y[n] = \exp\left(j\left(\phi[n] + 2\pi\varepsilon n\right)\right) \sum_{l=0}^{L-1} h_l x[n-l-v] + w[n] \tag{9.1}$$

where ε denotes the CFO (normalized by the sampling frequency), ϕ denotes the PN at the receiver, x represents the transmitted time-domain signal, h_l is the lth tap of the channel impulse response (CIR), L denotes the channel length, and w denotes the AWGN. Note that, for notational simplicity and without loss of generality, the effects of PA and PN at the transmitter have been omitted in the expression.

9.3.1 CP-OFDM

The CP-based OFDM (CP-OFDM) enjoys the simplest (de)modulation implementation among all the multicarrier waveforms. Its modulation can be expressed as

$$\mathbf{x} = [\mathbf{F}_{\mathrm{CP}}\,\mathbf{F}]^H\,\mathbf{s}, \tag{9.2}$$

where $\mathbf{s}$ denotes the $N \times 1$ column vector consisting of the subcarrier symbols, $\mathbf{F}$ denotes the $N \times N$ discrete Fourier transform (DFT) matrix whose element is given by $\exp(-j2\pi kn/N)/\sqrt{N}$ $(k, n = 0, \cdots, N-1)$, and $\mathbf{F}_{\mathrm{CP}}$ consists of the last N_{CP} columns of $\mathbf{F}$ (where N_{CP} denoting the CP length). Note that $\mathbf{F}_{\mathrm{CP}}\,\mathbf{s}$ forms the CP of the CP-OFDM symbol. The CP-OFDM demodulation matrix is given by $[\mathbf{0}\ \mathbf{F}]$, where $\mathbf{0}$ is an $N \times N_{\mathrm{CP}}$ zero matrix for CP removal.

9.3.2 W-OFDM

One can improve the OOB emission of the CP-OFDM by applying a windowing function to the transmitted time-domain signal, i.e., W-OFDM. Because of the windowing, the W-OFDM is also called weighted CP-OFDM (WCP-OFDM) [20]. By tapering the edges of the time-domain signal, the OOB emission can be reduced significantly. The tapering at the left edge causes no distortion of the useful signal thanks to the CP. To correct for the distortion due to the tapering at the right edge, the left edge of the received signal is added to the right edge, as explained in Chapter 5. This operation is called weighted overlap and add (WOLA) [9].

The raised-cosine filter is chosen as the windowing function in the simulator. The WOLA operation is implemented in the timing synchronization module at the W-OFDM receiver.

9.3.3 UF-OFDM

The UF-OFDM is also known as the universal filtered multicarrier (UFMC) [22]. It is achieved by first grouping the active subcarriers into B blocks of consecutive subcarriers, and then performing inverse discrete Fourier transform (IDFT) and filtering to each of the blocks. For the sake of convenience in the implementation, each block contains N_0 subcarriers. The filter bank is obtained by exponential modulation of a prototype filter of length N_v. The Dolph–Chebyshev filter is used as the prototype filter in the simulator. The UF-OFDM modulation can be expressed as

$$\mathbf{x} = \sum_{i=1}^{B} \mathbf{V}_i \mathbf{D}_i \mathbf{s}_i \tag{9.3}$$

where the subscript $_i$ is the block index, $\mathbf{s}_i = [s_i(0), \cdots, s_i(N_0 - 1)]^T$ consists of symbols and scattered pilots to be modulated on the N_0 subcarriers in the ith block, $\mathbf{D}_i$ consists of N_0 columns corresponding the subcarriers in the ith block submatrix of the IDFT matrix $\mathbf{F}^H$, and $\mathbf{V}_i$ is a Toeplitz matrix whose first column vector is given by $[v_i(0), \cdots, v_i(N_v), 0, \cdots, 0]^T$ where $v_i[n] = v[n]\exp(-j2\pi nk_i/N_v)$ with k_i denoting the center frequency of the ith block and $v[n]$ denoting the nth tap of the prototype filter. The UF-OFDM demodulation can be achieved by performing $2N_f$-point DFT of the zero-padded time-domain UF-OFDM symbol and then down sampling the frequency-domain signal by a factor of 2.

Note that a time-domain preprocessing has to be performed at the transmitter after the UF-OFDM modulation in order to remove the deterministic distortion introduced by the filter bank at the transmitter [22].

By performing filtering on groups of subcarriers, the OOB emission can be reduced significantly at the cost of increased complexity. There is also modest increase of symbol length due to the filters. Note that the performance of the UF-OFDM can be improved by introducing a CP in the presence of large delay spread [9] at the cost of increased overhead.

9.3.4 FBMC-OQAM

The FBMC-OQAM is also known as FBMC [3]. It applies a filter to each of the subcarriers. The filter length is K times of the number of subcarriers. K is the overlapping factor (which is a positive integer number). The filter bank can be implemented in the frequency domain using the frequency spreading method [3], which increases the DFT size from N to KN. Since the IDFT output and DFT input are overlapped in the time domain, a significant amount of redundancy is present in the computations of the frequency spreading method. In order to reduce this redundancy, the simulator implements the FBMC-OQAM in the time domain using the polyphase network (PPN) method [3].

Unlike the CP-OFDM, the neighboring subcarriers of the FBMC-OQAM are overlapped. The imaginary part of the impulse response of the subcarrier interference filter crosses the time axis at the integer multiples of the symbol period and the real part crosses the time axis at the odd multiples of half the symbol period [3]. The OQAM can be used in combination of the FBMC to avoid subcarrier interference without sacrificing the spectral efficiency. Specifically, this is achieved by doubling the symbol rate and using alternatively the real and the imaginary part for each subchannel. The FBMC-OQAM has good frequency localization (e.g., very low OOB emission) and high spectral efficiency (i.e., no CP overhead). Nevertheless, the FBMC-OQAM suffers from high complexity and long delay, i.e., $K - 1/2$ symbol durations. Another drawback is that it is not directly compatible with MIMO systems due to the staggered OQAM.

In the simulator, we adopt the PHYDYAS prototype filter with an (integer) overlapping factor K ranging from 2 to 4 [3]. A larger K results in better frequency localization, yet higher complexity and longer delay.

9.3.5 FBMC-QAM

To avoid the inconvenience caused by the OQAM, the QAM-based FBMC (FBMC-QAM) with dual filter banks was proposed [18]. A prototype filter in one filter bank is obtained by block interleaving the prototype filter in the other filter bank. The FBMC-QAM modulator performs filtering for even and odd subcarriers separately. The proposed filter banks in [18] ensure orthogonal subcarriers yet has higher OOB emission than that of the CP-OFDM. As a tradeoff between low OOB emission and high signal-to-interference ratio (SIR), two types of filters were proposed in [25]. Generally speaking, filters with lower OOB emission will result in lower SIR. The filter type I has a spectrum similar to that of the FBMC-OQAM with PHYDYAS filter (i.e., good frequency localization) yet poor SIR (10.6 dB). The filter type II has a wider spectrum, yet 9-dB higher SIR (19.6 dB) [25].

Like the FBMC-OQAM, the FBMC-QAM enjoys a high spectral efficiency (i.e., there is no CP overhead), yet suffers from high complexity and long delay. Unlike the FBMC-OQAM, the FBMC-QAM is compatible with MIMO systems at the cost of degradations in SIR and/or OOB emission.

In the simulator, the two types of filters in [25] and the filters in [18] are implemented for a fixed overlapping factor of $K = 4$. They are referred to as filter type I, II, and III, respectively.

9.4 SIMULATION EXERCISES

In this section, we compare the performances of different waveforms using the simulator. The simulator is written using object-oriented programming in MATLAB. To run the simulator, one needs to first direct to the subfolder **[.../waveformSimulator/ptplink]** in MATLAB. System parameters (waveform, symbol rate, subcarrier number, etc.) can be changed in a file named **sSysParDefault.m**. Different analog impairments (PN, CFO, AWGN, etc.) can be switched on and off in a file named **sAnaParDefault.m**. After all the parameters have been set, one needs to run **setpath.m** to add the paths of all the necessary folders. Finally, running **runSim.m**, it will perform simulation of the chosen waveform. Some simulation examples can be found in the subfolder **[.../ptplink/examples]**.

9.4.1 SPECTRAL REGROWTH

As mentioned in the previous section, certain waveforms (e.g., UF-OFDM and FBMC-OQAM) outperform the CP-OFDM in terms of OOB emissions (at the cost of increased complexity). Nevertheless, there can be spectral regrowth when waveforms are subject to PA nonlinearities. Here we show the spectral regrowths of CP-OFDM, UF-OFDM, W-OFDM, and FBMC-OQAM using the simulator.

The measurement setup is given below. We assume a DFT size of 512, out of which 158 are reserved as a guard band (i.e., 79 subcarriers in the beginning and end of each multicarrier symbol are set to zero), and QPSK modulation. In order to correctly include the OOB emission due to the PA nonlinearity, an interpolation with an oversampling factor of 5 and the Parks–McClellan optimal finite impulse response (FIR) filter as the anti-aliasing filter are used. For UF-OFDM modulation, we additionally assume Dolph–Chebyshev filter (with 64 filter taps and 40-dB stop band suppression) as the prototype filter and 16 subcarriers per filter. For W-OFDM modulation, we additionally assume a time-domain raised-cosine window (with 32 tapering length at the left and right soft edges). For FBMC-OQAM modulation, we assume an overlapping factor of 4 ($K = 4$) unless otherwise specified.

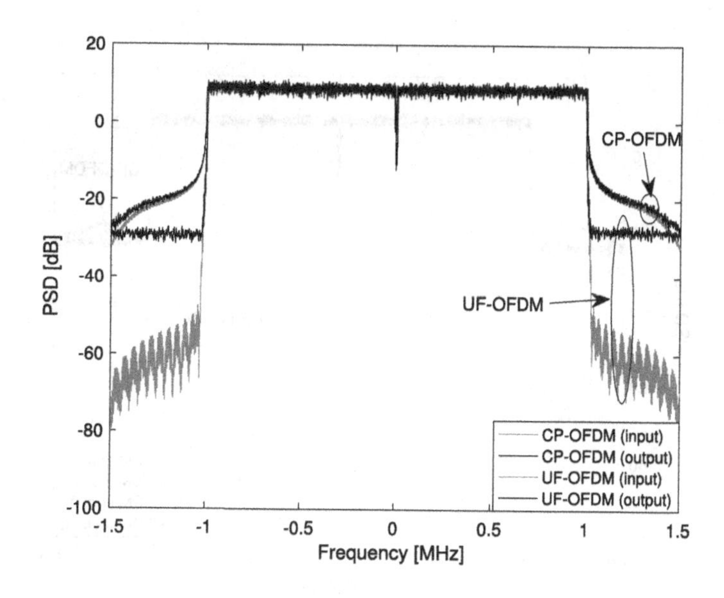

FIGURE 9.2

Spectral regrowths of CP- and UF-OFDM under PA nonlinearity: low input power (with 10-dB additional backoff).

Figs. 9.2–9.6 show the spectral regrowths of CP-, UF-, and W-OFDM under PA nonlinearities. As can be seen, the UF-OFDM has lower OOB emission as compared with the CP-OFDM, especially at low input power. Nevertheless, as the input power increases, the OOB emission of the UF-OFDM increases more than that of the CP-OFDM. That is, the OOB advantage of the UF-OFDM over the CP-OFDM diminishes at high input power. It is noted that the OOB emission of the CP-OFDM can be greatly improved by introducing a time-domain window to it, i.e., the W-OFDM. The OOB emissions of W- and UF-OFDM become comparable under PA nonlinearities. It is apparent that the FBMC-OQAM with an overlapping factor of 4 has much narrower OOB emission than the UF-OFDM does. Nevertheless, their spectra overlap under PA nonlinearities. It should also be noted that, in the absence of PA nonlinearities, the OOB emission of the FBMC-OQAM with an overlapping factor of 2 ($K = 2$) becomes larger than that with an overlapping factor of four (yet still smaller than that of the CP-OFDM). For the conciseness of the paper, the corresponding results are omitted here. All in all, it is shown that OOB emissions can be improved by filter banks; however, the improvements become insignificant when the PA nonlinearities are taken into account.

9.4.2 IMPAIRMENT OF CFO

The carrier frequencies generated by oscillators at the transmitter and the receiver are usually not identical. The difference between the two carrier frequency is referred to as the carrier frequency offset (CFO). The CFO destroys the orthogonality of multicarrier waveforms, resulting in performance degradations for different waveforms. In this exercise, we study the CFO impairment to different waveforms by simulations.

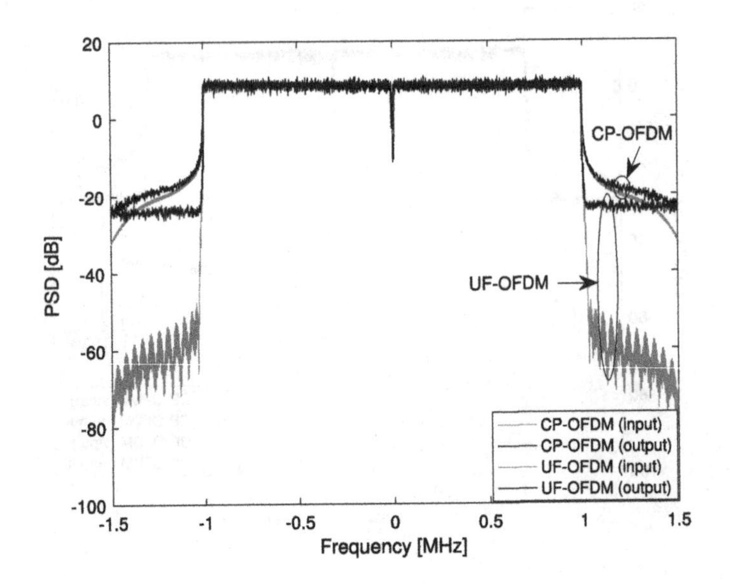

FIGURE 9.3

Spectral regrowths of CP- and UF-OFDM under PA nonlinearity: high input power.

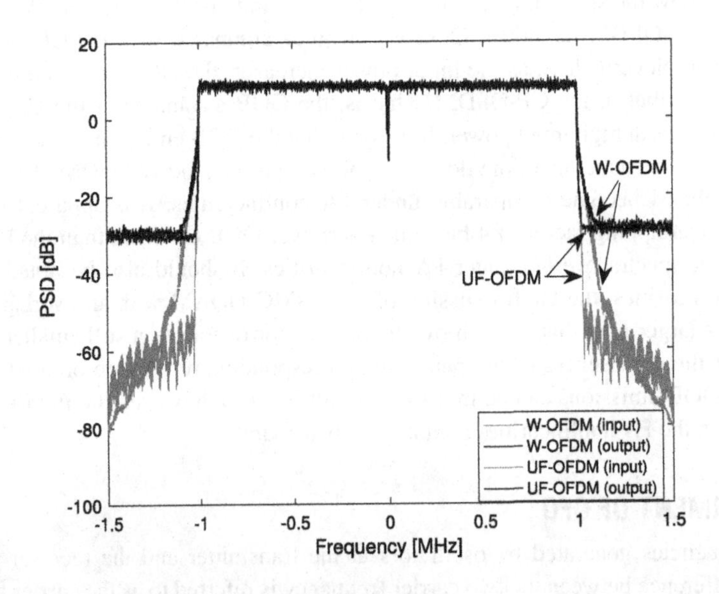

FIGURE 9.4

Spectral regrowths of W- and UF-OFDM with low input power.

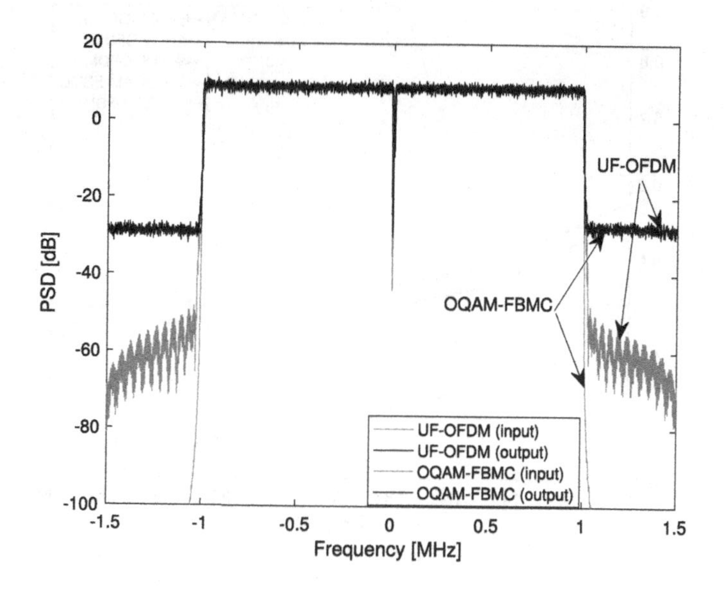

FIGURE 9.5

Spectral regrowths of UF-OFDM and FBMC-OQAM with low input power.

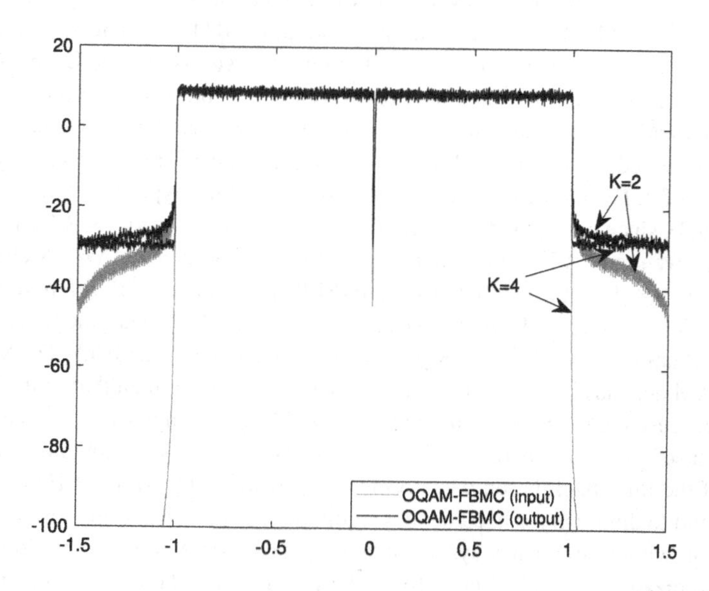

FIGURE 9.6

Spectral regrowths of FBMC-OQAM with different overlapping factors and low input power.

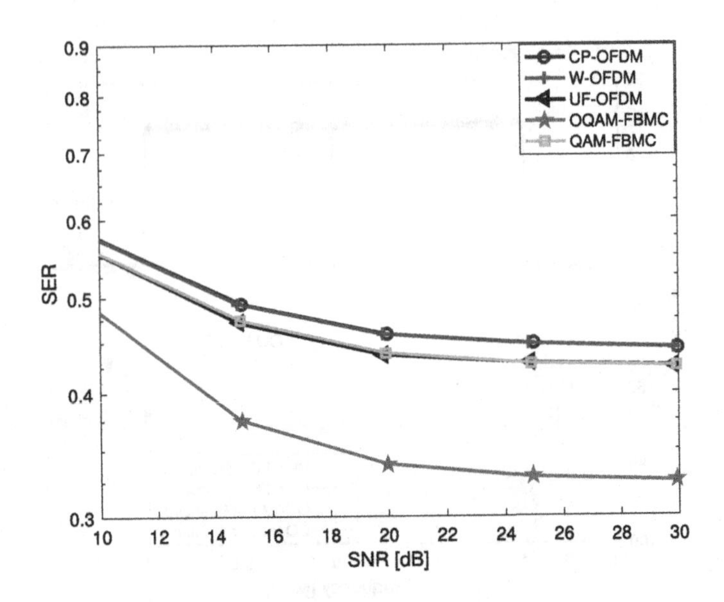

FIGURE 9.7

SER performances of different waveforms with 100 Hz CFO.

For fair comparisons of different waveforms, we disable the frequency synchronization module and assume 100 and 1000 Hz CFOs. In addition, we assume 1024 subcarriers, 100 MHz bandwidth, 16-QAM, and 100 multicarrier symbols. For the UF-OFDM, we assume the Dolph–Chebyshev filter (with 64 filter taps and 40-dB stop band suppression) as the prototype filter and 16 subcarriers per filter. For the W-OFDM, we assume a time-domain raised-cosine window with a tapering length of 32 at the left and right soft edges. For the FBMC-OQAM, we assume a PHYDYAS prototype filter with an overlapping factor of 4. For the FBMC-QAM, we assume the filter type II with an overlapping factor of 4 (cf. Section 9.3), since it is a good tradeoff between OOB emission and SIR [25]. In order to focus on the CFO effect, we ignore all the other hardware effects and assume an AWGN channel.

Fig. 9.7 and Fig. 9.8 show the symbol error rate (SER) performances of different waveforms under 100 and 1000 Hz CFOs, respectively. As can be seen, the FBMC-OQAM has the best SER performance and the UF-OFDM has the second best SER performance due to their filter bank. The SER performance of the FBMC-QAM depends on the value of the CFO and is no better than that of the UF-OFDM. The latter feature is because the filter type II in the FBMC-QAM is a compromise between OOB emission and SIR and because it does not guarantee orthogonality. Nevertheless, it should be noted that the SER improvements of the filter-bank-based waveforms (e.g., FBMC-OQAM and UF-OFDM) over that of the CP-OFDM tend to diminish (or become less significant) as the CFO increases.

As can be seen, without frequency synchronization, the system will not work. In practice, a coarse frequency synchronization (e.g., [23]) is performed prior to demodulation at the receiver. The residual CFO from the imperfect frequency synchronization can be further mitigated (together with the PN) using PN mitigation algorithms [8]. Since different waveforms apply different filters (which affect the signal powers of the waveforms), the exact SNR differs from the conventional E_S/N_0 (symbol energy

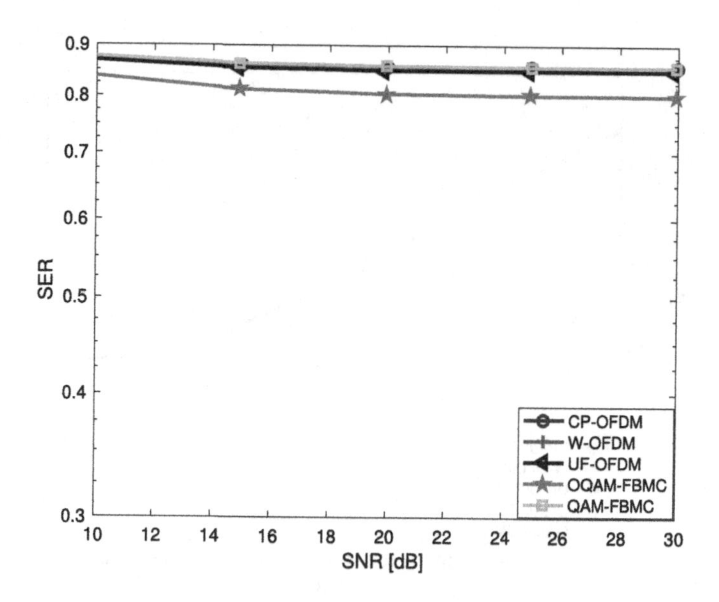

FIGURE 9.8

SER performances of different waveforms with 1000 Hz CFO.

over noise power density). As a result, we estimate the signal powers of different waveforms from their transmitted (time-domain) signals, respectively.

9.4.3 IMPAIRMENT OF PN

On top of the carrier frequency, the oscillator also produces random jitter frequency, which results in random PN. The PN causes random phase rotation of the signal and destroys the orthogonality of the multicarrier waveforms. In this exercise, we study the PN impairment to different waveforms by simulations.

As mentioned in Section 9.2.3, the PN effects in the frequency domain can be categorized as CPE and ICI. While the CPE can easily be estimated, the ICI is different to track. The advanced ICI correction algorithms (see, e.g., [19], [7]) are usually computationally expensive and waveform-specific, and may not always be feasible.

We assume perfect timing and frequency synchronization, 1024 subcarriers, 100 MHz bandwidth, 16-QAM, and 100 multicarrier symbols. For the UF-OFDM, we assume the Dolph–Chebyshev filter (with 64 filter taps and 40-dB stop band suppression) as the prototype filter and 16 subcarriers per filter. For the W-OFDM, we assume a time-domain raised-cosine window with 32 tapering length at the left and right soft edges. For the FBMC-OQAM, we use the PHYDYAS prototype filter with an overlapping factor of 4. For the FBMC-QAM, we assume the filter type II with an overlapping factor of four (cf. Section 9.3) since it is a good tradeoff between OOB emission and SIR [25]. In order to focus on the PN effect, we ignore all the other hardware effects and assume an AWGN channel.

Fig. 9.9 and Fig. 9.10 shows the SER performance of different waveforms in the presence of PN without and with CPE correction, respectively. As can be seen, the CPE corrections improve the SER

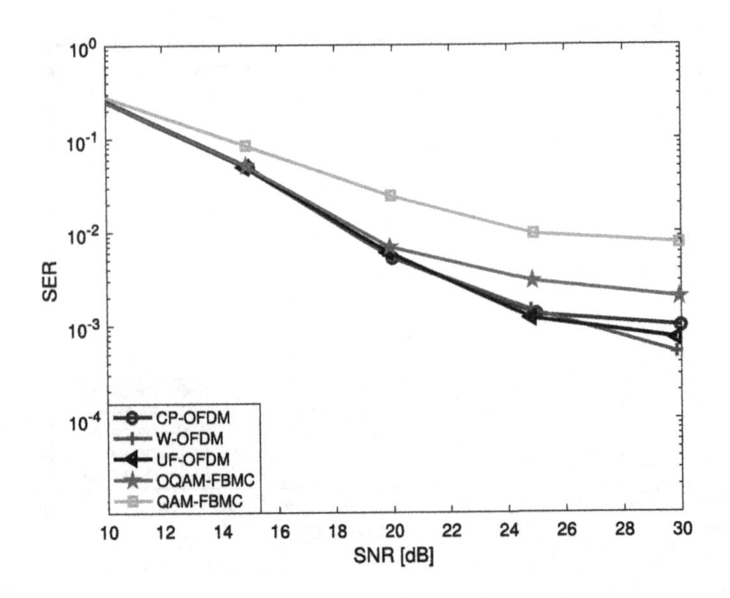

FIGURE 9.9

SER performances of different waveforms subject to PN without any receiver compensation.

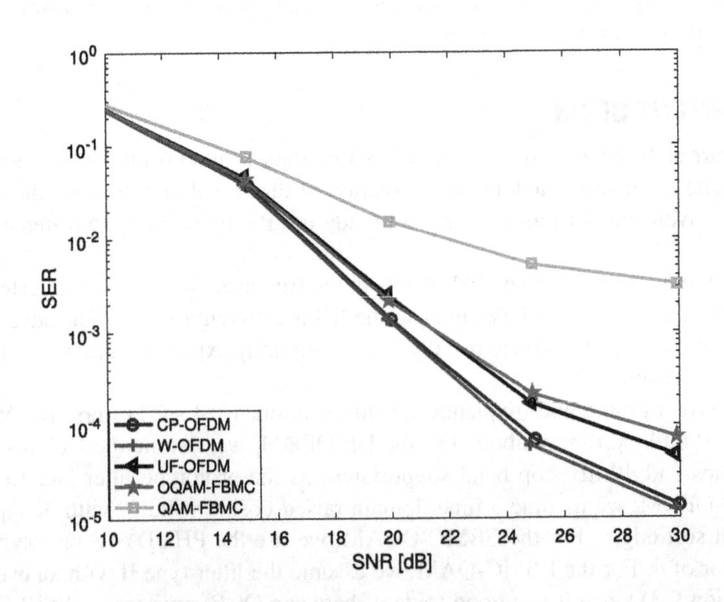

FIGURE 9.10

SER performances of different waveforms subject to PN with CPE correction.

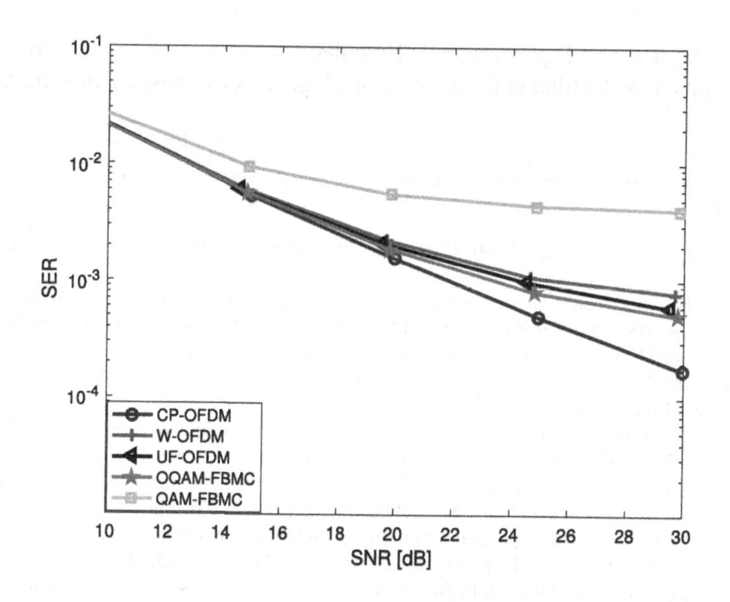

FIGURE 9.11

SER performances of different waveforms in time-varying fading channel with a user speed of 60 km/h at 6 GHz.

performances of the considered waveforms. Unlike the (constant) CFO, the PN arises from random jitter frequency. In this case, the W- and CP-OFDM have better SER performances than the other filter-bank-based waveforms, especially after the CPE removal. Due to the intrinsic SIR, the FBMC-QAM (with filter type II) has the worst SER performance under the PN.

9.4.4 IMPAIRMENT OF FADING CHANNEL

The multicarrier waveforms were introduced to cope with the multipath effect of the propagation channel in broadband communications. In this exercise, we compare SER performances of different waveforms in the multipath fading channel. We assume perfect timing and frequency synchronization, 512 subcarriers, 120 MHz bandwidth, and QAM. Like before, we assume the Dolph–Chebyshev filter (with 64 filter taps and 40-dB stop band suppression) as the prototype filter and 16 subcarriers per filter for UF-OFDM, a time-domain raised-cosine window with 32 tapering length at the left and right soft edges for W-OFDM, the PHYDYAS prototype filter with an overlapping factor of 4 for FBMC-OQAM, and the filter type II with an overlapping factor of 4 for FBMC-QAM. In order to focus on the multipath channel effect, we assume perfect channel estimation, ignore all the hardware effects, and resort to the QuaDRiGa channel model [16]. Unlike the WINNER channel model [1], the QuaDRiGa channel model can well emulate the time variation of the fading channel [14]. We assume a user speed of 60 km/h at 6 GHz. For each channel drop (realization), 50 multicarrier symbols were transmitted. In total, 60 channel drops were simulated. The SER performances of different waveforms are shown in Fig. 9.11. As can be seen, the CP-OFDM has the best SER performance in the time-varying fading channel; the FBMC-OQAM, UF-OFDM, and W-OFDM have similar SER performances; and the

FBMC-QAM has the worst SER performance. Note that the performance of the FBMC-QAM can be improved by using the type III filter at the expense of a higher OOB emission than the CP-OFDM [18].

REFERENCES

[1] P. Kyosti, J. Meinila, L. Hentila, et al., IST-4-027756 WINNER II D1.1.2 V1.2 WINNER II Channel Models: Part I Channel Models, 2007.
[2] S. Afsardoost, T. Eriksson, C. Fager, Digital predistortion using a vector-switched model, IEEE Transactions on Microwave Theory and Techniques (ISSN 0018-9480) 60 (4) (2012, April) 1166–1174, https://doi.org/10.1109/TMTT.2012.2184295.
[3] M. Bellanger, et al., FBMC physical layer: a primer, http://www.ict-phydyas.org, 2010, 06.
[4] G. Berardinelli, Generalized DFT-S-OFDM waveforms without cyclic prefix, IEEE Access 6 (2018) 4677–4689, https://doi.org/10.1109/ACCESS.2017.2781122.
[5] J.A.C. Bingham, Multicarrier modulation for data transmission: an idea whose time has come, IEEE Communications Magazine (ISSN 0163-6804) 28 (5) (1990, May) 5–14, https://doi.org/10.1109/35.54342.
[6] X. Chen, A. Wolfgang, Phase noise mitigation in OFDM-based backhaul in the presence of channel estimation and synchronization errors, in: 2016 IEEE 83rd Vehicular Technology Conference (VTC Spring), 2016, May, pp. 1–5.
[7] X. Chen, S. Zhang, Antenna mutual coupling effect on MIMO-OFDM system in the presence of phase noise, in: 2017 11th European Conference on Antennas and Propagation (EUCAP), 2017, March, pp. 653–657.
[8] X. Chen, A. Wolfgang, A. Zaidi, MIMO-OFDM for small cell backhaul in the presence of synchronization errors and phase noise, in: 2017 IEEE International Conference on Communications Workshops (ICC Workshops), 2017, May, pp. 1221–1226.
[9] A. Daher, E.H. Baghious, G. Burel, E. Radoi, Overlap-save and overlap-add filters: optimal design and comparison, IEEE Transactions on Signal Processing (ISSN 1053-587X) 58 (6) (2010, June) 3066–3075, https://doi.org/10.1109/TSP.2010.2044260.
[10] M. Faulkner, The effect of filtering on the performance of OFDM systems, IEEE Transactions on Vehicular Technology (ISSN 0018-9545) 49 (5) (2000, Sep.) 1877–1884, https://doi.org/10.1109/25.892590.
[11] K. Guan, G. Li, T. Kurner, A.F. Molisch, B. Peng, R. He, B. Hui, J. Kim, Z. Zhong, On millimeter wave and THz mobile radio channel for smart rail mobility, IEEE Transactions on Vehicular Technology (ISSN 0018-9545) 66 (7) (2017, July) 5658–5674, https://doi.org/10.1109/TVT.2016.2624504.
[12] H2020-ICT-671650-mmMAGIC/D5.1 2016 mmMAGIC Deliverable D5.1. Initial multi-node and antenna transmitter and receiver architectures and schemes.
[13] S. Haykin, Adaptive Filter Theory, Prentice-Hall, 1996.
[14] F.H.H. Institute, Quasi deterministic radio channel generator user manual and documentation, document revision: v1.4.8-571, 2016, Sep.
[15] J. Luo, A.A. Zaidi, J. Vihriälä, D. Giustiniano, et al., Preliminary radio interface concepts for mm-wave mobile communications. Deliverable D4.1, Millimetre-Wave Based Mobile Radio Access Network for Fifth Generation Integrated Communications (mmMAGIC), https://5g-mmmagic.eu/results, 2016.
[16] S. Jaeckel, L. Raschkowski, K. Borner, L. Thiele, Quadriga: a 3-d multi-cell channel model with time evolution for enabling virtual field trials, IEEE Transactions on Antennas and Propagation (ISSN 0018-926X) 62 (6) (2014, June) 3242–3256, https://doi.org/10.1109/TAP.2014.2310220.
[17] C. Kim, K. Kim, Y.H. Yun, Z. Ho, B. Lee, J.Y. Seol, QAM-FBMC: a new multi-carrier system for post-ofdm wireless communications, in: 2015 IEEE Global Communications Conference (GLOBECOM), 2015, Dec., pp. 1–6.
[18] H. Nam, M. Choi, C. Kim, D. Hong, S. Choi, A new filter-bank multicarrier system for QAM signal transmission and reception, in: 2014 IEEE International Conference on Communications (ICC), 2014, June, pp. 5227–5232.
[19] D. Petrovic, W. Rave, G. Fettweis, Effects of phase noise on OFDM systems with and without pll: characterization and compensation, IEEE Transactions on Communications (ISSN 0090-6778) 55 (8) (2007, Aug.) 1607–1616, https://doi.org/10.1109/TCOMM.2007.902593.
[20] D. Roque, C. Siclet, Performances of weighted cyclic prefix OFDM with low-complexity equalization, IEEE Communications Letters (ISSN 1089-7798) 17 (3) (2013, March) 439–442, https://doi.org/10.1109/LCOMM.2013.011513.121997.
[21] P. Siohan, C. Siclet, N. Lacaille, Analysis and design of OFDM/OQAM systems based on filterbank theory, IEEE Transactions on Signal Processing (ISSN 1053-587X) 50 (5) (2002, May) 1170–1183, https://doi.org/10.1109/78.995073.

[22] V. Vakilian, T. Wild, F. Schaich, S. ten Brink, J.F. Frigon, Universal-filtered multi-carrier technique for wireless systems beyond LTE, in: 2013 IEEE Globecom Workshops (GC Wkshps), 2013, Dec., pp. 223–228.
[23] J.J. van de Beek, M. Sandell, P.O. Borjesson, Ml estimation of time and frequency offset in OFDM systems, IEEE Transactions on Signal Processing (ISSN 1053-587X) 45 (7) (1997, Jul.) 1800–1805, https://doi.org/10.1109/78.599949.
[24] G. Yi, L. Gang, G. Jianhua, A novel time and frequency synchronization scheme for OFDM systems, IEEE Transactions on Consumer Electronics (ISSN 0098-3063) 54 (2) (2008, May) 321–325, https://doi.org/10.1109/TCE.2008.4560093.
[25] Y.H. Yun, C. Kim, K. Kim, Z. Ho, B. Lee, J.Y. Seol, A new waveform enabling enhanced QAM-FBMC systems, in: 2015 IEEE 16th International Workshop on Signal Processing Advances in Wireless Communications (SPAWC), 2015, June, pp. 116–120.

Index

Printed in the United States
By Bookmasters